Handbook of Analysis and Extraction Methods of Anthocyanins

Handbook of Analysis and Extraction Methods of Anthocyanins provides a comprehensive guide to learning about the properties of anthocyanins, which have gained increasing importance in recent years and have attracted widespread attention from industry, academia, and government, as well as the precise, applicable, and modern methods of their analysis developed to date. The first part of the book introduces the structure, biochemical properties, health effects, and high antioxidant capacity of anthocyanins based on scientific developments in recent years. The second part of the book is aimed at the technological use of anthocyanins in industry, focusing on the effects of food processing methods on anthocyanins, encapsulation, the possibilities of using them as colorants in industry, and their importance as a functional ingredient, as well as a healthy ingredient. The third part of the book presents in detail the extraction and purification methods of anthocyanins in different food products as well as sensitive, quantitative, applicable, and newly developed NMR, HPLC, UHPLC, GC/MS, UHPLC/MS, and LC/MS methods.

Key Features:

- Minimal Prerequisites: No prior functional food experience is needed, making the content accessible to a wide audience.

- New and Real-World Data: Learn with real-scientific information on anthocyanins, including all biological properties of anthocyanins, antioxidant activities, and new datasets on health protection, functional food development, encapsulation, and uses of anthocyanins as food colorant.

- Expanded Theoretical and Practical Data on Methods of Analysis of Anthocyanins: Includes deeper coverage of theory-based approaches of modern methods of analysis of anthocyanins, their connection with GC/MS, LC/MS etc.-based approaches, and a presentation of newly developed and formal aspects of these methods of anthocyanins.

Ideal for those interested in or looking to deepen their knowledge of functional foods and nutraceuticals, this edition provides a clear introduction to the antioxidant structure and health-protective properties of anthocyanins and current modern analysis methods.

Food Analysis & Properties

Series Editor
Leo M. L. Nollet

University College Ghent, Belgium

This CRC series **Food Analysis and Properties** is designed to provide a state-of-art coverage on topics to the understanding of physical, chemical and functional properties of foods: including (1) recent analysis techniques of a choice of food components; (2) developments and evolutions in analysis techniques related to food; (3) recent trends in analysis techniques of specific food components and/or a group of related food components.

Sequencing Technologies in Microbial Food Safety and Quality
Edited by Devarajan Thangadurai, Leo M.L. Nollet, Saher Islam, and Jeyabalan Sangeetha

Nanoemulsions in Food Technology: Development, Characterization, and Applications
Edited by Javed Ahmad and Leo M.L. Nollet

Mass Spectrometry in Food Analysis
Edited by Leo M.L. Nollet and Robert Winkler

Bioactive Peptides from Food: Sources, Analysis, and Functions
Edited by Leo M.L. Nollet and Semih Ötleş

Nutriomics: Well-being through Nutrition
Edited by Devarajan Thangadurai, Saher Islam, Leo M.L. Nollet, and Juliana Bunmi Adetunji

Analysis of Naturally Occurring Food Toxins of Plant Origin
Edited by Leo M. L. Nollet and Javed Ahmad

Analysis of Food Spices: Identification and Authentication
Edited by Leo M. L. Nollet, Javed Ahmad, and Javed Ahamad

Flavoromics: An Integrated Approach to Flavor and Sensory Assessment
Edited by Leo M. L. Nollet and Matteo Bordiga

Bioactive Compounds from Food: Benefits and Analysis
Edited by Leo M. L. Nollet and Javed Ahmad

Bioactive Compounds: Identification and Characterization of their Food and Pharmacological Potential
Edited by Mozaniel Santana de Oliveira and Leo M.L. Nollet

Handbook of Analysis and Extraction Methods of Anthocyanins
Edited by Leo M.L. Nollett and Semih Ötleş

For more information, please visit the Series Page: https://www.crcpress.com/Food-Analysis–Properties/book-series/CRCFOODANPRO

Handbook of Analysis and Extraction Methods of Anthocyanins

Edited by

Leo M.L. Nollet and Semih Ötleş

CRC Press
Taylor & Francis Group
Boca Raton London New York

CRC Press is an imprint of the
Taylor & Francis Group, an **informa** business

Designed cover image: Shutterstock

First edition published 2025
by CRC Press
2385 NW Executive Center Drive, Suite 320, Boca Raton FL 33431
and by CRC Press

4 Park Square, Milton Park, Abingdon, Oxon, OX14 4RN

CRC Press is an imprint of Taylor & Francis Group, LLC

ISBN: 978-1-032-58468-3 (hbk)
ISBN: 978-1-032-59161-2 (pbk)
ISBN: 978-1-003-45326-0 (ebk)

DOI: 10.1201/9781003453260

Typeset in Palatino
by KnowledgeWorks Global Ltd.

Contents

Section I General Introduction, Sources, and Chemistry

Section II Extraction Methods and Analysis of Anthocyanins

Series Preface

There will always be a need to analyze food compounds and their properties. Current trends in analyzing methods include automation, increasing the speed of analyses, and miniaturization. Over the years, the unit of detection has evolved from micrograms to pictograms.

A classical pathway of analysis is sampling, sample preparation, cleanup, derivatization, separation, and detection. At every step, researchers are working and developing new methodologies. A large number of papers are published every year on all facets of analysis. So, there is a need for books that gather information on one kind of analysis technique or on the analysis methods for a specific group of food components.

The scope of the CRC Series on *Food Analysis and Properties* aims to present a range of books edited by distinguished scientists and researchers who have significant experience in scientific pursuits and critical analysis. This series is designed to provide state-of-the-art coverage on topics such as the following:

1. Recent analysis techniques on a range of food components.
2. Developments and evolution in analysis techniques related to food.
3. Recent trends in analysis techniques for specific food components and/or a group of related food components.
4. The understanding of physical, chemical, and functional properties of foods.

The book *Handbook of Analysis and Extraction Methods of Anthocyanins* is volume number 26 of this series.

I am happy to be a series editor of such books for the following reasons:

- I am able to pass on my experience in editing high-quality books related to food.
- I get to know colleagues from all over the world more personally.
- I continue to learn about interesting developments in food analysis.

Much work is involved in the preparation of a book. I have been assisted and supported by a number of people, all of whom I would like to thank. I would especially like to thank the team at CRC Press/Taylor & Francis, with a special word of thanks to Steve Zollo, senior editor.

Many, many thanks to all the editors and authors of this volume and future volumes. I very much appreciate all their effort, time, and willingness to do a great job.

I dedicate this series to:

- My wife, for her patience with me (and all the time I spend on my computer).
- All patients suffering from prostate cancer; knowing what this means, I am hoping they will have some relief.

Dr. Leo M. L. Nollet (Retired)
University College Ghent
Ghent, Belgium

Preface

As the market share of functional foods increases in the world, interest in components with protective and therapeutic properties, especially antioxidant ones, has been increasing rapidly in both the scientific world and the nutraceutical industry in recent years. When examined from this perspective, anthocyanins stand out as a class of flavonoids with an important antioxidant character among the secondary metabolites of plants. Anthocyanins are pigments that give various fruits and vegetables their blue, red, purple, and black colors. In this book, the importance of anthocyanins is an important functional component. The studies conducted in recent years on their extraction technologies and analysis are discussed and a roadmap that can be considered for the future for both academia and industry is tried to be drawn.

The book is divided into two sections:

Section 1: General Introduction, Sources, and Chemistry, and Section 2: Extraction Methods and Analysis of Anthocyanins.

The book considers fundamental concepts, chemical properties, biosynthesis, bioavailability, stability in food products during food processing, encapsulation, antioxidant properties, health effects, uses as functional food, isolation and extraction methods, analysis (HPLC, UHPLC, LC/MS, NMR, GC/MS, and spectroscopy) of anthocyanins.

The book addresses food scientists, technologists, chemists, nutrition researchers, producers, and processors working in the whole food science and technology field as well as those who are interested in the development of innovative functional products.

We would like to acknowledge and thank all chapter authors for their fruitful collaboration in bringing together different topics of anthocyanins in one integral and valuable reference.

We would especially like to thank the team at CRC Press/Taylor & Francis, with a special word of thanks to Steve Zollo, senior editor.

Leo M.L. Nollet
Semih Ötleş

> When the going gets tough, the tough get going.

Biographies

Leo M. L. Nollet earned an M.S. (1973) and Ph.D. (1978) in biology from the Katholieke Universiteit Leuven, Belgium. He is an editor and associate editor of numerous books. He edited for M. Dekker, New York—now CRC Press of Taylor & Francis Publishing Group—the first, second, and third editions of *Food Analysis by HPLC* and *Handbook of Food Analysis*. The last edition is a two-volume book. Dr. Nollet also edited the *Handbook of Water Analysis* (first, second, and third editions) and *Chromatographic Analysis of the Environment*, third and fourth editions (CRC Press). With F. Toldrá, he coedited two books published in 2006, 2007, and 2017: *Advanced Technologies for Meat Processing* (CRC Press) and *Advances in Food Diagnostics* with another publisher. With M. Poschl, he coedited the book *Radionuclide Concentrations in Foods and the Environment*, also published in 2006 (CRC Press). Dr. Nollet has also coedited with Y. H. Hui and other colleagues on several books: *Handbook of Food Product Manufacturing; Handbook of Food Science, Technology, and Engineering* (CRC Press, 2005); *Food Biochemistry and Food Processing* (first and second editions; and the *Handbook of Fruit and Vegetable Flavors*. In addition, he edited the *Handbook of Meat, Poultry and Seafood Quality*, first and second editions.

From 2008 to 2011, he published five volumes on animal product-related books with F. Toldrá: *Handbook of Muscle Foods Analysis, Handbook of Processed Meats and Poultry Analysis, Handbook of Seafood and Seafood Products Analysis, Handbook of Dairy Foods Analysis* (2nd edition in 2021), and *Handbook of Analysis of Edible Animal By-Products*. Also, in 2011, with F. Toldrá, he coedited two volumes for CRC Press: *Safety Analysis of Foods of Animal Origin* and *Sensory Analysis of Foods of Animal Origin*. In 2012, they published the *Handbook of Analysis of Active Compounds in Functional Foods*. In a co-edition with Hamir Rathore, *Handbook of Pesticides: Methods of Pesticides Residues Analysis* was marketed in 2009; *Pesticides: Evaluation of Environmental Pollution* in 2012; *Biopesticides Handbook* in 2015; and *Green Pesticides Handbook: Essential Oils for Pest Control* in 2017. Other finished book projects include *Food Allergens: Analysis Instrumentation and Methods* (with A. van Hengel; CRC Press, 2011) and *Analysis of Endocrine Compounds in Food*. Dr. Nollet's recent projects include *Proteomics in Foods* with F. Toldrá and *Transformation Products of Emerging Contaminants in the Environment: Analysis, Processes, Occurrence, Effects, and Risks* with D. Lambropoulou.

In the series *Food Analysis and Properties*, he edited (with C. Ruiz-Capillas) *Flow Injection Analysis of Food Additives* (CRC Press, 2015) and *Marine Microorganisms: Extraction and Analysis of Bioactive Compounds* (CRC Press, 2016). With A.S. Franca, he coedited *Spectroscopic Methods in Food Analysis* (CRC Press, 2017), and with Horacio Heinzen and Amadeo R. Fernandez-Alba, he coedited *Multiresidue Methods for the Analysis of Pesticide Residues in Food* (CRC Press, 2017). Further volumes in the series *Food Analysis and Properties* are *Phenolic Compounds in Food: Characterization and Analysis* (with Janet Alejandra Gutierrez-Uribe, 2018), *Testing and Analysis of GMO-containing Foods and Feed* (with Salah E. O. Mahgoub, 2018), *Fingerprinting Techniques in Food Authentication and Traceability* (with K. S. Siddiqi, 2018), *Hyperspectral Imaging Analysis and Applications for Food Quality* (with N.C. Basantia, Leo M.L. Nollet, Mohammed Kamruzzaman, 2018), *Ambient Mass Spectroscopy Techniques in Food and the Environment* (with Basil K. Munjanja, 2019), *Food Aroma Evolution: During Food Processing, Cooking, and Aging* (with M. Bordiga, 2019), *Mass Spectrometry Imaging in Food Analysis* (2020), *Proteomics for Food Authentication* (with S. Ötleş, 2020), *Analysis of*

Nanoplastics and Microplastics in Food (with K.S. Siddiqi, 2020), *Chiral Organic Pollutants: Monitoring and Characterization in Food and the Environment* (with Edmond Sanganyado and Basil K. Munjanja, 2020), *Sequencing Technologies in Microbial Food Safety and Quality* (with Devarajan Thangadurai, Saher Islam, Jeyabalan Sangeetha, 2021), *Nanoemulsions in Food Technology: Development, Characterization, and Applications* (with Javed Ahmad, 2021), *Mass Spectrometry in Food Analysis* (with Robert Winkler, 2022), *Bioactive Peptides from Food: Sources, Analysis, and Functions* (with Semih Ötleş, 2022), and *Nutriomics: Well-being through Nutrition* (with Devarajan Thangadurai, Saher IslamJuliana Bunmi Adetunji, 2022). In 2023 he published with Javed Ahmad *Analysis of Naturally Occurring Food Toxins of Plant Origin*.

The last published books are as follows: *Biopesticides Handbook 2nd Ed.* (with S. Mir), *Handbook of Seafood and Seafood Products Analysis* (with F. Toldrá), *Flavoromics: An Integrated Approach to Flavor and Sensory Assessment* (with M. Bordiga), *Analysis of Food Spices* (with J. Ahmad and J. Ahamad), *Global Regulations of Medicinal, Pharmaceutical, and Food Products* (with F. Ali), *and Bioactive Compounds from Food Benefits and Analysis* (with J. Ahmad).

Semih Ötleş A native of Izmir, Türkiye, Professor Ötleş obtained his B.Sc. degree from the Department of Food Engineering (Ege University) in 1980. During his assistantship at Ege University, in 1985, he received an M.S. in Food Chemistry, and in 1989, after completing his thesis research on the instrumental analysis and chemistry of vitamins in foods, he received a Ph.D. in Food Chemistry from Ege University. In 1991–1992, he completed postdoctoral training on meat proteins, including OECD—Postdoctoral Fellowship, in the Research Center Melle at Gent University, Belgium. Afterward, he joined the Department of Food Engineering at Ege University as a scientist of Food Chemistry, being promoted to Associate Professor in 1993 and to Professor in 2000. He was Vice Dean at Engineering Faculty (2003–2009), Head of the Department of Nutrition and Dietetics (2008–2011), and Vice Rector in Ege University (2012–2016). The research activities of Professor Ötleş have been focused on instrumental methods of food analysis. Ötleş began a series of projects on the separation and instrumental analysis techniques, first for analysis of vitamins in foods, then protein chemistry, carbohydrates, carotenoids, proteomics, and most recently bioactive peptides. Other activities span the fields of instrumental food analysis like UPLC, GC, GC/MS, LC/MS/MS analysis, hyphenated techniques, soy chemistry, aromatics, medical and functional foods, and nutraceutical chemistry. Included are multiresidue analysis of various foods, n-3 fatty acids in fish oils, and medical and functional foods.

Section I

General Introduction, Sources, and Chemistry

1

State of the Art of Anthocyanins: Antioxidant Activity, Sources, Bioavailability, and Therapeutic Effect in Human Health

Noelia Tena, Julia Martín, and Agustín G. Asuero

1.1 Introduction

In recent years, the interest in plants and food containing antioxidant properties has increased. The chemical compounds present in vegetables and fruits with these capacities are: vitamins C and E, carotenoids, and flavonoids. The anthocyanins, which are the most important group of flavonoids in plants, are pigments with a flavylium cation (AH^+) structure that act as acids. This structure is directly related to its antioxidant activity. Most of the functional properties and the sensory quality of the anthocyanins can be explained by their chemical reactivity. The structures and properties of anthocyanins are dependent on different factors such as pH, temperature, and solvents which should be controlled to carry out antioxidant activity studies of these compounds [1–9].

Free radicals, reactive oxygen species (ROS), and/or reactive nitrogen species (RNS) are required for the proper performance of the human body and its organs. These radicals are on balance by a redox homeostasis in our body. However, the body may be occasionally affected by oxidative stress resulting from an off-balance state. This stress is important in the development of chronic degenerative diseases including coronary heart disease, cancer, and aging [2]. Anthocyanins have been described as compounds that prevent or inhibit, oxidation by scavenging free radicals and reducing oxidative stress. On a regular basis, anthocyanins act as H-atom donator or as single-electron transfer (SET). Different methods of analysis based on both mechanisms have been proposed to determine the antioxidant activity of anthocyanins. The antioxidant activity of these compounds depends on their total concentration, structure, and environment. A literature compilation about the concentrations of the most common anthocyanins in different foods is presented in this review in order to have an overview of the different sources of anthocyanins.

The beneficial properties attributed to the dietary ingestion of anthocyanin-rich foods (eye health, cardiovascular diseases, antiobesity, antidiabetic, antimicrobial, anticancer, or neuroprotective effect) have been deeply documented in studies carried out with experimental models. These health benefits contrast with the apparent small portion (<1–2%) of these compounds absorbed by our organism [3–5]. During the digestion process, anthocyanins undergo an intense variation in pH that together with the enzymatic and bacterial action can cause the hydrolysis and transformation of anthocyanins into metabolites, conjugated products, or simpler phenolic compounds [7, 10–12]. The question is: How

DOI: 10.1201/9781003453260-2

can anthocyanins be so influential in health? Are anthocyanins the only responsible for their beneficial effects? The last scientific developments highlight the potential synergy effect between parent compounds, metabolites (phases I and II), conjugated products, and microbe-generated metabolites to explain those biological events [4, 11–13].

Due to their particular physicochemical features, the bioavailability of anthocyanins is very difficult to assess. The first studies were performed analyzing blood and urine to determine the anthocyanin concentration levels after the ingestion of foods rich in anthocyanins [14–16]. However, the low absorption percentage obtained led to in vitro assays (mostly using cell culture systems) in order to facilitate the knowledge of their biochemical and chemical changes as well as the influence of the digestion steps. The last studies have emphasized the key role of the microbiota in the transformation of anthocyanins, which is not considered in in vitro assays but is still poorly considered in in vivo and ex vivo studies [14].

This review aims to highlight some aspects regarding the antioxidant activity of anthocyanins and their bioavailability after intake. The first part includes an exposition of the most common antioxidant bioassays used to determine in vitro the antioxidant activity of anthocyanins, being the advantages and disadvantages of each bioassay identified. Afterwards, the effect of the chemical structure and the environment on the ability of the anthocyanins to prevent oxidation is discussed and presented together with information about different sources and the range of concentrations of these compounds in food. The second part of the manuscript exposes the bioavailability and metabolism of anthocyanins as well as a summary including the main therapeutic effects of anthocyanins on different diseases.

1.2 Antioxidant Bioassays for Anthocyanins

Numerous antioxidant assays have been proposed to measure the ability of anthocyanins to prevent the oxidation process that naturally occurs. Depending on the source of the anthocyanins and their nature, in most of the cases, an extraction step before carrying out the antioxidant bioassay is needed. The extraction process is a critical step in the determination of the antioxidant activity bioassay, presenting a challenge due to the low stability of anthocyanins after extraction and their tendency to remain bound to the matrix of the sample. Multiple alternatives have been proposed in the literature for this procedure [17]. Results show that temperature, pH, solvent system, solvent-to-solid ratio, and number of extractions are factors that play an important role in the extraction efficiency and that should be optimized for each sample [18, 19]. Once the anthocyanins are in a liquid solution, the antioxidant activity could be determined by different bioassays. In general, two different mechanisms can be used to explain the antioxidant activity of anthocyanins: hydrogen atom donator (HAT) and SET. In HAT mechanism, the free radical $R^\bullet$ removes a hydrogen atom from the antioxidant (AH^+) converting the free radical to a more stable product. In the SET mechanism, the antioxidant (AH^+) donates an electron to the free radical reducing the oxidized intermediates into the stable form [20]. However, the difficulty in distinguishing between HAT and SET reactions is high. In most situations, these two reactions take place simultaneously, and the mechanism of the reaction is determined by the antioxidant structure, solubility, the partition coefficient, and solvent polarity [21]. Different antioxidant bioassays have been commonly used in the literature to determine the antioxidant activity of anthocyanins. A summary of the most common antioxidant bioassays used for anthocyanins is shown in Table 1.1.

TABLE 1.1

Commonly Used Methods for Measurement in Vitro of Antioxidant Activity

Bioassay	Reagents Involved in the Reaction	Detection	Method
DPPH (diphenyl-1-picrylhydrazyl) assay	Free radical (DPPH$^{\bullet+}$)	Decrease of Abs. at 515 nm	Spectrophotometric or colorimetric
ORAC Assay (oxygen radical absorbance capacity)	2,2′-Azobis(2-amidino-propane) dihydrochloride (AAPH) to produce free radical β-phycoerythrin or fluorescein or pyrogallol red	Decrease of fluorescence	Fluorescence spectroscopy
TRAP assay (total peroxyl radical trapping antioxidant parameter)	2,2′-Azobis(2-amidopropane) hydrochloride (ABAP) to produce free radical luminol	Decrease of luminescence	Chemiluminescence
FCT (ferric thiocyanate) assay	Ferrous chloride, formation of red ferric thiocyanate	Increase of Abs. at 500 nm	Spectrophotometric
FRAP (ferric reducing antioxidant power) assay	FeCl$_3$·6H$_2$O, formation of blue ferrous complexes	Increase of Abs. at 593 nm	Colorimetric
CUPRAC, cupric ion reducing antioxidant capacity	Cupric neocuproine, formation of Cu(I)-neocuproine	Increase of Abs. at 550 nm	Spectrophotometric
ABTS [2,2′-azino-bis (3-ehtylbenzothiazoline-6-sulfonic acid) diamonium salt] assay	Free radical (ABTS$^{\bullet+}$)	Decrease of Abs. at 415 nm	Colorimetric
Methods of inhibited autoxidation	Lipid molecules, azoinitiator	O$_2$ consumption/ hydroperoxide formation	Oxygen electrode, pressure gauge, detection of conjugated dienes

Note: Abs, absorbance.

Each one of these methods provides information about the antioxidant activity of the anthocyanins under specific conditions. In order to have more information about the reaction mechanism, a more comprehensive description of the procedure followed in each one of these bioassays are presented together with their main advantages and disadvantages.

1.2.1 DPPH (Diphenyl-1-Picrylhydrazyl) Assay

DPPH assay is a spectrophotometric method that can be applied for both solid and liquid samples, not being specific for any particular antioxidant. Thus, it can be applied to determine the total antioxidant capacity of the sample. This assay is based on the ability of the free radical (DPPH$^{\bullet}$) to react with hydrogen donor (AH$^+$). The free radical presents an intense abortion in the UV-vis spectral region at 515 nm, the absorbance at 515 nm decreases when the free radical is reduced.

The antioxidant activity of anthocyanins for DPPH has been assessed by several authors [22–26]. Reliable results can be obtained with this method in a fast and simple manner. Currently, a colorimetric alternative has been proposed to extend the application of this assay where a spectrophotometer is not available [27]. Furthermore, as the free radical is stable its production is not necessary every time that an analysis is carried out. The main disadvantages could be the complexity of the analysis, as it turns more complex when other compounds present in the sample absorb at 515 nm and many antioxidants are inert to DPPH.

1.2.2 ORAC Assay (Oxygen Radical Absorbance Capacity)

ORAC assay is a fluorescence method that combines the sample (AH^+) with a fluorescent compound, protein phycoerythrin (β-PE), and with a generator of free radicals, 2,2'-azobis(2-amidino-propane) dihydrochloride (AAPH). This assay is based on the loss of the fluorescent compound when it is oxidized by the generated free radicals. When antioxidants (AH^+) are involved in the reaction, the fluorescent compound is protected from oxidative degradation. Hence, the fluorescence signal remains. The fluorescence signal is monitored for 1 h at λem = 565 nm and λex = 540 nm, respectively. The antioxidant activity of anthocyanins has been assessed by ORAC assay by several authors [26, 28–30]. This is considered a good method to determine the antioxidant capacity of hydrophilic and hydrophobic samples, as it is easily adaptive to different samples changing the generator of free radicals. One disadvantage is the non-specificity of the fluorescence compounds that can react with the sample losing fluorescence even without the addition of a free radical generator.

1.2.3 TRAP Assay (Total Peroxyl Radical Trapping Antioxidant Parameter)

TRAP assay is a chemiluminescence method that consists of the following components:

  i. thermolabile azo-radical initiator (e.g., 2,2'-azobis(2-amidopropane) hydrochloride (ABAP)), which produces radicals ($R^\bullet$) that react rapidly with O_2 to give a peroxyl radicals ($ROO^\bullet$);
  ii. oxidizable compounds with chemiluminescence properties to monitor the reaction progress (e.g., luminol); and
  iii. the sample with the antioxidant properties (AH^+).

The antioxidant activity of anthocyanins has been assessed by TRAP assay by several authors [31–34]. This assay is sensitive to all known chain-breaking antioxidants. However, an important disadvantage to highlight is the difficulty comparing the results between laboratories due to the number of different endpoints that can be used. It is a relatively complex method, time-consuming, and requires a high degree of experience. Furthermore, such as in the ORAC assay, the oxidizable compounds are non-specific and other non-radical chain reactions could occur.

1.2.4 FCT (Ferric Thiocyanate) Assay

The FCT assay is a spectrophotometric method that consists of the oxidation of ferrous chloride to ferric ions by reacting with peroxide. The peroxides are formed during the reaction that takes place when the sample is mixed with ethanol, water, phosphate buffer solution (pH = 7), and linoleic acid. Then, ferrous chloride in hydrochloric acid is added to the reaction, and the ferric ion formed is combined with ammonium thiocyanate producing ferric thiocyanate, which is red. The absorbance of the sample is measured at 500 nm until the maximum value is reached.

This assay is used to measure the amount of peroxide produced during the initial stages of oxidation. In the case of anthocyanins, this method has been applied to determine the antioxidant activity avoiding the peroxidation of polyunsaturated fatty acid [35]. This assay is simple and reproducible. Nevertheless, results are not reliable in the case that compounds within the sample absorb around 500 nm. This drawback is common not only in this method but in other spectrophotometric assays.

1.2.5 FRAP (Ferric Reducing Antioxidant Power) Assay

FRAP assay is a colorimetric method that can be employed for the determination of the total antioxidant activity of anthocyanins. It is based on the reduction of complexes of 2,4,6-tripyridyl-s-triazine (TPTZ) with ferric chloride hexahydrate ($FeCl_3 \cdot 6H_2O$) under acidic conditions. The solution turns slightly brownish, forming blue ferrous complexes once the reduction is completed. The absorbance is measured at 593 nm against the blank.

This method has been extensively applied to determine the antioxidant activity of anthocyanins in different matrices: in elderberry [36]; in Roselle extract [37]; in raspberries, blackberries, red currants, gooseberries, and Cornelian cherries [38]; or in carrots, cabbage, cauliflower, potatoes, onions, asparagus, and eggplant [39]. However, the results of this method are in most of the cases compared to the results of other antioxidant assays [40]. Rapid and reproducible results are obtained with this method. However, some limitations related to FRAP assay should be considered. On the one hand, the samples must be aqueous and the pH value is critical. On the other hand, some compounds without antioxidant properties can reduce Fe^{3+} to Fe^{2+} inducing an overestimation of the antioxidant activity of the sample.

1.2.6 CUPRAC (Cupric Ion Reducing Antioxidant Capacity) Assay

CUPRAC assay is a spectrophotometric method similar to FRAP. In this method, the anthocyanins react with the CUPRAC reagent (cupric neocuproine) producing the Cu(I)-neocuproine which is a chromophore that absorbs at 450 nm. This method has been successfully applied to various food extracts [7].

The main advantages of this method are related to the positive characteristics of the CUPRAC reagent: availability and easy accessibility, rapidity, stability, low cost, sensitivity toward thiol-type antioxidants unlike FRAP, and responsiveness to both hydrophilic and lipophilic antioxidants. Nevertheless, the detection is based on the absorption at 450 nm which is non-specific enough and the presence of other compounds present in the sample could be interfered with the results.

1.2.7 ABTS (2,2′-Azino-bis (3-ehtylbenzothiazoline-6-sulfonic Acid) Diamonium Salt) Assay

The most recent ABTS assay method is based on decolorization techniques. It consists of the production of a stable radical, blue/green ABTS chromophore, by the reaction of ABTS with potassium persulfate. This stable radical has a maximum absorbance of 415 nm. A drop in absorbance of this compound occurs when the radical reacts with the antioxidant. This method can determine the antioxidant activity of mixtures of substances, helping to distinguish between additive and synergistic effects. The antioxidant activity is calculated relative to the reactivity of the Trolox standard under similar conditions.

This assay is frequently combined with the DPPH assay for the determination of the antioxidant activity of anthocyanins [41–43]. This method is simple and not a large sample volume and time of analysis are necessary. Nevertheless, a standard solution is required in order to obtain accurate results.

This assay is also described in the literature as TEAC [6-hydroxy-2,5,7,8-tetramethylchroman-2-carboxylic acid (Trolox)] equivalent antioxidant capacity. The same free radical is used in both TEAC and ABTS assays. However, different reagents have been proposed for the generation of the green-blue ABTS$^{\bullet+}$, resulting in different TEAC assays. This radical is produced by the oxidation of 2,2′-azinobis (3-ethylbenzothiazoline-6-sulfonic acid; ABTS). The oxidation can be reached in different ways: (i) in TEAC assay I, metmyoglobin reacted with H_2O_2 generating the ferrylmyoglobin radical, which then reacted with ABTS. (ii) In TEAC assay II, the $ABTS^{\bullet+}$ is formed by the reaction with manganese dioxide. (iii) In TEAC assay III, an enzymatic reaction using horseradish peroxidase is applied. Other proposals consist of applying electrochemical oxidation or using 2,2′-azobis-2-amidinopropane, dihydrochloride (AAPH), or potassium persulfate ($K_2S_2O_8$) as oxidants.

These alternatives are interchangeable when the appropriate solvent is selected. However, differences in assay conditions can be found, such as the reaction time or the wavelength used for the detection, for instance, sometimes in order to avoid interferences 734 nm is preferable to 415 nm.

These assays have been used to determine the antioxidant capacity of the anthocyanins, as a consequence some results have been reported in wine [44]; in corn [45]; in pomegranate juice [46]; or in blueberries [47].

The advantages of TEAC assay I are the simplicity, reproducibility, and flexibility to determine the antioxidant capacity in hydrophilic and lipophilic foods. However, the pre-addiction of the sample before the radical generation could result in an overestimation of the antioxidant capacity. Other possible disadvantages of this assay may be the fact that ABTS is not found naturally and that any compound with a redox potential lower than $ABTS^{\bullet+}$ may reacts with the radical.

Other alternatives applied to determine the antioxidant activity of anthocyanins involve the use of enzymes [48] or the use of chromatographic techniques. The latter alternative can not only extract the anthocyanins to determine their antioxidant activity [49] but also determine the total anthocyanins concentration (TAC), and the identification and quantification of the presence of each anthocyanin present in the sample [50, 51].

The in vitro assays exposed in this manuscript provide information about the antioxidant activity of the anthocyanins determined by their capacity to neutralize the initiators of the oxidation process (absorbing photons, neutralizing ROS, or chelating metals ion) stopping the initiation steps of the autoxidation process. Thus, these assays are based on the reaction of the anthocyanins with: (i) some colored persistent radicals such as the free radicals used in the DPPH test and TEAC test; or (ii) other oxidizing agents like Fe^{3+} ions used in the FRAP test or Cu^{2+} ions used in CUPRAC test. This neutralization capacity is measured under specific conditions of temperature, light, or a combination of both. Other assays to measure the antioxidant capacity of anthocyanins are based on a competitive probe reaction where the competitive reaction of radicals with the anthocyanins or with a probe is monitored by fluorimetric techniques in the case of the ORAC assay, among others. All of these methods offer information on the actual antioxidant activity. They trap free radicals and should be considered as chain-breaking antioxidants. This property can be measured properly only with methods based on inhibited autoxidation, while others (such as DPPH and ABTS) provide only a rough estimation. However, different specific studies have been carried out to determine the preventive antioxidant capacity [52]. Other specific assays to gather information about the inhibition of lipid substrate oxidation, where different factors and mechanisms of oxidation are studied simultaneously, can be found in the literature [53].

As it has been explained above, the assays presented in this manuscript follow different mechanisms of reaction to measure the antioxidant capacity. Therefore, the information provided by them and the interpretation of their results should be carefully considered. For example, that is the case of the results provided by FRAP and CUPRAC assays versus ORAC assay which consists of a radical-trapping reaction and directly measures the capacity to neutralize initiators.

Despite the fact that the methods mentioned above are the most commonly used to determine the antioxidant capacity of anthocyanins, there is currently a claim for the development of different alternatives in order to provide direct information about the capacity to prevent autoxidation in the biological system. Diverse alternatives have been proposed to measure the inhibition of autoxidation. That is the case exposed by Matera et al. [54], where the antioxidant activity is measured by studying the inhibited autoxidation in aqueous micelles. For that purpose, linoleic acid and Triton-X100 were mixed under controlled pH together with an initiator and the extracted anthocyanins. The reaction was tracked by monitoring the oxygen consumption with a Clark-type electrode. Another alternative to measuring the capacity of anthocyanins to inhibit autoxidation was proposed by Tisuda et al. [55], where the antioxidative activity was measured by using linoleic acid autoxidation in different systems; liposome, rabbit erythrocyte membrane, and rat liver microsomal. This type of study provides direct information about the relationship between the results of antioxidant assays and their biological activity in the prevention of autoxidation. Currently, the European Food Safety Authority (EFSA) has published guidance for the scientific requirements for health claims related to antioxidants, oxidative damage, and cardiovascular health. In this guidance, the European Health Claims Regulation has given recommendations to determine the biological effect of food supplements with purported antioxidant activity. According to this document, the antioxidant properties of foods are based on scavenging free radicals that are in vitro studied in model systems, and it is not established that this capability has a beneficial physiological effect in humans. Thus, the in vivo human studies in plasma using methods—such as TRAP, TEAC, FRAP, and ORAC—carried out to establish the antioxidant capacity of the plasma do not demonstrate that it has a beneficial physiological effect in humans. Consequently, the protection of cells from premature aging to promote healthy aging in relation to the antioxidant properties of food is not sufficiently defined because it is established by non-specific criteria. In order to provide information about the prevention of the autoxidation under specific criteria, the EFSA guidance recommends some assays for specific biological molecules. For example, it proposes: (i) the monitoring of F2-isoprostanes to determine the capacity for the protection of the lipids from oxidative damage; (ii) the analyses of 8-hydroxy-2-deoxy-guanosine in blood, tissue, and urine to assess oxidative damage to DNA; and (iii) the monitoring of protein carbonyls by ELISA, which can be applied to determine the protection of protein from oxidation damage [56].

All the assays explained in detail in this manuscript provide results that generally express the antioxidant activity as mmol Trolox equivalent per kg of fresh weight. However, these results do not report exactly the antioxidative activity of foods. For this purpose, information about the size portion of each food in the diet is also important to evaluate its antioxidant activity [57]. Due to the relevance of this type of results with regard to the antioxidant activity of anthocyanins, the USDA National Nutrient Database for Standard Reference has published a database where the total concentration of anthocyanins of different fruits and vegetables has been determined together with the service size that should be ingested from each fruit or vegetable [58].

1.3 Classification and Natural Sources of Anthocyanins

Health and therapeutic effects of anthocyanins are related to their chemical and biochemical reactivity, which are partially explained by their antioxidative activities [59, 60]. However, the antioxidative activity of anthocyanins does not necessarily transfer to biological activity because any actions on the body depend both on bioavailability and cellular molecular targets [61]. Furthermore, not all the blue, red, and purple fruits, vegetables, and flowers have the same composition and concentration of anthocyanins and in consequence the same antioxidative activity. The fruits with the highest concentration of anthocyanins are berries, currants, grapes, and some tropical fruits. In the group of edible vegetables, leafy vegetables, grains, roots, and tubers show the highest concentration of anthocyanins as well [62]. Furthermore, the presence of anthocyanins can be detected in different parts of the plant such as the stem, leaves, and storage organs.

In nature, these pigments are commonly present as anthocyanin, which is in the form of glycoside, and as anthocyanidin also known as aglycone (one or more saccharides bonded with the aglycone). The base structure of anthocyanins is shown in Figure 1.1. Nowadays the number of anthocyanins identified in nature is higher than 600 [6]. Among these anthocyanins, glycoside forms of delphinidin, cyanidin, petunidin, peonidin, malvidin, and pelargonidin are the most abundant [54, 55, 62–65]. Table 1.2 shows the most common anthocyanins that have been identified in different fruits, vegetables, and edible flowers and the numerical code that has been assigned to each anthocyanin to refer them in the manuscript.

In supplementary information, Tables S1–S6 show the most common anthocyanins identified in different fruits, vegetables, and flowers, grouped according to their chemical structure: delphinidin and its derivatives (Table S1), cyanidin and its derivatives (Table S2), petunidin (Table S3), peonidin (Table S4), malvidin (Table S5), and pelargonidin and its derivatives (Table S6). Each table also indicates the natural source of the anthocyanins, the type of extraction that was applied, the chromatographic method used to identify and quantify them, and the antioxidant assays applied to determine the antioxidant capacity.

The main conclusion from Table S1 is that delphinidin 3-glucoside (code 4) is the most common and abundant delphinidin in fruits and edible flowers. However, this group of anthocyanins is not so frequent in vegetables, being delphinine the most common. The results showed in Table S2 pointed out that the most common cyanindin in fruit and edible flower is cyanidin 3-glucoside (code 11), being also abundant cyanidin 3-galactoside (code 10) in fruits. In vegetables, cyanidin (code 8) is the most common and abundant but cyanidin 3-glucoside (code 11) is abundant in some grains as well. Table S3 shows that the following anthocyanins can be identified in fruits: petunidin 3-arabinoside, petunidin 3-galactoside, and petunidin 3-glucoside (code 30, 31, 32) in almost all the studied fruits,

Name	R_1	R_2	R_3
Delphinidin	OH	OH	H
Petunidin	OH	OCH_3	H
Malvidin	OCH_3	OCH_3	H
Cyanidin	OH	H	H
Peonidin	OCH_3	H	H
Pelargonidin	H	H	H

FIGURE 1.1
Structure of anthocyanins R_3 = sugar, and anthocyanidins R_3 = H.

TABLE 1.2

Anthocyanins Identified in Different Fruits, Vegetables, and Edible Flowers and Their Codes

Code	Anthocyanin	Code	Anthocyanin
1	Delphinidin	30	Petunidin 3-arabinoside
2	Delphinidin 3-arabinoside	31	Petunidin 3-galactoside
3	Delphinidin 3-galactoside	32	Petunidin 3-glucoside
4	Delphinidin 3-glucoside	33	Petunidin 3-halactoside
5	Delphinidin 3,5-diglucoside	34	Petunidin 3-rutinoside
6	Delphinidin 3-rutinoside	35	Peonidin
7	Delphinidin 3-*O*-(6″-*p*-coumaroyl-glucoside)	36	Peonidin 3-galactoside
8	Cyanidin	37	Peonidin 3-glucoside
9	Cyanidin 3-arabidoside	38	Peonidin 3-rutinoside
10	Cyanidin 3-galactoside	39	Peonidin 3-(6′-malonylglucoside)
11	Cyanidin 3-glucoside	40	Peonidin 3-(3″,6″-dimalonylglucoside)
12	Cyanidin 3,5-diglucoside	41	Peonidin 3-glucoside/malvidin 3-galactoside
13	Cyanidin 3-rutinoside	42	Peonidin 3-arabinoside/malvidin 3-glucoside
14	Cyanidin 3-(6′-malonylglucoside)	43	Peonidin 3-*O*-sophoroside-5-*O*-glucoside
15	Cyanidin 3-(3″,6″-dimalonylglucoside)	44	Peonidin 3-*p*-hydroxybenzoylsophoroside-5-glucoside
16	Cyanidin 3-xyloside	45	Peonidin 3-caffeoylsophoroside-5-glucoside
17	Cyanidin 3-xylosylrutinoside	46	Peonidin 3-dicaffeoylsophoroside-5-glucoside
18	Cyanidin 3-dioxaloylglucoside	47	Peonidin 3-caffeoyl-*p*-hydroxybenzoylsophoroside-5-glucoside
19	Cyanidin 3-halavtoside	48	Peonidin 3-caffeoy-feruloylsophoroside-5-glucoside
20	Cyanidin 3-*O*-sophoroside	49	Malvidin
21	Cyanidin 3-sophoroside-5-rhamnoside	50	Malvidin 3-arabinoside
22	Cyanidin 3-sambubioside	51	Malvidin 3-galactoside
23	Cyanidin 3-sambubioside-5-rhamnoside	52	Malvidin 3-glucoside
24	Cyanidin-3-*p*-hydroxybenzoylsophoroside-5-glucoside	53	Malvidin 3,5-diglucoside
25	Cyanidin-3-caffeoylsophoroside-5-glucoside	54	Pelargonidin
26	Cyanidin-3-caffeoyl-*p*-hydroxybenzoylsophoroside-5-glucoside	55	Pelargonidin 3-glucoside
27	Cyanidin 3-(*p*-coumaroyl)-diglucoside-5-glucoside	56	Pelargonidin 3-rutinoside
28	Cyanidin 3-(*p*-coumaroyl)-diglucoside-5-glucoside	57	Pelargonidin 3,5-diglucoside
29	Petunidin		

with petunidin 3-glucoside (code 32) being the most abundant and the only petunidin identified in the flowers. Petunidin (code 29) is the only anthocyanin of this group identified in the studied vegetables. Table S4 reveals that no peonidin was identified in flowers. In fruits, peonidin 3-glucoside (code 37) was the most abundant but also peonidin 3-galactoside (code 36) was identified in most of them. In vegetables, the most common anthocyanin of this group was peonidin (code 35). From the group of malvidin (Table S5), it can be concluded that the most abundant in fruits is malvidin 3-galactoside (code 51) followed by malvidin 3-glucoside (code 52). The last stage was also the most abundant in flowers followed by malvidin 3,5-diglucoside (code 53). In vegetables, the malvidin was the only anthocyanin quantified of this group. Table S6 shows that pelargonidin 3-glucoside (code 55) was the most abundant in fruits followed by pelargonidin 3-rutinoside (code 56). The pelargonidin 3,5-diglucoside (code 57) was the most abundant in flowers, and the pelargonidin (code 54) was the only anthocyanin of this group identified in the studied vegetables.

The protection of these pigments against the oxidation process depends on their structures. Not all of them possess the same activities to scavenge diverse reactive oxygen or nitrogen species. The antioxidant ability of anthocyanins depends on the ring orientation since it will determine the willingness to donate a proton and the capacity to transfer and electron. The number of free hydroxyls around the pyrone ring and their positions also play a key role in the antioxidant activity [2]. The presence of other types of radicals in the main structure has an important role in the antioxidant activity as well. Hence, anthocyanins, chalcones, and quinoidal bases with a double bond conjugated to the keto group are efficient antioxidants at scavenging free radicals. Also, the glycosylated B-ring structure of anthocyanins contributes to the high antioxidant activity, where orthohydroxylation and methoxylation substantially increase the antioxidant activity. Furthermore, anthocyanidins have higher antioxidant activity in comparison with anthocyanins, which has been reported in the literature. The reason may be the lower stability of the anthocyanidin compared to the anthocyanin due to its structure, which consequently makes anthocyanidin highly reactive [66]. Acylation of anthocyanin with one or more phenolic acids has a significant increase in antioxidant activity [54, 67], but glycosylation leads to a reduction in the activity [66, 68].

The efficacy of scavenging diverse free radicals differs from one anthocyanin to the other. Pelargonidin-3-glucoside, cyanidin-3-glucoside, and delphinidin-3-glucoside and their standard aglycones have strong antioxidative activity in a liposomal system and reduced formation of malondialdehyde by UVB irradiation [55, 69]. Furthermore, the results pointed out the highest inhibitory effect on lipid peroxidation and $O_2^{\bullet}$ scavenging activity of delphinidin and delphinidin-3-glucoside followed by cyanidin and pelargonidin [2]. On the contrary, pelargonidin had the highest inhibitory effect on hydroxyl radical scavenging activity [66]. Moreover, a study demonstrates the highest inhibitory effect on copper (II)-induced low-density lipoprotein (LDL) oxidation of cyanidin and cyanidin-3-glucoside compared with other phenolic acids, anthocyanins, and anthocyanin aglycones, whereas delphinidin has intermediate efficacy [70]. Another study analyzed the antioxidant activity of malvidin-3-glucoside and the result showed that the quinoidal-base and pseudo-base of malvidin-3-glucoside significantly inhibited peroxidation of linoleate compared with catechin, malvidin, and resveratrol [71]. However, the oxidation activity assigned to these anthocyanins is dependent on the type of reactive species and in consequence on the type of antioxidant assay carried out for the determination of the antioxidant activity. Thus, FRAP and TEAC assays have reported a significant reduction of the antioxidant activity by the metoxilation in the position 5 or 3 and 5 in petunidin and malvidin monoglucoside, respectively [72]. Another factor implicated in the reactivity of the anthocyanins and also in their antioxidative activity is the pH. In the literature different studies have demonstrated the effect of the pH on the antioxidant capacity of the anthocyanins from different sources; Roselle [26], wine [71, 72], black rice complexed with cycloamylose [73, 74], palm juice [75], and *Hibiscus acetosella* [76]. The pH is an important factor that should be controlled in order to determine the reactivity of the anthocyanins. The acid nature of the anthocyanin structure is shown in Figure 1.2. This acid nature

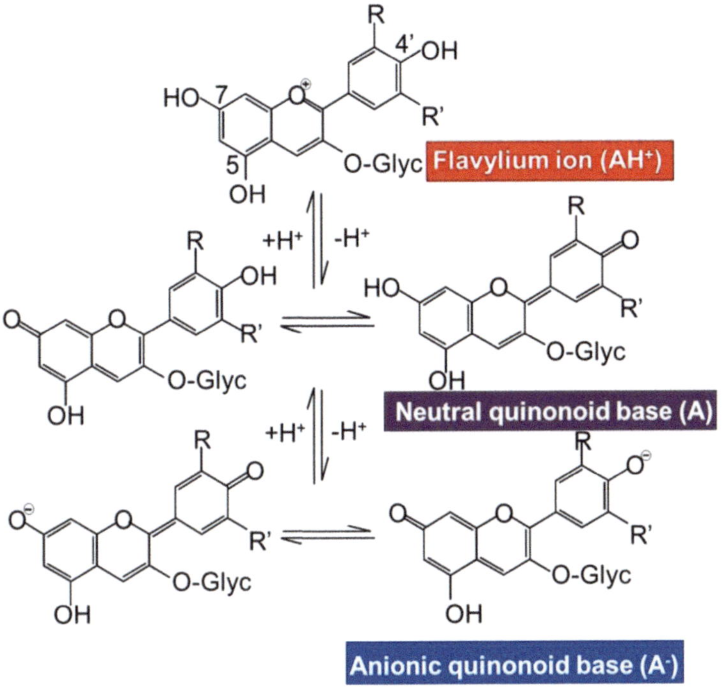

FIGURE 1.2
Effect of pH on the structure and color of anthocyanins.

is due to the conjugation of the double bonds in the rings of the main structure and the hydroxyl groups at C4′, C5, and C7, respectively. The hydroxyl group at C7 is the strongest acid. The deprotonation can be produced at acid pH~4 yielding a neutral quinonoid base stabilized by tautomerization with the hydroxyl group at C5. The hydroxyl group at C4 is also susceptible to being deprotonated at higher pH~7 yielding the anionic base. If the pH level is still rising to the basic pH, higher than 8, the deprotonation is produced in the C5 yielding the dianionic base which can lead to the chalcone anion [1]. Hence, the pH of the solution controls the proportions of protonated and deprotonated hydrated and isomeric forms of anthocyanin, affecting its reactivity. Therefore, the pH should be controlled during the extraction process of anthocyanins and also during the antioxidative bioassays because their results are pH-dependent [2, 26].

Once the identification of the most common anthocyanins in different foods has been carried out and the influence of the structure and the environment in the capacity of anthocyanins to prevent the oxidation has been discussed, the next step was to determine the bioavailability of these anthocyanins and their implication, enhancing human health.

1.4 Bioavailability of Anthocyanins

The daily intake of anthocyanins can be estimated via food databases and can range from a few to hundreds of milligrams per person due to the methodological differences in the assessment, together with the influence of nutritional, cultural, and social differences of the investigated populations [8]. The pattern followed by anthocyanins after oral dispensation is unique and different from other flavonoids [77]. Anthocyanins have a markedly low bioavailability, only 1–2% of the ingested anthocyanins maintain their parent C6–C3–C6 structure in the organism. Food digestion is a pH-dependent process and, therefore, anthocyanins are subjected to transformations in addition to hydrolyzation by several enzymes in the small intestine [12, 78]. A portion of the ingested anthocyanins reaches the large intestine, where they are metabolized into low-molecular-weight catabolites, which can be excreted in the feces within 2–4 h (up to 8 h) or absorbed again. Active transporters through either gastric or intestinal cell barrier play an important role in their transfer and absorption within the liver, kidney, brain, or other organs and tissues, besides the stomach [13, 79]. In a recent review on tissue bioavailability in animals, Sandoval-Ramírez et al. [80] concluded that the TAC absorbed was 2.17×10^5 pmol/g in mice kidney, 1.73×10^5 pmol/g in liver, 3.6×10^3 pmol/g in heart, and 1.16×10^5 pmol/g in lung; and 6.08×10^3 pmol/g in pig brain. In the wall of the intestine and then in the liver, anthocyanins and their catabolites undergo phase 2 enzymatic metabolism being also transformed into their glucuronidated, sulfated, and methylated forms [10, 12–14, 78, 81–83]. The presence of microbial catabolites at many sites of the body, at higher concentrations than the native form, has suggested that part of the biological activities attributed to anthocyanins is related to the synergetic effect of their colonic catabolites [13, 84]. Anthocyanin metabolites and transformation products have been characterized and quantified by several authors [85–89]. Ferrars et al. [88] identified a wide variety of anthocyanin phenolic metabolites, including 11 novel metabolites, in post-menopausal women after 12 weeks of elderberry intake, at concentration levels higher than their anthocyanin native forms. There are many critical factors affecting the fate of anthocyanins and their metabolites in our organism: the ability to cross membranes, pH, digestive enzymes, microbiota, biliary acids, or food matrix. The use of radiolabeled (^{14}C) or stable-isotope-labeled (^{13}C) tracer studies provides useful information about to which extent anthocyanins are metabolized to phenolic acid derivatives. In this sense, Czank et al. [90] investigated the fate of anthocyanins in eight male participants after the ingestion of ^{13}C-cyanidin-3-O-glucoside (500 mg). The relative mean bioavailability was 12.38% (5.37% excreted in urine and 6.91% in breath). The authors found a maximum serum concentration 42-fold higher for ^{13}C-labeled metabolites than their respective native compound ^{13}C-cyanidin-3-glucoside. Up to 49 metabolites were detected including among others: phase II conjugates of cyanidin-3-glucoside and cyanidin (cyanidin-glucuronide, methyl cyanidin-glucuronide, and methyl cyanidin-3-glucoside-glucuronide); degradation products (protocatechuic acid, phloroglucinaldehyde, and phloroglucinaldehyde); phase II conjugates of protocatechuic acid, phenylacetic acids, phenylpropenoic acids, and hippuric acid.

The mechanisms through which anthocyanins may exert their bioactivity are not fully understood as it is not clear whether their activity is linked to native forms, their derivatives, or both. The distinction of their different biological roles is a very challenging task. Some comparative studies have been conducted on the antioxidant activity of anthocyanin metabolites [91]. Recently, Kim et al. [92] provided basic information on the chemical changes of cyanidin glycosides during in vitro gastrointestinal digestion. Cyanidin-3-O-galactoside was degraded into caffeoylquinic acid, which was not found after in vitro digestion of cyanidin-3-O-glucoside. The bioactivity (DPPH) of the anthocyanin metabolites decreased in the intestinal fraction. However, the bioactivity increased after simulated colonic digestion, possibly because of the newly formed colonic metabolites. Furthermore, anthocyanin metabolites from the chokeberry extract exhibited higher DPPH radical activities than those from the mulberry extract. In another study, α-glucosidase inhibitory

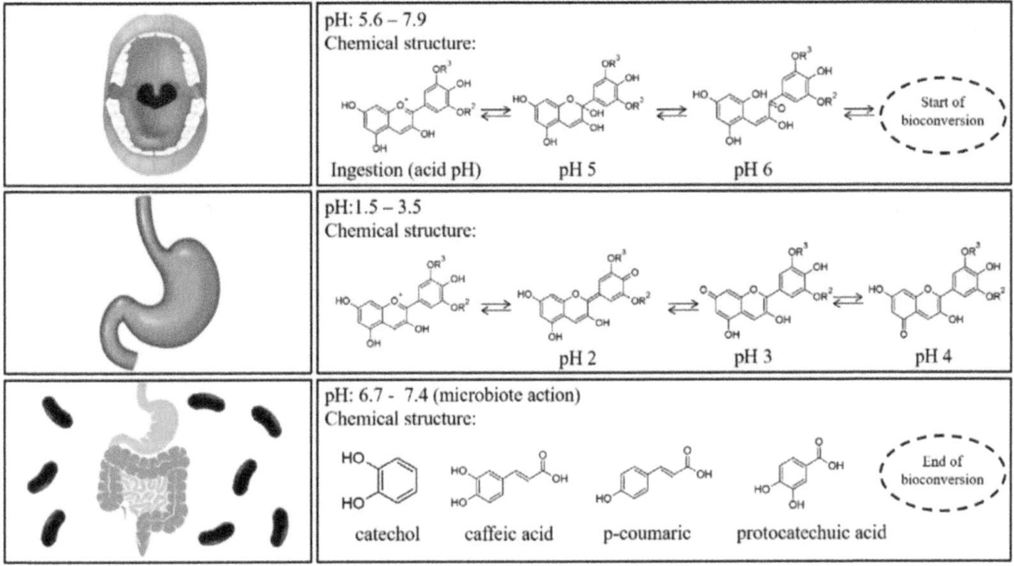

FIGURE 1.3
Schematic representation of the anthocyanins chemical structures influenced by the digestion process steps (r_2 and r_3 = h or methyl) (Taken from Braga et al. [14], with permission of Elsevier).

activity and ROS scavenging activities of conjugated-pelargonidin-3-*O*-glucoside samples were potentially increased after gastrointestinal digestion [93].

A scheme of the physicochemical reactions observed during the three main stages of the human digestion process can be observed in Figure 1.3 [11]. Biotransformation reactions start in the oral cavity through salivary amylase (pH 5.6–7.9). Once in the stomach at pH 1.5–3.5, anthocyanins exist in multiple ionic forms being mainly present as red flavylium cations and quinoidal blue species. Finally, in the intestinal step (pH 6.7–7.4), anthocyanins are present as colorless carbinol (with limited absorption) and occur the biotransformation into low-molecular-weight molecules such as phenolic acids or catechol (gallic acid, vanillic acid, protocatechuic acid, 4-hydroxybenzoic acid, and syringic acid have been identified as the main degradation products of delphinidin-3-*O*-glucoside, peonidin-3-*O*-glucoside, cyanidin-3-*O*-glucoside, pelargonidin-3-*O*-glucoside, and malvidin-3-*O*-gluco-side, respectively) [13].

For a better understanding of the anthocyanin bioavailability, different in vivo and in vitro models simulating digestion have been proposed [11]. Gowd et al. [94] assessed the phenolic profile of blackberry anthocyanin extract followed by human gut microbiota fermentation at different time intervals (0–48 h). Authors revealed the formation of gut metabolites enhances the high glucose plus palmitic-acid-induced ROS, mitochondrial membrane collapse, and glutathione depletion in HepG2 cells. Several studies have also reported that after anthocyanin colonic fermentation occurs an increase of beneficial bacteria (*Bifidobacterium* spp., *Actinobacteria*, *Bacteroidetes*, *Lactobacillus*/*Enterococcus* spp., *Akkermansia*) [95–100]. Intestinal microbiota possesses β-glucosidase activity, allowing the release of glucose from the aglycone and providing energy to support bacterial growth. A study recently carried out by Zhou et al. [95] suggests that the consumption of blueberry and its extracts could exert prebiotic activity and a modulatory effect on the composition and abundance of human intestinal microbiota. Anthocyanins could enhance human

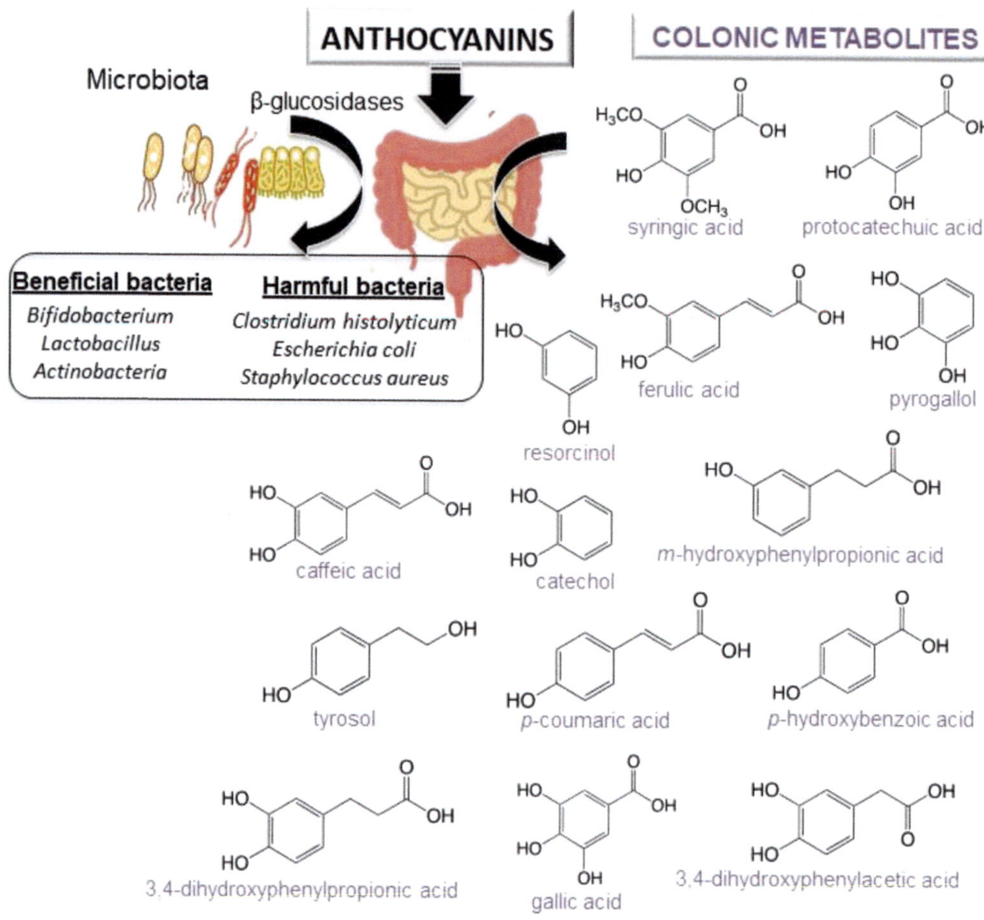

FIGURE 1.4
Scheme of anthocyanins metabolites from colonic metabolism and colon microbiota alteration.

health by modulating gut microorganisms, which are often related to different diseases [95, 101]. Nevertheless, it is important to note that anthocyanin derivatives can also reduce some harmful bacteria such as *C. histolyticum* after colonic fermentation [101, 102]. A summary of anthocyanins colonic metabolism metabolites and colon microbiota alteration is shown in Figure 1.4 [11].

In any case, there are still many doubts about the mechanisms involved and which factors have a crucial impact on bioavailability [7]. Anthocyanins with an efficient effect for one individual may not have the same effect for another [12], and there is still a high variability in the results obtained. Based on the literature and recent reviews [11, 14], this variability is due to the lack of homogeneity introduced at three levels: (i) food matrix and food processing; (ii) enzymatic levels (affected by genetic factors and diet, age, and sex); and (iii) microbiota functionality. Reported data considering inter- or intra-individual variability is very scarce and bioavailability methods are not standardized making it very difficult to reach firm conclusions. On the one hand, in vitro methods (cell-based assays) fail to consider the role of the individual microbiota present in the human body; while, on the other hand, in in vivo trials (human trials and animal studies) each subject has their own microbiota [14].

It is also important to note that the incorporation of anthocyanins into food and medical products is a challenging task due to their high instability and susceptibility to degradation. In this sense, the use of nano/microencapsulation with natural polymers is one of the best strategies to improve the stability of sensitive substances in in vitro simulated gastrointestinal digestion and colonic fermentation [103, 104]. According to a recent review on this topic [104] different techniques have been tested to encapsulate anthocyanins including spray-drying > freeze-drying > gelation > lipid-based particles > electrohydrodynamic processes. The first one is the most economical, simplest, and the most applied method (80–90%) [105, 106]. The use of other techniques still remains poorly explored probably due to the hydrophilic nature of anthocyanins, being therefore a promising area of future research [104, 107–109].

Blackberry anthocyanins encapsulated with β-cyclodextrin [101, 110] or gum arabic [111] helped to delay the release of anthocyanins during in vitro simulated gastrointestinal digestion. The stability of anthocyanins can be also influenced by the type of wall material. Recently, Wu et al. [103] evaluated the effect of four different wall materials during in vitro simulated digestion and colonic fermentation. The encapsulation technique enhanced significantly the colonic accessibility and delayed the release of anthocyanins, especially for soy protein. Degradation products of anthocyanins such as syringic acid produced during colonic fermentation by the action of gut microbiota were indicative of their benefits for host health.

1.5 Therapeutic Effects of Anthocyanins

Available scientific studies prove the beneficial effects of the presence of anthocyanins in fruits and vegetables in the prevention of diseases [60, 66, 81, 112]. Even after the ingestion of high doses of anthocyanin and derivatives, no negative effects have been observed [113]. This section covers the main health benefits of anthocyanins in different types of pathologies including eye health, cardiovascular disease, antiobesity, antidiabetic, antimicrobial effects, anticancer activities, and neurodegenerative disorders. A summary of the positive effects of anthocyanins is shown in Table 1.3 [114–160], and their mechanisms of action in disease prevention are discussed below.

Eye health: Since the first report in 1966 about the positive effects of anthocyanins on vision in humans, anthocyanin-rich extracts have been worldwide utilized as a popular supplement for ocular health [161, 162]. Oral dispensation of blackcurrant anthocyanins may be a promising supplement for patients with open-angle glaucoma, being also effective for antiglaucoma medication, while anthocyanin-rich bilberry extract has a protective effect on vision during retinal inflammation [115]. It has also been confirmed that cyanidin helps the regeneration of rhodopsin and smooth muscle relaxation in rats [116]. Results have also shown that bilberry extracts were able to suppress the photooxidation of pyridinium disretinoid A2E, an auto-fluorescence pigment that accumulates in retinal epithelial cells with age and can cause light-induced damage to the cell. In a comparative study, a significant improvement in nocturnal visual function and an improved contrast sensitivity levels in subjects with myopia versus placebo group was observed [114]. Anthocyanins act also in inhibiting transient myopia, reducing eye fatigue or enhancing retinal blood flow with glaucoma [118, 121, 161, 163].

Cardiovascular diseases: It is especially important the role of anthocyanins in preventing myocardial infarction and cardiovascular disease related to mortality. Extracts of

TABLE 1.3

Health Benefits of Anthocyanins

Eye Health	Administration	References
Improvement of vision in patients with open-angle glaucoma	Oral capsule	[114]
Protective effect during retinal inflammation	IV in rats	[115]
Regeneration of rhodopsin and smooth muscle relaxation	IV in mouse model	[116]
Improvement of dark adaptation	Oral capsule	[117]
Prevention of cataractogenesis of diabetic cataract	Incubation of Enucleated rat lenses	[118]
Antiapoptotic effects against oxidative damage of lens epithelial cell	Cell studies	[119]
Prevention of retinal degeneration induced by N-methyl-N-nitrosourea	Oral solution	[120]
Increase of ocular blood flows	Oral capsule	[121]
Cardiovascular diseases		
Inhibition of platelet aggregation (in vitro antithrombotic properties)	Cell studies	[122]
Increase of high-density lipoprotein cholesterol levels and decrease of low-density lipoprotein cholesterol levels	Oral capsule	[123]
Lower risk of nonfatal myocardial infarction	Oral intake	[124]
Vasorelaxation properties in isolated coronary artery rings in pigs	Cell studies	[125]
Decrease of susceptibility to ischemia-reperfusion injury and infarct size	Rodent food	[126]
Improvement of lipid profile and platelet function	Oral capsule	[127]
Antiobesity effects		
Improvement of weight gain and lipid profile on obese rats	Fat-diet-induced mouse model	[128]
Suppression of body weight gain and improve blood lipid profile in rats	Fat-diet-induced mouse model	[129]
Reduction of sugar concentration in urine and plasma in rats	Intraperitoneal and intragastric administration	[130]
Ameliorated obesity in high-fat-fed mice	Cell studies	[131]
Upregulation of adipocytokine secretion and gene expression in rat adipocytes	Cell studies	[132]
Suppression of fat tissue gain, weight gain, and other metabolic disorders	Fat-diet-induced mouse model	[133]
Antidiabetic effects		
Amelioration of hyperglycemia and insulin sensitivity in diabetic mice	Fat-diet-induced mouse model	[134]
Improvement of dyslipidemia, enhancement of antioxidant capacity, and prevention of insulin resistance in human with type 2 diabetes	Oral capsule	[135]
Alleviation of glomerular angiogenesis of diabetic kidneys in mice	Cell studies	[136]
Inhibition of DPP IV activity (a protease that regulates blood glucose levels via degradation of incretins)	Computational studies	[137]
Amelioration of renal apoptosis in diabetic nephropathy mice	Oral solution	[138]
Activation of adipose tissue-derived adiponectin to defend against diabetes-related endothelial dysfunction in mice	Diet-induced mouse model	[139]

(Continued)

TABLE 1.3 *(Continued)*

Health Benefits of Anthocyanins

Eye Health	Administration	References
Antimicrobial effects		
Induction of cell damage by destroying the cell wall, membrane, and intercellular matrix	Cell studies	[140]
Highest sensitivity to *Aeromonas hydrophila* and *Listeria innocua*	Microbial strains	[141]
Antibacterial effects toward *Enterococcus faecium* resistant to vancomycin, *Pseudomonas aeruginosa*, *Staphylococcus aureus*, and *Escherichia coli*	Microbial strains	[142]
Inhibition of Gram-negative bacteria	Microbial strains	[143]
Anticancer effects		
Suppression of cell proliferation, inflammation, and angiogenesis and induction of apoptosis in esophageal tissue of rats	Diet-induced rat model	[144]
Anti-invasive potential in breast cancer cell lines	Cell studies	[145]
Anticancer effect on BALB/c nude mice bearing MDA-MB-453 cell xenografts and breast cancer cell lines	Cell studies	[146]
Inhibition of cell migration and invasion, suppression of activation of rapidly accelerated fibrosarcoma, mitogen-activated protein kinase and c-Jun N-terminal kinase, and downregulation of secretion of matrix metalloproteinase 2	Cell studies	[147]
Inhibition of growth of human HT-29 colon cancer cells, increase of expression of tumor suppression genes, and decrease of cyclooxygenase-2 gene expression	Cell studies	[148]
Reduction of colonic aberrant crypt foci, colonic cellular proliferation, and COX-2 mRNA expression in rats	Diet-induced rat model	[149]
Suppression of formation of aberrant crypt foci in colons of CF-1 mice	Cell studies and diet-induced rat model	[150]
Promotion of apoptosis in benign prostatic hyperplasia rats	Oral doses in rat model	[151]
Anti-invasive effect on human hepatoma Hep3B cells and inhibition of matrix metalloproteinase MMP-2 and MMP-9 gene expression	Cell studies	[152]
Inhibition of Akt-mTOR signaling thereby inducing maturation of acute myeloid leukemia cells, besides inducing apoptotic players such as TRAIL in cancer systems	Cell studies	[153]
Neurodegenerative diseases		
Neuroprotective activity by suppression of dopaminergic cell death in Parkinson's disease	Cell studies	[154]
Improvement of learning and memory ability in mice. Higher antioxidant enzyme activity and less lipid oxidation in both brain and liver	Diet-induced mouse model	[155]
Regulation of cholinergic neurotransmission to restore Na+, K$^+$-ATPase, and Ca^{2+}-ATPase activities and to prevent memory deficits in rats	Oral and injected rat models	[156]
Neuroprotective effect: memory and synaptic dysfunction	Oral rat models	[157]
Improvement of its free radical scavenging capabilities via p38/JNK pathway against Abeta1-42-induced oxidative stress	Cell studies	[158]
Enhancement of neuroprotection against Abeta1-42-induced neuroinflammation and neurodegeneration	Oral mouse model and cell studies	[159]
Enhancement of the neuroprotection in an Abeta1-42 mouse model of Alzheimer's disease	Oral mouse model and cell studies	[160]

anthocyanins have been used to inhibit platelet aggregation being preventive in the initial stage of thrombi; in the treatment of problems with poor micro-circulation resulting from capillary fragility; and also to prevent the LDL oxidation [122, 164–166]. In a placebo-controlled trial in dyslipidemia patients (40–65 years), the intake of berry-derived anthocyanins improved lipoprotein profile through cholesteryl ester transfer protein inhibition [123]. Authors observed a greater increase in high-density lipoprotein (HDL) cholesterol levels and in the cellular cholesterol efflux to serum as well as a decrease in LDL cholesterol levels in the anthocyanin group in contrast to the placebo group. Similar results were reported by Álvarez Suárez et al. [127] in an in vivo study using healthy volunteers supplemented with strawberries (500 g). Daily consumption improved the lipid profile reducing total cholesterol, LDL cholesterol, and triglyceride levels, while HDL cholesterol remained unchanged. This increased antihemolytic defenses and platelet function in the subjects. In another attempt, higher intakes of fruit-based anthocyanins were associated with a lower risk of nonfatal myocardial infarction (14%) and ischemic stroke in a prospective cohort study of men over 24 years [124]. A meta-analysis of 45 randomized controlled trials stated that the consumption of berries and purified anthocyanins (2.2–1230-mg anthocyanins/day) increases significantly HDL-cholesterol and reduces LDL-cholesterol, triglycerides, systolic blood pressure, and diastolic blood pressure as well as the inflammatory markers CRP and TNFα [167]. The analysis also suggested that some individuals are more susceptible to the protective effects of anthocyanin consumption: (i) those overweight; (ii) those over 50 years; and (iii) those with an increased risk of cardiovascular disease. Another meta-analysis of 99 randomized controlled trials showed that the consumption of anthocyanin-rich products decreased significantly both systolic and diastolic blood pressure regardless of the health status of the participants [168].

In in vitro assays, anthocyanins have also shown inhibition of the porcine pancreatic elastase [169], an enzyme that plays a significant function in pathologies such as arteriosclerosis, emphysema, or rheumatoid arthritis by attacking fibers and collagen. Moreover, acceleration in the cicatrization process due to anthocyanin-rich extract has been demonstrated, showing preventive and curative activity against gastroduodenal ulcers induced in rats [7]. Their influence on the biosynthesis of mucopolysaccharides provably improves the efficacy of the gastric mucous layer and increases the base substance of the connective tissue and of the capillaries [170].

Antiobesity and antidiabetic effects: Anthocyanins have shown antiobesity effects through multiple mechanisms such as inhibiting lipid absorption, regulating lipid metabolism, increasing energy expenditure, suppressing food intake, and regulating gut microbiota, which suggests anthocyanins are promising candidates in antiobesity therapies [171]. Kwon et al. [128] observed that an anthocyanins-added diet from black soybean in rats decreases body weight gains, being significantly lowered in the rats fed with a high-fat diet plus black soybean anthocyanins compared with the rats fed with a high-fat diet without black soybean. Anthocyanins also improved the lipid profile and suppressed the high-fat-diet-induced weight gain in the liver intermediately and decreased the weights of epididymal and perirenal fat pads.

In addition, type 2 diabetes is closely related to obesity [66]. Anthocyanins can alleviate complications in type 2 diabetes by inhibiting intestinal glucose absorption, inducing pancreatic insulin secretion, upregulating glucose transporter type 4, and suppressing hepatic gluconeogenesis [172]. After the supplementation of a high-fat diet for 13 weeks with different berries in mice, Heyman et al. [173] observed that those supplemented mice gained lesser body weight and presented lower fasting insulin levels than the control group as well as mediated positive effects on glucose homeostasis. Jankowski et al. [130] described a substantial decrease in the sugar concentration in urine and blood serum after streptozotocin injection in fed rats with grapes. The mechanisms of anthocyanins suggested by the authors were the reduction of the biosynthesis of collagen, lipoproteins, and glycoproteins, as well as the reduction of the activity of elastase and adenosine deaminase (both high in diabetic patients). Treatment with cherries in rats resulted in a significant reduction of blood glucose and urinary microalbumin and an increase of the creatinine secretion level in urea [174]. The pulp, seed, and skin from "red chilto" (a red fruit from Argentina) had a hypoglycemic effect and acted increasing glucose absorption, decreasing glucose diffusion rate, and promoting glucose transport across the cell membrane [175] in an in vitro simulated gastroduodenal digestion. Consumption of blueberries and apples/pears in humans was also associated with a lower risk of type 2 diabetes [176].

Antimicrobial effects: The antimicrobial activity of anthocyanins against a wide range of microorganisms is also well documented. Possible mechanisms induced cell damage by destroying the cell wall, membrane, and intercellular matrix [66, 140, 177]. Blackberry extracts have antibacterial activity with the highest sensitivity to *Aeromonas hydrophilia* and *Listeria innocua* [141]. Cranberry extracts have antibacterial activity toward *Enterococcus faecium* resistant to vancomycin, *Pseudomonas aeruginosa*, *Staphylococcus aureus*, and *Escherichia coli* [142]. Different types of berry extracts inhibit Gram-negative bacteria but not Gram-positive bacteria [143] probably because Gram-negative bacteria act as a preventive barrier against hydrophobic compounds but not against hydrophilic compounds [178].

Anticancer activity: Possible mechanisms of the anticancer activity of anthocyanins have been described by many authors: antimutagenic activity; inhibition of oxidative DNA damage and carcinogen activation; induction of phase II enzymes for detoxification; cell cycle arrest; inhibition of cyclooxygenase-2 enzymes; as well as induction of apoptosis and antiangiogenesis [179–184].

In breast cancer, anthocyanins cause the inhibition of key modulators that promote its progression and development by acting directly in the DNA fragmentation and promoting the death of MCF-7 cancer cells [185, 186]. In addition, the studies indicate that anthocyanins exert extensive in vitro anti-invasive and in vivo antimetastatic activities. For example, delphinidin can act as a potential antimetastatic agent that suppresses PMA-induced cancer cell invasion through the specific inhibition of NF-κB-dependent MMP-9 gene expression [187, 188]. In lung cancer, the treatment of cyanidin-3-glucoside and cyanidin 3-rutinoside, isolated from mulberry, inhibits the migration and invasion of A549 cells and also decreases MMP-2 and uPA and enhances TIMP-2 and PAI. Anthocyanins also inhibit the growth of carcinogenic cells that provoke colon cancer, induce the apoptosis effect, and are even able to act as modulators of the macrophages in the immune response [180]. Forester et al. [189] also reported the positive effect of anthocyanin metabolites decreasing cell viability and causing cell cycle arrest and apoptosis in colon cancer. In oral and cervical cancer, the invasion of SCC-4 cells and HeLa cells was diminished by the treatment of peonidin 3-glucoside and cyanidin-3-glucoside [190].

It is also important to note that the structures of anthocyanins have a considerable influence on their biological activities [191–193]. In this sense, the type of aglycones, sugars, and acylated acids, and the position and degree of glycosylation and acylation seem to be the main factors influencing the anticancer property [191]. Jing et al. [192] compared the anticancer properties of anthocyanin-rich extracts using human colon cancer HT29 cell line. Authors reported the following growth inhibitory activity rates: purple corn > chokeberry and bilberry > purple carrot and grape > radish and elderberry. Those non-acylated monoglycosylated anthocyanins had greater anticancer properties than those with pelargonidin, triglycoside, and/or acylation with cinnamic acid.

Neurodegenerative diseases: Anthocyanins are also uniquely suited for the treatment of neurodegenerative diseases such as Alzheimer's, Parkinson's, or amyotrophic lateral sclerosis. Their main mechanisms include antioxidant pathways, calcium homeostasis, inflammation, protein homeostasis, and the balance of pro-survival and pro-apoptotic signaling [194, 195].

In a primary cell model of Parkinson's disease, dopaminergic cell death elicited by rotenone was suppressed by extracts prepared from blueberries, grape seed, hibiscus, blackcurrant, and mulberry [154]. Moreover, Strathearn et al. [154] observed that those extracts rich in anthocyanins and proanthocyanidins exhibited greater neuroprotective activity than extracts rich in other polyphenols.

The oral dispensation of anthocyanins (200 mg/kg) in rats was able to regulate cholinergic neurotransmission, to restore Na^+, K^+-ATPase, and Ca^{2+}-ATPase activities, and prevent memory deficits caused by scopolamine dispensation [156]. Rehman et al. [157] showed the neuroprotective effect of anthocyanins based on an artificial aging model using D-galactose to induce oxidative stress and inflammatory response. The potential mechanisms of their action included: decreased expression of the receptor for advanced glycation end product, reduced level of ROS, and lipid peroxidation. Shih et al. [155] observed that mice fed with anthocyanin-rich mulberry extracts demonstrated significantly less amyloid β protein and showed improvement in learning and memory ability in avoidance response tests. The fed mice also showed a higher antioxidant enzyme activity and less lipid oxidation in both the brain and liver, as compared to the control mice. Besides, the treatment with anthocyanin-rich mulberry extract has been proven to decrease the levels of serum aspartate aminotransferase, alanine aminotransferase, triglyceride, and total cholesterol that increase with aging.

Furthermore, the therapeutic profile of anthocyanins can be improved by encapsulation [158–160]. For instance, in Alzheimer's disease Amin et al. [158] showed that encapsulated nanoparticles loaded with anthocyanins are rapidly taken up by cells enhancing their neuroprotective profile against amyloid beta toxicity above that of anthocyanins alone. Similar activity was also observed in in vivo studies in mice [158, 160].

1.6 Conclusions

In order to establish the antioxidant activity of anthocyanins and how their intake affects human health, many factors should be taken into account. Firstly, an evaluation of the antioxidant activity from a multiparametric perspective is required as the total concentration, the structure, the nature of the sample, the pH, and the mechanism of the reaction play an important role in their effect. On a regular basis, more than one antioxidant assay—one for each mechanism—should be carried out. Besides identification and quantification of the anthocyanins is also highly recommended to establish an accurate value for the antioxidant activity of a sample. Secondly, it is required to know the bioavailability of these compounds after their intake. Many studies have demonstrated the benefits of anthocyanin-rich extracts in the prevention of diseases. Nonetheless, it is important to note that their efficacy depends on their bioavailability. Along the digestion process, anthocyanins are metabolized into various conjugates, which then ultimately metabolize into phenolic acid degradation products as well. The accumulated evidence suggests the synergy effect between all possible forms to explain their attributed health-promoting properties. An inter- and intra-individual variability in anthocyanins absorption, metabolism, distribution, and excretion is also evident. Among the main factors that probably affect this variability are: food matrix and processing, enzymatic levels, and microbiota functionality. Attention should be paid in different lines: (i) to perform well-designed standardized methods when evaluating bioavailability; (ii) to consider inter or intra-individual variability in anthocyanin metabolism; (iii) to assess the antioxidant activity of anthocyanin metabolites during gastrointestinal digestion; and, since anthocyanin might exert different biological activities, (iv) to draw a more accurate characterization profile.

Supplementary Materials

The following are available online at http://www.mdpi.com/2076-3921/9/5/451/s1, Table S1: Natural source, type of extraction, chromatographic method, and antioxidant assay applied to identify and quantify delphinidin and its derivatives identified with codes, Table S2: Natural source, type of extraction, chromatographic method, and antioxidant assay applied to identify and quantify cyanidin and its derivatives identified with codes, Table S3: Natural source, type of extraction, chromatographic method, and antioxidant assay applied to identify and quantify petunidin and its derivatives identified with codes, Table S4: Natural source, type of extraction, chromatographic method, and antioxidant assay applied to identify and quantify peonidin and its derivatives identified with codes, Table S5: Natural source, type of extraction, chromatographic method, and antioxidant assay applied to identify and quantify malvidin and its derivatives identified with codes, Table S6: Natural source, type of extraction, chromatographic method, and antioxidant assay applied to identify and quantify pelargonidin and its derivatives identified with codes.

Author Contributions

Investigation, N.T., J.M., and A.G.A.; Project administration, A.G.A.; Supervision, A.G.A.; Writing—original draft, N.T. and J.M.; Writing—review and editing, N.T. and J.M. All authors have read and agreed to the published version of the manuscript.

Funding

This research received no external funding.

Acknowledgments

The authors would like to thank the Vicerrectorado de Investigación de la Universidad de Sevilla for its support. We would also like to thank the reviewers for their helpful comments that have greatly contributed to the improvement of the document. We are also grateful for the very helpful revision by Irene García Sáez.

Conflicts of Interest

The authors declare no conflict of interest.

References

1. Dangles, O.; Fenger, J.A. The chemical reactivity of anthocyanins and its consequences in food science and nutrition. *Molecules.* 2018, *23*, 1970.
2. Miguel, M.G. Anthocyanins: Antioxidant and/or anti-inflammatory activities. *J. Appl. Pharm. Sci.* 2011, *1*, 7–15.
3. Kay, C.D.; Pereira-Caro, G.; Ludwig, I.A.; Clifford, M.N.; Crozier, A. Anthocyanins and flavanones are more bioavailable than previously perceived: A review of recent evidence. *Annu. Rev. Food Sci. Technol.* 2017, *8*, 155–180.
4. Lila, M.A.; Burton-Freeman, B.; Grace, M.; Kalt, W. Unraveling anthocyanin bioavailability for human health. *Annu. Rev. Food Technol.* 2016, *7*, 375–393.
5. Fang, J. Bioavailability of anthocyanins. *Drug Metab. Rev.* 2014, *46*, 508–520.
6. Lingua, M.S.; Fabani, M.P.; Wunderlin, D.A.; Baroni, M.V. From grape to wine: Changes in phenolic composition and its influence on antioxidant activity. *Food Chem.* 2016, *208*, 228–238.
7. Martín, J.; Kuskoski, E.M.; Navas, M.J.; Asuero, A.G.. Antioxidant Capacity of Anthocyanin Pigments. In *Flavonoids—From Biosynthesis to Human Health*; Justino, J., Ed.; Science, Technology and Medicine Open Access Publisher: Rijeka, Croatia, 2017; Chapter 11; pp. 205–255.
8. Martín Bueno, J.; Sáez-Plaza, P.; Ramos-Escudero, F.; Jímenez, A.M.; Fett, R.; Asuero, A.G. Analysis and antioxidant capacity of anthocyanin pigments. Part II: Chemical structure, color, and intake of anthocyanins. *Crit. Rev. Anal. Chem.* 2012, *42*, 126–151.
9. Navas, M.J.; Jiménez-Moreno, A.M.; Martín Bueno, J.; Sáez-Plaza, P.; Asuero, A.G. Analysis and antioxidant capacity of anthocyanin pigments. Part IV: Extraction of anthocyanins. *Crit. Rev. Anal. Chem.* 2012, *42*, 313–342.
10. Gamel, T.H.; Wright, A.J.; Tucker, A.J.; Pickard, M.; Rabalski, I.; Podgorski, M.; Di Ilio, N.; O'Brien, C.; Abdel-Aal, E.M. Absorption and metabolites of anthocyanins and phenolic acids after consumption of purple wheat crackers and bars by healthy adults. *J. Cereal Sci.* 2019, *86*, 60–68.
11. Cavalcante Braga, A.R.; Murador, D.C.; Mendes de Souza Mesquita, L.; Vera de Rosso, V. Bioavailability of anthocyanins: Gaps in knowledge, challenges and future research. *J. Food Compos. Anal.* 2018, *68*, 31–40.
12. Fernandes, I.; Faria, A.; Calhau, C.; de Freitas, V.; Mateus, N. Bioavailability of anthocyanins and derivatives. *J. Funct. Foods.* 2014, *7*, 54–66.
13. Fernandes, I.; Faria, A.; de Freitas, V.; Calhau, C.; Mateus, N. Multiple-approach studies to assess anthocyanin bioavailability. *Phytochem. Rev.* 2015, *14*, 899–919.
14. Eker, M.E.; Aaby, K.; Budic-Leto, I.; Rimac Brnčic, S.; El, S.N.; Karakaya, S.; Simsek, S.; Manach, C.; Wiczkowski, W.; de Pascual-Teresa, S. A review of factors affecting anthocyanin bioavailability: Possible implications for the inter-individual variability. *Foods.* 2020, *9*, 2.
15. Mülleder, U.; Murkovic, M.; Pfannhauser, W. Urinary excretion of cyanidin glycosides. *J. Biochem. Biophys. Methods.* 2002, *53*, 61–66.
16. Bub, A.; Watzl, B.; Heeb, D.; Rechkemmer, G.; Briviba, K. Malvidin-3-glucoside bioavailability in humans after ingestion of red wine, dealcoholized red wine and red grape juice. *Eur. J. Nutr.* 2001, *40*, 113–120.
17. Hong, H.T.; Netzel, M.E.; O'Hare, T.J. Optimization of extraction procedure and development of LC–DAD–MS methodology for anthocyanin analysis in anthocyanin-pigmented corn kernels. *Food Chem.* 2020, *319*, 126515.

18. Castañeda-Ovando, A.; Pacheco-Hernandez, M.L.; Paez-Hernandez, M.E.; Rodriguez, J.A.; Galan-Vidal, C.A. Chemical studies of anthocyanins: A review. *Food Chem.* 2009, *113*, 859–871.

19. Fernandez-Aulis, F.; Hernandez-Vazquez, L.; Aguilar-Osorio, G.; Arrieta-Baez, D.; Navarro-Ocan, A. Extraction and identification of anthocyanins in corn cob and corn husk from cacahuacintle maize. *J. Food Sci.* 2019, *84*, 954–962.

20. Siti Azima, A.M.; Noriham, A.; Manshoor, N. Anthocyanin content in relation to the antioxidant activity and colour properties of Garcinia mangostana peel, *Syzigium cumini* and *Clitoria ternatea* extracts. *Int. Food Res. J.* 2014, *21*, 2369–2375.

21. Liang, N.; Kitts, D.D. Antioxidant property of coffee components: Assessment of methods that define mechanisms of action. *Molecules.* 2014, *19*, 19180–19208.

22. Shalaby, E.A.; Shanab, S.M.M. Antioxidant compounds, assays of determination and mode of action. *Afr. J. Pharm. Pharmacol.* 2013, *7*, 528–539.

23. Goupy, P.; Bautista-Ortin, A.-B.; Fulcrand, H.; Dangles, O. Antioxidant activity of wine pigments derived from anthocyanins: Hydrogen transfer reactions to the DPPH radical and inhibition of the heme-induced peroxidation of linoleic acid. *J. Agric. Food Chem.* 2009, *57*, 5762–5770.

24. Molyneux, P. The use of the stable free radical diphenylpicrylhydrazyl (DPPH) for estimating antioxidant activity. Songklanakarin. *J. Sci. Technol.* 2004, *26*, 211–219.

25. Sudheeran, P.K.; Feygenberg, O.; Maurer, D.; Alkan, N. Improved cold tolerance of Mango fruit with enhanced anthocyanin and flavonoid contents. *Molecules.* 2018, *23*, 1832.

26. Wu, H.Y.; Yang, K.M.; Chiang, P.Y. Roselle anthocyanins: Antioxidant properties and stability to heat and pH. *Molecules.* 2018, *23*, 1357.

27. Akar, Z.; Kucuk, M.; Dogan, H. A new colorimetric DPPH scavenging activity method with no need for a spectrophotometer applied on synthetic and natural antioxidants and medicinal herbs. *J. Enzym. Inhib. Med. Chem.* 2017, *32*, 640–647.

28. Ehlenfeldt, M.K.; Prior, R.L. Oxygen radical absorbance capacity (ORAC) and phenolic and anthocyanin concentrations in fruit and leaf tissues of highbush blueberry. *J. Agric. Food Chem.* 2001, *49*, 2222–2227.

29. Bellido, G.G.; Beta, T. Anthocyanin composition and oxygen radical scavenging capacity (ORAC) of milled and pearled purple, black, and common barley. *J. Agric. Food Chem.* 2009, *57*, 1022–1028.

30. Zheng, Y.; Wang, S.Y.; Wang, C.Y.; Zheng, W. Changes in strawberry phenolics, anthocyanins, and antioxidant capacity in response to high oxygen treatments. *LWT.* 2007, *40*, 49–57.

31. Denev, P.; Ciz, M.; Ambrozova, G.; Lojek, A.; Yanakieva, I.; Kratchanova, M. Solid-phase extraction of berries' anthocyanins and evaluation of their antioxidative properties. *Food Chem.* 2010, *123*, 1055–1061.

32. Rossetto, M.; Vanzani, P.; Lunelli, M.; Scarpa, M.; Mattivi, F.; Rigo, A. Peroxyl radical trapping activity of anthocyanins and generation of free radical intermediates. *Free Radic. Res.* 2007, *41*, 854–859.

33. Dresch, R.R.; Kreinecker-Dresch, M.T.; Biegelmeyer, R.; Fretes-Argenta, D.; Fagundes da Rocha, R.; Ferreira-Teixeira, H.; Fonseca-Moreira, J.C.; Henriques, A.T. Potential use of secondary products of the agri-food industry for topical formulations and comparative analysis of antioxidant activity of grape leaf polyphenols. *Nat. Prod. Res.* 2018, *32*, 486–492.

34. Philpott, M.; Lim, C.C.; Ferguson, L.R. Dietary protection against free radicals: A case for multiple testing to establish structure-activity relationships for antioxidant potential of anthocyanic plant species. *Int. J. Mol. Sci.* 2009, *10*, 1081–1103.

35. Su, M.S.; Chien, P.J. Antioxidant activity, anthocyanins, and phenolics of rabbiteye blueberry (*Vaccinium ashei*) fluid products as affected by fermentation. *Food Chem.* 2007, *104*, 182–187.

36. Özgen, M.; Scheerens, J.C.; Neil Reese, R.; Miller, R.A. Total phenolic, anthocyanin contents and antioxidant capacity of selected elderberry (*Sambucus canadensis* L.) accessions. *Pharm. Mag.* 2010, *6*, 198–203.

37. Tsai, P.J.; Intosh, J.M.; Pearce, P.; Camden, B.; Jordan, B.R. Anthocyanin and antioxidant capacity in roselle (*Hibiscus sabdariffa* L.) extract. *Food Res. Int.* 2002, *35*, 351–356.

38. Pantelidis, G.E.; Vasilakakis, M.; Manganaris, G.A.; Diamantidis, G. Antioxidant capacity, phenol, anthocyanin and ascorbic acid contents in raspberries, blackberries, red currants, gooseberries and Cornelian cherries. *Food Chem.* 2007, *102*, 777–783.

39. Li, H.; Deng, Z.; Zhu, H.; Hu, C.; Liu, R.; Young, J.C.; Tsao, R. Highly pigmented vegetables: Anthocyanin compositions and their role in antioxidant activities. *Food Res. Int.* 2012, *46*, 250–259.

40. Moon, J.K.; Shibamoto, T. Antioxidant assays for plant and food components. *J. Agric. Food Chem.* 2009, *57*, 1655–1666.

41. Ge, Q.; Ma, X. Composition and antioxidant activity of anthocyanins isolated from Yunnan edible rose (*An ning*). *Food Sci. Hum. Wellness.* 2013, *2*, 68–74.

42. Frond, A.D.; Iuhas, C.I.; Stirbu, I.; Leopold, L.; Socaci, S.; Andreea, S.; Ayvaz, H.; Andreea, S.; Mihai, S.; Diaconeasa, Z.; et al. Phytochemical characterization of five edible purple-reddish vegetables: Anthocyanins, flavonoids, and phenolic acid derivatives. *Molecules* 2019, *24*, 1536.

43. Pasko, P.; Barton, H.; Zagrodzki, P.; Gorinstein, S.; Folta, M.; Zachwieja, Z. Anthocyanins, total polyphenols and antioxidant activity in amaranth andquinoa seeds and sprouts during their growth. *Food Chem.* 2009, *115*, 994–998.

44. Villaño, D.; Fernández-Pachón, M.S.; Troncoso, A.M.; García-Parrilla, M.C. The antioxidant activity of wines determined by the ABTS (+) method: Influence of sample dilution and time. *Talanta.* 2004, *64*, 501–509.

45. Hu, Q.P.; Xu, J.G. Profiles of carotenoids, anthocyanins, phenolics, and antioxidant activity of selected color waxy corn grains during maturation. *J. Agric. Food Chem.* 2011, *59*, 2026–2033.

46. Pala, C.U.; Toklucu, A.K. Effect of UV-c light on anthocyanin content and other quality parameters of pomegranate juice. *J. Food Compos. Anal.* 2011, *24*, 790–795.

47. Bunea, A.; Rugina, D.O.; Pintea, A.M.; Conta, Z.; Bunea, C.I.; Socaciu, C. Comparative polyphenolic content and antioxidant activities of some wild and cultivated blueberries from Romania. *Not. Bot. Horti Agrobo.* 2011, *39*, 70–76.

48. Moniruzzaman, M.; Khalil, M.I.; Sulaiman, S.A.; Gan, S.H. Advances in the analytical methods for determining the antioxidant properties of honey: A review. *Afr. J. Tradit. Complement. Altern. Med.* 2012, *9*, 36–42.

49. Degenhardt, A.; Knapp, H.; Winterhalter, P. Separation and purification of anthocyanins by high-speed countercurrent chromatography and screening for antioxidant activity. *J. Agric. Food Chem.* 2000, *48*, 338–343.

50. Rubinskiene, M.; Jasutiene, I.; Venskutonis, P.R.; Viskelis, P. HPLC determination of the composition and stability of blackcurrant anthocyanins. *J. Chromatogr. Sci.* 2005, *43*, 478–482.

51. Welch, C.R.; Wub, Q.; Simon, J.E. Recent advances in anthocyanin analysis and characterization. *Curr. Anal. Chem.* 2008, *4*, 75–101.

52. Amorati, R.; Valgimigli, L. Methods to measure the antioxidant activity of phytochemicals and plant extracts. *J. Agric. Food Chem.* 2018, *66*, 3324–3329.

53. Félix, R.; Valentão, P.; Andrade, P.B.; Félix, C.; Novais, S.C.; Lemos, M.F.L. Evaluating the in vitro potential of natural extracts to protect lipids from oxidative damage. *Antioxidants.* 2020, *9*, 231.

54. Matera, R.; Gabbanini, S.; Berretti, S.; Amorati, R.; De Nicola, G.R.; Iori, R.; Valgimigli, L. Acylated anthocyanins from sprouts of *Raphanus sativus* cv. Sango: Isolation, structure elucidation and antioxidant activity. *Food Chem.* 2015, *166*, 397–406.

55. Tsuda, T.; Watanabe, M.; Ohshima, K.; Norinobu, S.; Choi, S.W.; Kawakishi, S.; Osawa, T. Antioxidative activity of the anthocyanin pigments cyanidin 3-o-β-D-glucoside and cyanidin. *J. Agri. Food Chem.* 1994, *42*, 2407–2410.

56. EFSA Guidance for the scientific requirements for health claims related to antioxidants, oxidative damage and cardiovascular health (revision 1). *EFSA J.* 2018, *16*, 5136.

57. Prior, R.L.; Go, L.; Wu, X.; Jacob, R.A.; Sotoudeh, G.; Kader, A.A.; Cook, R.A. Plasma antioxidant capacity changes following a meal as a measure of the ability of a food to alter in vivo antioxidant status. *J. Am. Coll. Nutr.* 2007, *26*, 170–181.

58. Comert, E.D.; Mogol, B.A.; Gokmen, V. Relationship between color and antioxidant capacity of fruits and vegetables. *Curr. Res. Food Sci.* 2020, *2*, 1–10.

59. Bakuradze, T.; Tausend, A.; Galan, J.; Groh, I.A.M.; Berry, D.; Tur, J.A.; Marko, D.; Richling, E. Antioxidative activity and health benefits of anthocyanin-rich fruit juice in healthy volunteers. *Free Radic. Res.* 2019, *53*, 1045–1055.

60. Khoo, H.E.; Lim, S.M.; Azlan, A. Evidence-based therapeutic effects of anthocyanins from foods. *Pak. J. Nutr.*. 2019, *18*, 1–11.

61. Williamson, G. The role of polyphenols in modern nutrition. *Nutr. Bull.* 2017, *42*, 226–235.

62. Chaiyasut, C.; Sivamaruthi, B.S.; Pengkumsri, N.; Sirilun, S.; Peerajan, S.; Chaiyasut, K.; Kesika, P. Anthocyanin profile and its antioxidant activity of widely used fruits, vegetables, and flowers in Thailand Asian. *J. Pharm. Clin. Res.* 2016, *9*, 218–224.

63. Ogawa, K.; Sakakibara, H.; Iwata, R.; Ishii, T.; Sato, T.; Goda, T.; Shimoi, K.; Kumazawa, S. Anthocyanin composition and antioxidant activity of the crowberry (Empetrum nigrum) and other berries. *J. Agric. Food Chem.* 2008, *56*, 4457–4462.

64. Zykin, P.A.; Andreeva, E.A.; Lykholay, A.N.; Tsvetkova, N.V.; Voylokov, A.V. Anthocyanin composition and content in rye plants with different grain color. *Molecules.* 2018, *23*, 948.

65. De Morais, J.S.; Sant'Ana, A.S.; Dantas, A.M.; Silva, B.S.; Lima, M.S.; Borges, G.C.; Magnani, M. Antioxidant activity and bioaccessibility of phenolic compounds in white, red, blue, purple, yellow and orange edible flowers through a simulated intestinal barrier. *Food Res. Int.* 2020, *131*, 109046.

66. Khoo, H.E.; Azlan, A.; Tang, S.T.; Lim, S.M. Anthocyanidins and anthocyanins: Colored pigments as food, pharmaceutical ingredients, and the potential health benefits. *Food Nutr. Res.* 2017, *61*, 1361779.

67. Tamura, H.; Yamagami, A. Antioxidative activity of monoacylated anthocyanins isolated from Muscat Bailey A grape. *J. Agric. Food Chem.* 1994, *42*, 1612–1615.

68. Stintzing, F.C.; Stintzing, A.S.; Carle, R.; Frei, B.; Wrolstad, R.E. Color and antioxidant properties of cyanidin-based anthocyanin pigments. *J. Agric. Food Chem.* 2002, *50*, 6172–6181.

69. Tsuda, T.; Shiga, K.; Ohshima, K.; Kawakishi, S.; Osawa, T. Inhibition of lipid peroxidation and the active oxygen radical scavenging effect of anthocyanin pigments isolated from *Phaseolus vulgaris* L. *Biochem. Pharmacol.* 1996, *52*, 1033–1110.

70. Brown, J.E.; Kelly, M.F. Inhibition of lipid peroxidation by anthocyanins, anthocyanidins and their phenolic degradation products. *Eur. J. Lipid Sci. Technol.* 2007, *109*, 66–71.

71. Lapidot, T.; Harel, S.; Akiri, B.; Granit, R.; Kanner, J. pH-dependent forms of red wine anthocyanins as antioxidants. *J. Agric. Food Chem.* 1999, *47*, 67–70.

72. Muselík, J.; García-Alonso, M.; Martín-López, M.P.; Žemlic̆ka, M.; Rivas-Gonzalo, J.C. Measurement of antioxidant activity of wine catechins, procyanidins, anthocyanins and pyranoanthocyanins. *Int. J. Mol. Sci.* 2007, *8*, 797–809.

73. Jung, Y.K.; Joo, K.S.; Rho, S.J.; Kim, Y.R. pH-dependent antioxidant stability of black rice anthocyanin complexed with cycloamylose. *LWT Food Sci. Technol.* 2020, *129*, 109474.

74. Sui, X.; Dong, X.; Zhou, W. Combined effect of pH and high temperature on the stability and antioxidant capacity of two anthocyanins in aqueous solution. *Food Chem.* 2014, *163*, 163–170.

75. Ghosh, S.; Chakraborty, R.; Raychaudhuri, U. Determination of pH-dependent antioxidant activity of palm (*Borassus flabellifer*) polyphenol compounds by photoluminol and DPPH methods: A comparison of redox reaction sensitivity. *3 Biotech.* 2015, *5*, 633–640.

76. Março, P.H.; Poppi, R.J.; Scarminio, I.S.; Tauler, R. Investigation of the pH effect and UV radiation on kinetic degradation of anthocyanin mixtures extracted from *Hibiscus acetosella*. *Food Chem.* 2011, *125*, 1020–1102.

77. Manach, C.; Williamson, G.; Morand, C.; Scalbert, A.; Remesy, C. Bioavailability and bioefficacy of polyphenols in humans. I. Review of 97 bioavailability studies. *Am. J. Clin. Nutr.* 2005, *81*, 230–242.

78. Felgines, C.; Talavéra, S.; Texier, O.; Gil-Izquierdo, A.; Lamaison, J.-L.; Remesy, C. Blackberry anthocyanins are mainly recovered from urine as methylated and glucuronidated conjugates in humans. *J. Agric. Food Chem.* 2005, *53*, 7721–7727.

79. Passamonti, S.; Vanzo, A.; Vrhovsek, U.; Terdoslavich, M.; Cocolo, A.; Decorti, G.; Mattivi, F. Hepatic uptake of grape anthocyanins and the role of bilitranslocase. *Food Res. Int.* 2005, *38*, 953–960.

80. Sandoval-Ramírez, B.A.; Catalanń, U.; Fernanńdez-Castillejo, S.; Rubió, L.; Maciá, A.; Sola, R. Anthocyanin tissue bioavailability in animals: Possible implications for human health. A systematic review. *J. Agric. Food Chem.* 2018, *66*, 11531–11543.

81. Lucioli, S. Anthocyanins: Mechanism of action and therapeutic efficacy. In *Medicinal Plants as Antioxidant Agents: Understanding Their Mechanism of Action and Therapeutic Efficacy*; Capasso, A., Ed.; Research Signpost: Kerala, India, 2012; pp. 27–57.

82. Jin, Y.A.D.; George, T.G.M.; Lovegrove, J.A.A. Randomised trial to investigate the effects of acute consumption of a blackcurrant juice drink on markers of vascular reactivity and bioavailability of anthocyanins in human subjects. *Eur. J. Clin. Nutr.* 2011, *65*, 849–857.

83. Wu, X.; Cao, G.; Prior, R.L. Absorption and metabolism of anthocyanins in elderly women after consumption of elderberry or blueberry. *J. Nutr.* 2002, *132*, 1865–1871.

84. Williamson, G.; Clifford, M.N.. Colonic metabolites of berry polyphenols: The missing link to biological activity? *Br. J. Nutr.* 2010, *104* (Suppl. S3), S48–S66.

85. Kaiser, M.; Müller-Ehl, L.; Passon, M.; Schieber, A. Development and validation of methods for the determination of anthocyanins in physiological fluids via UHPLC-MS^n. *Molecules.* 2020, *25*, 518.

86. Ludwig, I.A.; Mena, P.; Calani, L.; Borges, G.; Pereira-Caro, G.; Bresciani, L.; Del Rio, D.; Lean, M.E.J.; Crozier, A. New insights into the bioavailability of red raspberry anthocyanins and ellagitannins. *Free Radic. Biol. Med.* 2015, *89*, 758–769.

87. Ferrars, R.M.; Cassidy, A.; Curtis, P.; Kay, C.D. Phenolic metabolites of anthocyanins following a dietary intervention study in post-menopausal women. *Mol. Nutr. Food Res.* 2014, *58*, 490–502.

88. Ferrars, R.M.; Czank, C.; Saha, S.; Needs, P.W.; Zhang, Q.; Raheem, K.S.; Botting, N.P.; Kroon, P.A.; Kay, C.D. Methods for isolating, identifying, and quantifying anthocyanin metabolites in clinical samples. *Anal. Chem.* 2014, *86*, 10052–10058.

89. Nurmi, T.; Mursu, J.; Heinonen, M.; Nurmi, A.; Hiltunen, R.; Voutilainen, S. Metabolism of berry anthocyanins to phenolic acids in humans. *J. Agric. Food Chem.* 2009, *57*, 2274–2281.

90. Czank, C.; Cassidy, A.; Zhang, Q.; Morrison, D.J.; Preston, T.; Kroon, P.A.; Botting, N.P.; Kay, C.D. Human metabolism and elimination of the anthocyanin, cyanidin-3-glucoside: A 13C-tracer study. *Am. J. Clin. Nutr.* 2013, *97*, 995–1003.

91. Novotny, J.A.. Anthocyanin Bioavailability: Past Progress and Current Challenges. In *Emerging Trends in Dietary Components for Preventing and Combating Disease*; ACS Symposium Series; American Chemical Society: Washington, DC, USA, 2012; Chapter 32; pp. 559–568.

92. Kim, I.; Moon, J.K.; Hur, S.J.; Lee, J. Structural changes in mulberry (*Morus microphylla* Buckl) and chokeberry (*Aronia melanocarpa*) anthocyanins during simulated in vitro human digestion. *Food Chem.* 2020, *318*, 126449.

93. Gowd, V.; Karim, N.; Xie, L.; Shishir, M.R.I.; Xu, Y.; Chen, W. In vitro study of bioaccessibility, antioxidant, and α-glucosidase inhibitory effect of pelargonidin-3-*O*-glucoside after interacting with beta-lactoglobulin and chitosan/pectin. *Int. J. Biol. Macromol.* 2020, *154*, 380–389.

94. Gowd, V.; Bao, T.; Chen, W. Antioxidant potential and phenolic profile of blackberry anthocyanin extract followed by human gut microbiota fermentation. *Food Res. Int.* 2019, *120*, 523–533.

95. Zhou, L.; Xiec, M.; Yanga, F.; Liu, J. Antioxidant activity of high purity blueberry anthocyanins and the effects on human intestinal microbiota. *Food Sci. Technol.* 2020, *117*, 628101.

96. Zhu, Y.; Sun, H.; He, S.; Lou, Q.; Yu, M.; Tang, M.; Tu, L. Metabolism and prebiotics activity of anthocyanins from black rice (*Oryza sativa* L.) in vitro. *PLoS ONE.* 2018, *13*, 0195754.

97. Hidalgo, M.; Oruna-Concha, M.J.; Kolida, S.; Walton, G.E.; Kallithraka, S.; Spencer, J.P.E.; Gibson, G.R.; de Pascual-Teresa, S. Metabolism of anthocyanins by human gut microflora and their influence on gut bacterial growth. *J. Agric. Food Chem.* 2012, *60*, 3882–3890.

98. Boto-Ordóñez, M.; Urpi-Sarda, M.; Queipo-Ortuño, M.I.; Tulipani, S.; Tinahones, F.J.; Andres-Lacueva, C. High levels of bifidobacteria are associated with increased levels of anthocyanin microbial metabolites: A randomized clinical trial. *Food Funct.* 2014, *5*, 1932–1938.

99. Garcia-Mazcorro, J.F.; Lage, N.N.; Mertens-Talcott, S.; Talcott, S.; Chew, B.; Dowd, S.E.; Kawas, J.R.; Noratto, G.D. Effect of dark sweet cherry powder consumption on the gut microbiota, short-chain fatty acids, and biomarkers of gut health in obese db/db mice. *Peer J.* 2018, *2018*, 1–31.

100. Lacombe, A.; Li, R.W.; Klimis-Zacas, D.; Kristo, A.S.; Tadepalli, S.; Krauss, E.; Young, R.; Wu, V.C.H. Lowbush wild blueberries have the potential to modify gut microbiota and xenobiotic metabolism in the rat colon. *PLoS ONE.* 2013, *8*, 67497.

101. Flores, G.; Ruiz del Castillo, M.L.; Costabile, A.; Klee, A.; Bigetti Guergoletto, K.; Gibson, G.R. In vitro fermentation of anthocyanins encapsulated with cyclodextrins: Release, metabolism and influence on gut microbiota growth. *J. Funct. Foods.* 2015, *16*, 50–57.

102. Guergoletto, K.B.; Costabile, A.; Flores, G.; Garcia, S.; Gibson, G.R. In vitro fermentation of juçara pulp (*Euterpe edulis*) by human colonic microbiota. *Food Chem.* 2016, *196*, 251–258.

103. Wu, Y.; Han, Y.; Tao, Y.; Li, D.; Xie, G.; Show, P.L.; Lee, S.Y. In vitro gastrointestinal digestion and fecal fermentation reveal the effect of different encapsulation materials on the release, degradation and modulation of gut microbiota of blueberry anthocyanin extract. *Food Res. Int.* 2020, *132*, 109098.

104. Sharif, N.; Khoshnoudi-Nia, S.; Jafari, S.M. Nano/microencapsulation of anthocyanins; A systematic review and meta-analysis. *Food Res. Int.* 2020, *132*, 109077.

105. Yousuf, B.; Gul, K.; Wani, A.A.; Singh, P. Health benefits of anthocyanins and their encapsulation for potential use in food systems: A review. *Crit. Rev. Food Sci. Nutr.* 2016, *56*, 2223–2230.

106. Mahdavi, S.A.; Jafari, S.M.; Ghorbani, M.; Assadpoor, E. Spray-drying microencapsulation of anthocyanins by natural biopolymers: A review. *Dry Technol.* 2014, *32*, 509–518.

107. Robert, P.; Fredes, C. The encapsulation of anthocyanins from berry-type fruits. Trends in foods. *Molecules.* 2015, *20*, 5875–5888.

108. Cavalcanti, R.N.; Santos, D.T.; Meireles, M.A.A. Non-thermal stabilization mechanisms of anthocyanins in model and food systems—An overview. *Food Res. Int.* 2011, *44*, 499–509.

109. Fang, Z.; Bhandari, B. Encapsulation of polyphenols—A review. *Trends Food Sci. Technol.* 2010, *21*, 510–523.

110. Fernandes, A.; Rocha, M.; Santos, L.; Brás, J.; Oliveira, J.; Mateus, N.; de Freitas, V. Blackberry anthocyanins: β-cyclodextrin fortification for thermal and gastro- intestinal stabilization. *Food Chem.* 2018, *245*, 426–431.

111. Huang, Y.; Zhou, W.B. Microencapsulation of anthocyanins through two-step emulsification and release characteristics during in vitro digestion. *Food Chem.* 2019, *278*, 357–363.

112. De Pascual-Teresa, S.; Sanchez-Ballesta, M.T. Anthocyanins: From plant to health. *Phytochem. Rev.* 2008, *7*, 281–299.

113. Clifford, M.N. Anthocyanins—Nature, occurrence and dietary sources. *J. Sci. Food Agric.* 2000, *80*, 1063–1072.

114. Shim, S.H.; Kim, J.M.; Choi, C.Y.; Kim, C.Y.; Park, K.H. Ginkgo biloba extract and bilberry anthocyanins improve visual function in patients with normal tension glaucoma. *J. Med. Food.* 2012, *15*, 818–823.

115. Ahmad, A.; Kaleem, M.; Ahmed, Z.; Shafiq, H. Therapeutic potential of flavonoids and their mechanism of action against microbial and viral infections—A review. *Food Res. Int.* 2015, *77*, 221–235.

116. Miyake, S.; Takahashi, N.; Sasaki, M.; Takahashi, N.; Ozawa, Y. Vision preservation during retinal inflammation by anthocyanin – Rich bilberry extract: Cellular and molecular mechanism. *Lab. Investig.* 2012, *92*, 102–109.

117. Lee, J.; Lee, H.K.; Kim, C.Y.; Hong, Y.J.; Choe, C.M.; You, T.W.; Seong, G.J. Purified high-dose anthocyanoside oligomer administration improves nocturnal vision and clinical symptoms in myopia subjects. *Br. J. Nutr.* 2005, *93*, 895–899.

118. Thiraphatthanavong, P.; Wattanathorn, J.; Muchimapura, S.; Wipawee, T.M.; Wannanon, P.; Terdthai, T.U.; Suriharn, B.; Lertrat, K. Preventive effect of *Zea mays* L. (purple waxy corn) on experimental diabetic cataract. *BioMed Res. Int.* 2014, *2014*, 507435.

119. Mok, J.W.; Chang, D.J.; Joo, C.K. Antiapoptotic effects of anthocyanin from the seed coat of black soybean against oxidative damage of human lens epithelial cell induced by H_2O_2. *Curr. Eye Res.* 2014, *39*, 1090–1098.

120. Paik, S.S.; Jeong, E.; Jung, S.W.; Ha, T.J.; Kang, S.; Sim, S.; Jeon, J.H.; Chun, M.H.; Kim, I.B. Anthocyanins from the seed coat of black soybean reduce retinal degeneration induced by N-methyl-N-nitrosourea. *Exp. Eye Res.* 2012, *97*, 55–62.

121. Ohguro, H.; Ohguro, I.; Katai, M.; Tanaka, S. Two-year randomized, placebo-controlled study of black currant anthocyanins on visual field in glaucoma. *Ophthalmologica*. 2012, *228*, 26–35.

122. Yang, Y.; Shi, Z.; Reheman, A.; Jin, W.; Li, C.; Zhu, G.; Wang, Y.; Freedman, J.J.; Ling, W.; Ni, H. Anthocyanins inhibit platelet activation and attenuate thrombus growth in both human and murine thrombosis models. *Blood*. 2010, *116*, 3197.

123. Qin, Y.; Xia, M.; Ma, J.; Hao, Y.; Liu, J.; Mou, H.; Cao, L.; Ling, W. Anthocyanin supplementation improves serum LDL-and HDL-cholesterol concentrations associated with the inhibition of cholesteryl ester transfer protein in dyslipidemic subjects. *Am. J. Clin. Nutr.* 2009, *90*, 485–492.

124. Cassidy, A.; Bertoia, M.; Chiuve, S.; Flint, A.; Forman, J.; Rimm, E.B. Habitual intake of anthocyanins and flavanones and risk of cardiovascular disease in men. *Am. J. Clin. Nutr.* 2016, *104*, 587–594.

125. Bell, D.R.; Gochenaur, K. Direct vasoactive and vasoprotective properties of anthocyanin-rich extracts. *J. Appl. Physiol.* 2006, *100*, 1164–1170.

126. Toufektsian, M.C.; De Lorgeril, M.; Nagy, N.; Salen, P.; Donati, M.B.; Giordano, L.; Mock, H.-P.; Peterek, S.; Matros, A.; Petroni, K.; et al. Chronic dietary intake of plant-derived anthocyanins protects the rat heart against ischemia-reperfusion injury. *J. Nutr.* 2008, *138*, 747–752.

127. Alvarez-Suarez, J.M.; Giampieri, F.; Tulipani, S.; Casoli, T.; di Stefano, G. One month strawberry-rich anthocyanin supplementation ameliorates cardiovascular risk, oxidative stress markers and platelet activation in humans. *J. Nutr. Biochem.* 2014, *25*, 289–294.

128. Kwon, S.H.; Ahn, I.S.; Kim, S.O.; Kong, C.S.; Chung, H.Y.; Do, M.S.; Park, K.Y. Anti-obesity and hypolipidemic effects of black soybean anthocyanins. *J. Med. Food.* 2007, *10*, 552–556.

129. Wu, T.; Yu, Z.; Tang, Q.; Song, H.; Gao, Z.; Chen, W.; Zheng, X. Honeysuckle anthocyanin supplementation prevents diet-induced obesity in C57BL/6 mice. *Food Funct.* 2013, *4*, 1654–1661.

130. Jankowski, A.; Jankowska, B.; Niedworok, J. The effects of anthocyanin dye from grapes on experimental diabetes. *Folia Med. Cracov.* 2000, *41*, 5–15.

131. Jayaprakasam, B.; Vareed, S.K.; Olson, L.K.; Nair, M.G. Insulin secretion by bioactive anthocyanins and anthocyanidins present in fruits. *J. Agric. Food Chem.* 2005, *53*, 28–31.

132. Tsuda, T.; Ueno, Y.; Aoki, H.; Koda, T.; Horio, F.; Takahashi, N.; Kawada, T.; Osawa, T. Anthocyanin enhances adipocytokine secretion and adipocyte-specific gene expression in isolated rat adipocytes. *Biochem. Biophys. Res. Commun.* 2004, *316*, 149–157.

133. Tsuda, T.; Horio, F.; Uchida, K.; Aoki, H.; Osawa, T. Dietary cyanidin 3-O-β-D-glucoside-rich purple corn color prevents obesity and ameliorates hyperglycemia in mice. *J. Nutr.* 2003, *133*, 2125–2130.

134. Takikawa, M.; Inoue, S.; Horio, F.; Tsuda, T. Dietary anthocyanin-rich bilberry extract ameliorates hyperglycemia and insulin sensitivity via activation of AMP-activated protein kinase in diabetic mice. *J. Nutr.* 2010, *140*, 527–533.

135. Li, D.; Zhang, Y.; Liu, Y.; Sun, R.; Xia, M. Purified anthocyanin supplementation reduces dyslipidemia, enhances antioxidant capacity, and prevents insulin resistance in diabetic patients. *J. Nutr.* 2015, *145*, 742–748.

136. Kang, M.K.; Lim, S.S.; Lee, J.Y.; Yeo, K.M.; Kang, Y.-H. Anthocyanin-rich purple corn extract inhibit diabetes-associated glomerular angiogenesis. *PLoS ONE*. 2013, *8*, 79823.

137. Fan, J.; Johnson, M.H.; Lila, M.A.; Yousef, G.; de Mejia, E.G. Berry and citrus phenolic compounds inhibit dipeptidyl peptidase IV: Implications in diabetes management. *Evid. Based Complement. Alternat. Med.* 2013, *2013*, 479505.

138. Koh, E.S.; Lim, J.H.; Kim, M.Y.; Chung, S.; Shin, S.J.; Choi, B.S.; Kim, H.W.; Hwang, S.Y.; Kim, S.W.; Park, C.W.; et al. Anthocyanin-rich seoritae extract ameliorates renal lipotoxicity via activation of AMP-activated protein kinase in diabetic mice. *J. Transl. Med.* 2015, *13*, 203.

139. Liu, Y.; Li, D.; Zhang, Y.; Sun, R.; Xia, M. Anthocyanin increases adiponectin secretion and protects against diabetes-related endothelial dysfunction. *Am. J. Physiol. Endocrinol. Metab.* 2014, *306*, E975–E988.

140. Pojer, E.; Mattivi, F.; Johnson, D.; Stockley, C.S. The case for anthocyanin consumption to promote human health: A review. *Compr. Rev. Food Sci. Food Saf.* 2013, *12*, 483–508.

141. Genskowsky, E.; Puente, L.A.; Perez-Alvarez, J.A.; Fernández-López, J.; Muñoz, L.A.; Viuda-Martos, M. Determination of polyphenolic profile, antioxidant activity and antibacterial properties of maqui [*Aristotelia chilensis* (Molina) Stuntz] a Chilean blackberry. *J. Sci. Food Agr.* 2016, *96*, 4235–4242.

142. Côté, J.; Caillet, S.; Doyon, G.; Ng, K. Antimicrobial effect of cranberry juice and extracts. *Food Cont.* 2011, *22*, 1413–1418.

143. Puupponen-Pimiä, R.; Nohynek, L.; Meier, C.; Kähkönen, M.; Heinonen, M.; Hopia, A.; Oksman-Caldentey, K.M. Antimicrobial properties of phenolic compounds from berries. *J. Appl. Microbiol.* 2001, *90*, 494–507.

144. Wang, L.S.; Hecht, S.S.; Carmella, S.G.; Yu, N.; Larue, B.; Henry, C.; McIntyre, C.; Rocha, C.; Lechner, J.F.; Stoner, G.D. Anthocyanins in black raspberries prevent esophageal tumors in rats. *Cancer Prev. Res.* 2009, *2*, 84–93.

145. Faria, A.; Pestana, D.; Teixeira, D.; de Freitas, V.; Mateus, N.; Calhau, C. Blueberry anthocyanins and pyruvic acid adducts: Anticancer properties in breast cancer cell lines. *Phytother. Res.* 2010, *24*, 1862–1869.

146. Hui, C.; Bin, Y.; Xiaoping, Y.; Long, Y.; Chunye, C.; Mantian, M.; Wenhua, L. Anticancer activities of an anthocyanin-rich extract from black rice against breast cancer cells in vitro and in vivo. *Nutr. Cancer.* 2010, *62*, 1128–1136.

147. Chen, X.Y.; Zhou, J.; Luo, L.P.; Han, B.; Li, F.; Chen, J.Y.; Zhu, Y.F.; Chen, W.; Yu, X.P. Black rice anthocyanins suppress metastasis of breast cancer cells by targeting RAS/RAF/MAPK pathway. *BioMed Res. Int.* 2015, *2015*, 414250.

148. Malik, M.; Zhao, C.; Schoene, N.; Guisti, M.M.; Moyer, M.P.; Magnuson, B.A. Anthocyanin-rich extract from *Aronia meloncarpa* E. induces a cell cycle block in colon cancer but not normal colonic cells. *Nutr. Cancer.* 2003, *46*, 186–196.

149. Lala, G.; Malik, M.; Zhao, C.; He, J.; Kwon, Y.; Giusti, M.M.; Magnuson, B.A. Anthocyanin-rich extracts inhibit multiple biomarkers of colon cancer in rats. *Nutr. Cancer.* 2006, *54*, 84–93.

150. Lim, S.; Xu, J.; Kim, J.; Chen, T.Y.; Su, X.; Standard, J.; Carey, E.; Griffin, J.; Herndon, B.; Katz, B.; et al. Role of anthocyanin-enriched purple-fleshed sweet potato p40 in colorectal cancer prevention. *Mol. Nutr. Food Res.* 2013, *57*, 1908–1917.

151. Jang, H.; Ha, U.S.; Kim, S.J.; Yoon, B.I.; Han, D.S.; Yuk, S.M.; Kim, S.W. Anthocyanin extracted from black soybean reduces prostate weight and promotes apoptosis in the prostatic hyperplasia-induced rat model. *J. Agric. Food Chem.* 2010, *58*, 12686–12691.

152. Shin, D.Y.; Lee, W.S.; Kim, S.H.; Kim, M.J.; Yun, J.W.; Lu, J.N.; Lee, S.J.; Tsoy, I.; Kim, H.J.; Ryu, C.H.; et al. Anti-invasive activity of anthocyanins isolated from *Vitis coignetiae* in human hepatocarcinoma cells. *J. Med. Food.* 2009, *12*, 967–972.

153. Bontempo, P.; de Masi, L.; Carafa, V.; Rigano, D.; Scisciola, L.; Iside, C.; Grassi, R.; Molinari, A.M.; Aversano, R.; Nebbioso, A.; et al. Anticancer activities of anthocyanin extract from genotyped *Solanum tuberosum* L. "Vitelotte". *J. Funct. Foods* 2015, *19*, 584–593.

154. Strathearn, K.E.; Yousef, G.G.; Grace, M.H.; Roy, S.L.; Tambe, M.A.; Ferruzzi, M.G.; Wu, Q.-L.; Simon, J.E.; Lila, M.A.; Rochet, J.-C. Neuroprotective effects of anthocyanin and proanthocyanidin-rich extracts in cellular models of Parkinson's disease. *Brain Res.* 2014, *1555*, 60–77.

155. Shih, P.-H.; Chan, Y.-C.; Liao, J.-W.; Wang, M.-F.; Yen, G.-C. Antioxidant and cognitive promotion effects of anthocyanin-rich mulberry (*Morus atropurpurea* L.) on senescence-accelerated mice and prevention of Alzheimer's disease. *J. Nutr. Biochem.* 2010, *21*, 598–605.

156. Gutierres, J.M.; Carvalho, F.B.; Schetinger, M.R.C.; Agostinho, P.; Marisco, P.C. Neuroprotective effect of anthocyanins on acetylcholinesterase activity and attenuation of scopolamine-induced amnesia in rats. *Int. J. Dev. Neurosci.* 2014, *33*, 88–97.

157. Rehman, S.U.; Shah, S.A.; Ali, T.; Chung, J.I.; Kim, M.O. Anthocyanins reversed D-galactose-induced oxidative stress and neuroinflammation mediated cognitive impairment in adult rats. *Mol. Neurobiol.* 2017, *54*, 255–271.

158. Amin, F.U.; Shah, S.A.; Badshah, H.; Khan, M.; Kim, M.O. Anthocyanins encapsulated by PLGA@PEG nanoparticles potentially improved its free radical scavenging capabilities via p38/JNK pathway against Abeta1-42-induced oxidative stress. *J. Nanobiotechnol.* 2017, *15*, 12.

159. Kim, M.J.; Rehman, S.U.; Amin, F.U.; Kim, M.O. Enhanced neuroprotection of anthocyanin-loaded PEG-gold nanoparticles against Abeta1-42-induced neuroinflammation and neurodegeneration via the NF-KB/JNK/GSK3beta signaling pathway. *Nanomedicine.* 2017, *13*, 2533–2544.

160. Ali, T.; Kim, M.J.; Rehman, S.U.; Ahmad, A.; Kim, M.O. Anthocyanin-loaded PEG-gold nanoparticles enhanced the neuroprotection of anthocyanins in an Abeta1-42 mouse model of Alzheimer's disease. *Mol. Neurobiol.* 2017, *54*, 6490–6506.

161. Nomi, Y.; Iwasaki-Kurashige, K.; Matsumoto, H. Therapeutic effects of anthocyanins for vision and eye health. *Molecules.* 2019, *24*, 3311.

162. Ghosh, D.; Konishi, T. Anthocyanins and anthocyanin-rich extracts: Role in diabetes and eye function. *Asian Pac. J. Clin. Nutr.* 2007, *16*, 200–208.

163. Tsuda, T. Anthocyanins as functional food factors: Chemistry, nutrition and health promotion. *Food Sci. Technol. Res.* 2012, *18*, 315–324.

164. Smeriglio, A.; Monteleone, D.; Trombetta, D. Health effects of *Vaccinium myrtillus* L.: Evaluation of efficacy and technological strategies for preservation of active ingredients. *Mini Rev. Med. Chem.* 2014, *14*, 567–584.

165. Liu, L.K.; Lee, H.J.; Shih, Y.W.; Chyau, C.C.; Wang, C.J. Mulberry anthocyanins extracts inhibit LDL oxidation and macrophage-derived foam cell formation induced by oxidative LDL. *J. Food Sci.* 2008, *73*, 4113–4121.

166. Rechner, A.R.; Kroner, C. Anthocyanins and colonic metabolites of dietary polyphenols inhibit platelet function. *Thromb. Res.* 2005, *116*, 327–334.

167. Luís, Â; Domingues, F.; Pereira, L. Association between berries intake and cardiovascular diseases risk factors: A systematic review with meta-analysis and trial sequential analysis of randomized controlled trials. *Food Funct.* 2018, *9*, 740–757.

168. García-Conesa, M.T.; Chambers, K.; Combet, E.; Pinto, P.; Garcia-Aloy, M.; Andresń-Lacueva, C.; dePascual-Teresa, S.; Mena, P.; Konic Ristic, A.; Hollands, W.J.; et al. Meta-analysis of the effects of foods and derived products containing ellagitannins and anthocyanins on cardiometabolic biomarkers: Analysis of factors influencing variability of the individual responses. *Int. J. Mol. Sci.* 2018, *19*, 694.

169. Ksonzékova, P.; Mariychuk, R.; Eliasova, A.; Mudronova, D.; Csank, T.; Kiraly, J.; Marcincakova, D.; Pistl, J.; Tkacikova, L. In vitro studies of biological activities of anthocyanin-rich berry extracts on porcine intestinal epithelial cells. *J. Sci. Food Agric.* 2016, *96*, 1093–1100.

170. Bone, K.; Mills, S.. *Principles and Practice of Phytotherapy: Modern Herbal Medicine*; Churchill Livingstone: Edinburg, TX, USA, 2013.

171. Xie, L.; Su, H.; Sun, C.; Zheng, X.; Chen, W. Recent advances in understanding the anti-obesity activity of anthocyanins and their biosynthesis in microorganisms. *Trends Food Sci. Technol.* 2018, *72*, 13–24.

172. Rupasinghe, H.P.V.; Arumuggam, N. Health benefits of anthocyanins. *Food Chem. Func. Anal.* 2019, *2019*, 123–158.

173. Heyman, L.; Axling, U.; Blanco, N.; Sterner, O.; Holm, C.; Berger, K. Evaluation of beneficial metabolic effects of berries in high-fat fed C57BL/6J mice. *J. Nutr. Metab.* 2014, *2014*, 403041.

174. Lachin, T.; Reza, H. Anti diabetic effect of cherries in alloxan induced diabetic rats. *Recent Pat. Endocr. Metab. Immune Drug Discov.* 2012, *6*, 67–72.

175. Orqueda, M.E.; Torres, S.; Zampini, I.C.; Cattaneo, F.; Fernandez Di Pardo, A.; Valle, E.M.; Jimenez-Aspee, F.; Schmeda-Hirschmann, G.; Isla, M.I. Integral use of *Argentinean solanum betaceum* red fruits as functional food ingredient to prevent metabolic syndrome: Effect of in vitro simulated gastroduodenal digestión. *Heliyon.* 2020, *6*, 03387.

176. Wedick, N.M.; Pan, A.; Cassidy, A.; Rimm, E.B.; Sampson, L.; Rosner, B.; Willett, W.; Hu, F.B.; Sun, Q.; van Dam, R.M. Dietary flavonoid intakes and risk of type 2 diabetes in US men and women. *Am. J. Clin. Nutr.* 2012, *95*, 925–933.

177. Cisowska, A.; Wojnicz, D.; Hendrich, A.B. Anthocyanins as antimicrobial agents of natural plant origin. *Nat. Prod. Commun.* 2011, *6*, 149–156.

178. Helander, I.M.; Alakomi, H.L.; Latva-Kala, K.; Mattila-Sandholm, T.; Pol, I.; Smid, E.J.; Gorris, L.G.M.; von Wright, A. Characterization of the action of selected essential oil components on Gram-negative bacteria. *J. Agric. Food Chem.* 1998, *46*, 3590–3595.

179. Riaz, M.; Zia-Ul-Haq, M.; Saad, B. *Anthocyanins and Human Health: Biomolecular and Therapeutical Aspects*; Springer: New York, 2016.

180. Wallace, T.C.; Giusti, M.M.. *Anthocyanins in Health and Disease*; CRC Press: Boca Raton, FL, 2014.

181. Fraga, C.G. (Ed.) *Plant Phenolics and Human Health: Biochemistry, Nutrition and Pharmacology*; Wiley: New York, 2010.

182. He, J.; Giusti, M.M. Anthocyanins: Natural colorants with health-promoting properties. *Annu. Rev. Food Sci. Technol.* 2010, *1*, 163–187.

183. Galvano, F.; Salamone, F.; Nicolosi, A.; Vitaglione, P. Anthocyanins-based drugs for colon cancer treatment: The nutriotionist's point of view. *Cancer Chemother. Pharmacol.* 2009, *64*, 431–432.

184. Webb, M.R.; Min, K.; Ebeler, S.E. Anthocyanins interactions with DNA: Intercalation, topoisomerase i inhibition and oxidative reactions. *J. Food Biochem.* 2008, *32*, 176–196.

185. Fernandes, I.; Faria, A.; Azevedo, J.; Soares, S.; Calhau, C.; De Freitas, V.; Mateus, N. Influence of anthocyanins, derivative pigments and other catechol and pyrogallol-type phenolics on breast cancer cell proliferation. *J. Agric. Food Chem.* 2010, *58*, 3785–3792.

186. Sun, J.; Hai Liu, R. Cranberry phytochemical extracts induce cell cycle arrest and apoptosis in human MCF-7 breast cancer cells. *Cancer Lett.* 2006, *241*, 124–134.

187. Hoshyar, R.; Mahboob, Z.; Zarban, A. The antioxidant and chemical properties of Berberis vulgaris and its cytotoxic effect on human breast carcinoma cells. *Cytotechnology.* 2015, *68*, 1207–1213.

188. Im, N.K.; Jang, W.J.; Jeong, C.H.; Jeong, G.S. Delphinidin suppresses PMA-induced MMP-9 expression by blocking the NF-kappaB activation through MAPK signaling pathways in MCF-7 human breast carcinoma cells. *J. Med. Food.* 2014, *17*, 855–861.

189. Forester, S.C.; Choy, Y.Y.; Waterhouse, A.L.; Oteiza, P.I. The anthocyanin metabolites gallic acid, 3-*o*-methylgallic acid, and 2,4,6- trihydroxybenzaldehyde decrease human colon cancer cell viability by regulating pro-oncogenic signals. *Mol. Carcinog.* 2014, *53*, 432–439.

190. Chen, P.N.; Kuo, W.H.; Chiang, C.L.; Chiou, H.L.; Hsieh, Y.S.; Chu, S.C. Black rice anthocyanins inhibit cancer cells invasion via repressions of MMPs and u-PA expression. *Chem. Biol. Interact.* 2006, *163*, 218–229.

191. Li, D.; Wang, P.; Luo, Y.; Zhao, M.; Chen, F. Health benefits of anthocyanins and molecular mechanisms: Update from recent decade. *Crit. Rev. Food Sci. Nutr.* 2017, *57*, 1729–1741.

192. Jing, P.; Bomser, J.A.; Schwartz, S.J.; He, J.; Magnuson, B.A.; Giusti, M.M. Structure–function relationships of anthocyanins from various anthocyanin-rich extracts on the inhibition of colon cancer cell growth. *J. Agric. Food Chem.* 2008, *56*, 9391–9398.

193. Zhao, C.; Giusti, M.M.; Malik, M.; Moyer, M.P.; Magnuson, B.A. Effects of commercial anthocyanin-rich extracts on colonic cancer and nontumorigenic colonic cell growth. *J. Agric. Food Chem.* 2004, *52*, 6122–6128.

194. Winter, A.N.; Bickford, P.C. Anthocyanins and their metabolites as therapeutic agents for neurodegenerative disease. *Antioxidants.* 2019, *8*, 333.

195. Klinkenberg, I.; Blokland, A. The validity of scopolamine as a pharmacological model for cognitive impairment: A review of animal behavioral studies. *Neurosci. Biobehav. Rev.* 2010, *34*, 1307–1350.

2

Stability of Anthocyanins during Food Processing

Maria Josiane Macedo, Juliana Dara Rabêlo Silva, and Mário Roberto
Maróstica Júnior

2.1 Introduction

Anthocyanins are a predominant and extensively used category of water-soluble natural colorants. They have emerged as a viable alternative to traditional synthetic colorants in response to the growing demand for additives of natural origin and increased awareness of the potential health and environmental risks associated with the use of chemicals in the production of synthetic pigments (Yao et al. 2021; Chen et al. 2022). These factors underline the potential of anthocyanins as value-added ingredients or promising additives for applications in the food industry.

The intense coloration of anthocyanins, which ranges from red to purple, blue and in-between, is attributed to their chemical structure. Specifically, the presence of conjugated double bonds allows anthocyanins to absorb light at an approximate wavelength of 500 nm (Echegaray et al. 2022). However, anthocyanins are highly unstable pigments and susceptible to degradation. A variety of intrinsic and extrinsic factors can influence the stability of anthocyanins, including relative humidity, light, pH, temperature, sugars (acylated and non-acylated), vitamin C, oxygen, sulfur dioxide or sulfites, enzymes, co-pigments, and metal ions (Enaru et al. 2021; Yao et al. 2021; Chen et al. 2022; Cunha et al. 2023). It is therefore essential to improve the physicochemical stability of anthocyanins in order to develop processing strategies that preserve their integrity.

Therefore, the purpose of this chapter is to compile scientific information on the main factors that influence the stability of anthocyanins and to discuss fundamental methods and strategies for preserving their characteristics during food processing.

2.2 Factors Affecting the Stability of Anthocyanins

Anthocyanins, natural pigments found in many fruits and vegetables, are renowned for their antioxidant properties and health benefits. However, their stability during food processing and storage poses a significant concern in the food industry. Several factors can influence the degradation of anthocyanins. Thus, this chapter reports the role of these factors in the stability of these compounds.

DOI: 10.1201/9781003453260-3

2.2.1 pH

The stability of anthocyanin pigments is directly linked to changes in their structure in response to pH variations. The predominant forms of anthocyanins are determined by the equilibrium constant (K) and are dependent on the pH of the medium, which also influences their color (Trouillas et al. 2016). This phenomenon is associated with the values of equilibrium constants, specifically deprotonation (K_a) and hydration (K_h). If K_h is greater than K_a, the equilibrium shifts to hemiacetal or pseudobase forms, and conversely, the equilibrium shifts to quinoidal bases, which have a blue color (Mercadante 2018; Oancea 2021; Gull et al. 2022).

As shown in Figure 3.1, in acidic conditions (pH 1–2), the prevalence is of the flavylium cation, which is in the protonated form, making them stable and red in color. As the pH increases, approaching neutrality, deprotonation and hydration reactions of the flavylium cation may occur (Trouillas et al. 2016).

The flavylium cation, when hydrated, results in the creation of a carbinol pseudobase, which appears colorless at pH conditions between 4 and 5, or undergoes deprotonation between pH 6 and 7, giving rise to the quinoidal base form with a blue color. Under highly alkaline conditions (pH > 9), tautomerization occurs through ring opening, resulting in a chalcone with a pale-yellow color, and subsequently, anthocyanins may undergo degradation (Albuquerque et al. 2020).

It has been well-elucidated that anthocyanins and their coloration are more stable in acidic pH (pH < 3.0); therefore, due to their instability at different pH levels, their application in foods becomes quite limited (Oancea 2021). Liu et al. (2019) investigated the effect of pH on anthocyanin emulsions, noting a color shift from violet at pH 7.0 to salmon pink at pH 3.0. Hou et al. (2013) found that pH strongly influenced the degradation of anthocyanins in black rice, where increasing the pH from 1.0 to 6.0 accelerated the degradation of anthocyanins. They were more stable in acidic conditions (pH 1.0–3.0), whereas a significant reduction in stability of anthocyanins was observed at pH > 3.0.

Anthocyanins from purple sweet potatoes were applied as colorants in ice pops, and the best conditions were found by Rodríguez-Mena et al. (2023), who observed that the pH of the extraction solvent influenced the obtaining of anthocyanins, with pH 2.5 showing the best color properties.

The pH control is an essential parameter for minimizing the loss of anthocyanins during food processing. Furthermore, the color change can directly affect the appearance and acceptance of food by consumers, especially when the color reflects the product's quality, therefore representing a major challenge in the food industry due to the high instability of these pigments at different pH.

2.2.2 Temperature

The thermal degradation of anthocyanins can be influenced by the pH of the medium, with varying degrees of degradation observed during heating. Brouillard and Delaporte (1977) initially proposed mechanisms for this degradation, which have been recently reviewed and updated by Oancea (2021). Under high temperature and different pH conditions, may occur irreversible oxidative reactions leading to the degradation of anthocyanins into aldehyde and benzoic acid derivatives (Figure 2.1) (Oancea 2021).

At pH 1.0, multiple hydrolyses of glycosidic bonds occur, resulting in the formation of anthocyanidin, followed by the cleavage of rings A and B, yielding final derivatives such as phenolic aldehyde and benzoic acid. Similarly, at pH 3.5, hydrolytic opening of the

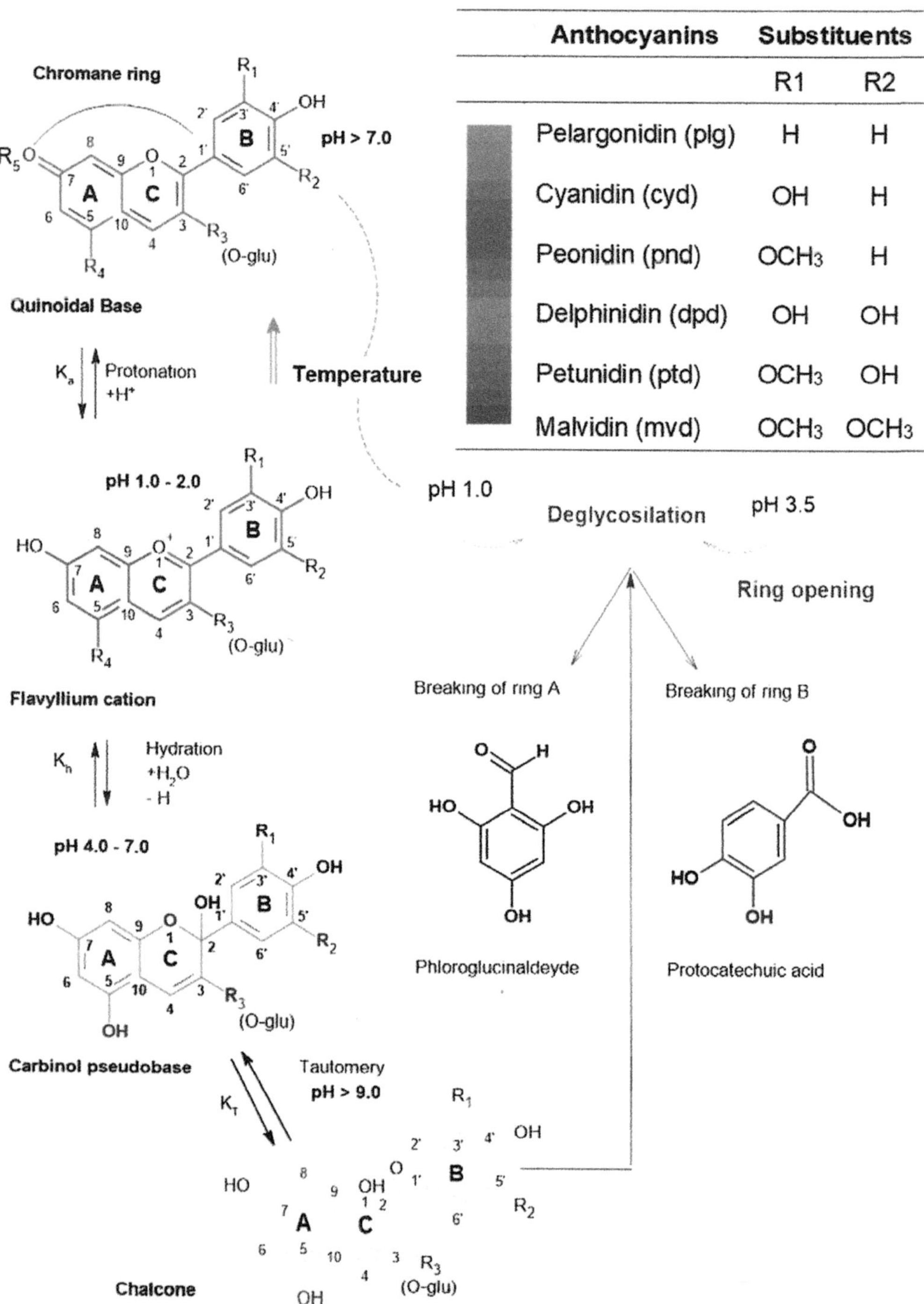

Anthocyanins	Substituents	
	R1	R2
Pelargonidin (plg)	H	H
Cyanidin (cyd)	OH	H
Peonidin (pnd)	OCH₃	H
Delphinidin (dpd)	OH	OH
Petunidin (ptd)	OCH₃	OH
Malvidin (mvd)	OCH₃	OCH₃

FIGURE 2.1

Interconversion and thermal degradation of anthocyanins.

pyrrolic ring precedes the formation of a chalcone, which then undergoes further degradation of rings A and B, producing similar final derivatives. Cleavage of C3–C4 leads to the formation of phloroglucinaldehyde, while cleavage of C2–C3 bonds generates phenolic acids. This process likely involves a combination of hydrolytic and autooxidative reactions (Mercadante 2018; Oancea 2021).

The degradation of anthocyanins during isothermal heating follows a first-order kinetic model, where a lower kinetic rate constant (k) indicates better stability of anthocyanins, as demonstrated in previous studies (Patras et al. 2010; Hou et al. 2013; De Marchi et al. 2024).

In a study conducted by Loypimai et al. (2016), the authors found that increasing the temperature from 60°C to 100°C significantly increased the values of k for anthocyanins and led to the darkening of pigments. Moreover, under the influence of a higher pH (pH 5.0), they observed a greater sensitivity to temperature changes, resulting in a deterioration of anthocyanin stability. Similar results were previously reported by Hou et al. (2013), who found that elevated pH (pH 5.0–6.0) combined with high temperature (90°C–100°C) resulted in high degradation of anthocyanin pigments.

Various thermal processing methods like blanching, pasteurization, and baking are commonly employed in food processing to ensure safety and prolong shelf life, often involving high temperatures (>60°C) (Oancea 2021). However, it's crucial to consider that exposure time and heating techniques directly influence the stability of compounds, potentially causing pigment discoloration (Kara & Erçelebi 2013).

When analyzing the processing of blueberry puree, which involved the use of high temperature (95°C blanching) combined with pasteurization, Brownmiller et al. (2008) found a significant reduction of 43% in the total anthocyanin content compared to fresh fruits. In a comprehensive review, thermal processing effects on the stability of pigments obtained from fruits and their by-products were well elucidated (Rawson et al. 2011).

Žilić et al. (2016), while investigating the impact of temperature on the production of cookies using anthocyanin-rich corn flours, observed a significant reduction, between 3.5 and 5 times, in anthocyanin levels after baking at 200°C. In the same study, by adjusting the pH of the cookie dough with citric acid (from pH 7.86–8.05 to pH 4.59–3.72) and reducing the temperature to 150°C, the authors noted a positive influence on anthocyanin stability and coloration, with values of anthocyanins 2.5–3.4 times higher, highlighting the influence of pH and temperature on anthocyanin stability during cookie processing.

These findings highlight the critical role of temperature in food production with anthocyanins, as it can impact the content, color, and flavor, affecting sensory characteristics and consumer acceptance (Sendri et al. 2023). Furthermore, the extent of anthocyanin degradation during thermal food processing also depends on both the specific composition of anthocyanins and the characteristics of the food matrix. For example, the presence of sucrose, reducing sugars, and Maillard reaction products, as well as ascorbic acid and phenolic compounds, can alter the degradation kinetics of these compounds (Žilić et al. 2016).

2.2.3 Light

The light is responsible for favoring the biosynthesis process in the plant; however, after the extraction of anthocyanins and their incorporation into foods, the pigments tend to undergo constant degradation during processing, storage, and transportation (Giusti et al. 2023).

The degradation rates of anthocyanins from Roselle (*Hibiscus* L.) were investigated by Askar et al. (2015). The authors observed that degradation occurred more rapidly when exposed to sunlight compared to those kept in the dark. Additionally, they found that degradation was directly related to pH, with higher rates in basic medium than in acidic medium. According to the authors, the rate of photodiscoloration increased with increasing pH, possibly due to the formation of more hydroxyl radicals, which are induced by the exposure of solutions to sunlight.

In the study conducted by Aramwit et al. (2010), it was observed that after 10 hours of exposure to fluorescent light at room temperature, there was a significant reduction in both the total content of anthocyanins obtained from mulberries (*Morus alba*) and in antioxidant activity. It was found that the anthocyanin content and antioxidant activity decreased proportionally with the increase in exposure time of the extract to light.

Rigolon et al. (2024) investigated the stability of sports drinks made from jabuticaba peel and blueberry pulp over 91 days under varying storage conditions. They observed that exposure to light led to gradual color degradation and a reduction in anthocyanin content in both beverages, regardless of the beverage matrix.

Therefore, exposure to light impacts the stability of anthocyanins in food products. Although essential for their formation in plants, light accelerates their degradation in processed foods, especially under sunlight and basic pH conditions. Therefore, careful handling and packaging are crucial to preserve the anthocyanins in food products.

2.2.4 Oxygen

Anthocyanins are easily oxidized by oxygen, making them susceptible to oxidative degradation during various stages of food processing and storage. This can be attributed to the reaction of anthocyanins to hinder radical activity in high concentrations of oxygen, resulting in the depletion of antioxidant pigments and the formation of colorless or brown compounds (Markakis 1982).

Kim et al. (2021) found in their study that the thermal processing of blueberry puree in the absence of oxygen prevented anthocyanin oxidation. All anthocyanins in samples heated in the absence of oxygen showed higher levels of anthocyanins and achieved greater color retention compared to those processed in the presence of oxygen.

The degradation of anthocyanins can be accelerated by the presence of oxygen, mediated by oxidizing enzymes such as polyphenol oxidase (PPO), which catalyzes the oxidation of chlorogenic acid (CG) to the corresponding o-quinone (chlorogenoquinone, CGQ). The interaction of this quinone with anthocyanins results in the formation of brown-colored condensation products (Patras et al. 2010). Thus, the presence of oxygen not only amplifies the effects of other anthocyanin degradation processes but also facilitates and accelerates their enzymatic oxidation.

2.3 Interaction with Other Food Components

2.3.1 Enzymes

During the processing of raw materials, activated enzymes are largely responsible for the degradation of anthocyanins, due to their high specificity and efficiency. β-glucosidase is thought to initiate the degradation of anthocyanins by catalyzing the cleavage of the covalent bond between the glycosyl residue and the aglycone of the anthocyanin pigment. Consequently, the highly labile anthocyanidin undergoes degradation, diminishing its solubility and transforming it into colorless compounds, thereby diminishing the vibrancy of the pigments' coloration (Cavalcanti et al. 2011; Cai et al. 2022).

Peroxidases and phenolases, which encompass phenol oxidases and PPOs, are prevalent enzymes that contribute to the breakdown of anthocyanins. These enzymes, naturally occurring in fruits and berries, interact directly with anthocyanins, leading to their degradation via a co-oxidation reagent of enzymatically produced o-quinones and/or subsequent quinone oxidation derivatives (Cavalcanti et al. 2011; Enaru et al. 2021; Cai et al. 2022).

Nonetheless, research carried out by Fang et al. (2007) elucidated that the rate of cyanidin 3-glucoside decolorization remained nearly constant, irrespective of the presence or absence of gallic acid (GA), and even when the enzyme extract was deactivated in the presence of GA. This suggests that bayberry PPO was unable to interact directly with cyanidin 3-glucoside.

2.3.2 Ascorbic Acid

Ascorbic acid, an essential component in food processing, is a reliable indicator of food quality, as any fluctuation in its level signals a deterioration in food quality after processing and during storage. This compound, present in a variety of fruits and vegetables, is often incorporated as an antioxidant in many foods with the aim of increasing their nutritional value (Enaru et al. 2021).

In an environment containing anthocyanins, the presence of ascorbic acid accelerates the degradation of these pigments and the subsequent loss of color. The mechanisms proposed for this phenomenon include the direct condensation of ascorbic acid with anthocyanins and the formation of hydrogen peroxide, which leads to the oxidative cleavage of the pyrithione ring (Cavalcanti et al. 2011; Enaru et al. 2021; Cai et al. 2022). Direct condensation occurs at carbon four of the anthocyanin molecule, resulting in the loss of both compounds. Conversely, the second mechanism involves the loss of anthocyanin pigments through oxidative cleavage of the pyrithione ring, which operates through a free radical mechanism. In this process, ascorbic acid serves as a catalyst for molecular oxygen, generating free radicals (Cavalcanti et al. 2011; Enaru et al. 2021; Cai et al. 2022).

This occurrence poses a substantial hurdle to the utilization of colorants derived from anthocyanin in the food sector, particularly in beverages and juices frequently fortified with vitamin C (Farr & Giusti 2018). This was corroborated by Pacheco-Palencia et al. (2007), where the enrichment with ascorbic acid expedited the degradation of anthocyanin in clarified juices at all storage temperatures.

The influence of ascorbic acid on anthocyanins is multifaceted and depends on a number of factors, including the food matrix (He & Schwartz 2008). To illustrate, in a context where hydrogen peroxide (H_2O_2) is present, it was observed that ascorbic acid concentrations between 60 and 80 mg/L contributed to the stabilization of the anthocyanins present in pomegranate juice. In contrast, in cherry juice, an identical concentration of 80 mg/L accelerated the degradation of anthocyanins (Mehmet & Zkan 2002). Thus, it is important to note that in situations where oxygen and ascorbic acid levels are high, the degrading effect becomes more prominent.

2.3.3 Sugar

Sugars, abundantly found in various food products like juices, wines, and preserves, play a pivotal role in anthocyanin coloration. An elevated sugar concentration leads to diminished water activity, thereby decelerating the pseudobases formation and subsequently safeguarding the color of anthocyanins. Conversely, lower sugar concentrations, along with their degradation by-products, can compromise the stability of anthocyanins. Various sugars, such as sucrose, fructose, glucose, and xylose, may lead to the unanticipated breakdown of anthocyanins. Furfural, a common by-product of sugar degradation, has been proven to hasten the degradation of anthocyanins more than methylfurfural, culminating in the creation of brown pigment polymers (Cavalcanti et al. 2011; Cai et al. 2022).

Tsai et al. (2005) illustrated the influence of sucrose on the anthocyanin and antioxidant potential of blackberry extract under heat. Their observations included an escalation in polymeric and co-pigmented anthocyanins, coupled with a reduction in monomeric anthocyanins. Sucrose was found to exert a hyperchromic effect on anthocyanins before heating, and the browning by-products could amplify the discoloration of anthocyanins during the initial heating phase. Similarly, Nikkhah et al. (2007) reported that moderate sugar concentrations, such as 20%, had a protective effect on anthocyanin content and color quality in different extracts of chilled berries. However, this protective effect diminished with higher sugar concentrations.

2.3.4 Sulfites

Sulfur-containing compounds, specifically sulfites and sulfates, find widespread use in preserving fruit-derived foods. However, these compounds can lead to the fading of anthocyanin pigments, resulting in discoloration. This effect arises from the formation of colorless structures derived from sulfur when these compounds are introduced at positions 2 or 4 of the anthocyanins. The process is reversible: heating the anthocyanins releases sulfur dioxide, allowing the color to regenerate. Similarly, acidification to a low pH also liberates sulfur dioxide and restores the anthocyanins. However, high concentrations of sulfites, greater than 10 g/kg, cause permanent anthocyanin degradation (Cavalcanti et al. 2011; Enaru et al. 2021).

Sulfur dioxide (SO_2), for example, is widely used in the fruit and vegetable preservation industry as an inhibitor of microbial growth and enzymatic and nonenzymatic browning. Additionally, SO_2 has demonstrated its efficacy as an excellent choice for low-pH foods, such as wine, owing to its antiseptic properties against yeasts and bacteria, as well as its antioxidant effects. This makes it one of the most versatile and effective additives, capable of whitening pigments, eliminating unpleasant odors from oxidation and enabling the extraction of solid grape components such as stalks, seeds, and skins (Enaru et al. 2021). This additive is also used in three other stages of winemaking: (1) During the fermentation process, in the grapes or must, with the main aim of preventing oxidation; (2) Following the completion of fermentation and before the aging or storage stages, it limits microbial growth that could impact the wines; (3) Finally, just before bottling, the wines are stabilized with SO_2 to prevent any alterations or accidents in the bottles.

Nevertheless, stringent regulation governs the utilization of this compound due to its potential to induce sensory alterations in the end product. High concentrations of SO_2 can lead to undesirable flavors, affecting sulfur gas and reducing compounds, including hydrosulfates and mercaptans (Enaru et al. 2021). It is therefore crucial to continue research to optimize the use of these additives, guaranteeing the quality of the final product and consumer safety.

2.4 Processing Factors

2.4.1 Extraction

Anthocyanins are located within the vacuoles of plant cells and necessitate effective tissue rupture for extraction, a process facilitated by solvent penetration followed by solubilization.

The careful selection of solvents, taking into account polarity, is paramount. Solid-liquid extraction, utilizing solvents such as methanol, ethanol, ethyl acetate, or acetone, represents a widely adopted conventional method (Silva et al. 2017; Muhlack et al. 2018; Romero-Díez et al. 2019). This approach is often complemented by pH adjustment through the addition of acids in low concentrations to reduce pH, thereby enhancing stability and facilitating diffusion into the solvent (Gull et al. 2022).

Wang et al. (2016) observed that specific extraction conditions, involving different solvents, favor distinct groups of anthocyanins. The authors found that blueberry extracts showed better extraction with methanol, resulting in a significant increase in monomeric anthocyanins compared to other solvents such as ethanol and acetone. In contrast, cherries obtained better results using ethanol.

The authors noted significant effects of solvent type and extraction method on anthocyanin composition. Methanol yielded higher extraction efficiency for delphinidin-3-arabinoside, petunidin-3-glucoside, malvidin-3-galactoside (Mv-3-gal), peonidin-3-arabinoside (Pn-3-ara), and malvidin-3-glucoside (Mv-3-glu) in blueberries, while acetone was more effective for cyanidin-3-galactoside (Cy-3-gal), Mv-3-gal, and Mv-3-glu. In cherries, methanol favored cyanidin-3-rutinoside, and ethanol yielded higher cyanidin-3-galactoside levels in red pear peel.

Such results reinforce the relationship between the different polarity of each anthocyanin compound concerning the solvent. In this regard, the authors demonstrate that extraction conditions can selectively alter the compositions of anthocyanins in the extract, necessitating a rigorous selection of ideal conditions based on the compounds of interest to be investigated.

Currently, a novel group of solvents, Deep Eutectic Solvents (DES), known for their eco-friendliness, is gaining traction in industries. DES, forming a supramolecular structure, extract phytochemical compounds due to enhanced solubility. Recent research, including a review by Foroutani et al. (2024), shows DES-based extracts are more stable than traditional solvents. By combining hydrogen bond acceptors (HBAs) with donors (HBDs), DES create mixtures with lower melting points, yielding natural deep eutectic solvents (NADES), safer and eco-friendly (Benvenutti, Zielinski & Ferreira 2022; Foroutani et al. 2024).

Zannou et al. (2020) investigated the recovery of anthocyanins from Roselle (*Hibiscus sabdariffa* L.) using DES, employing sodium acetate as a HBA and formic acid acting as a HBD in a molar ratio of 1:2. They observed higher polyphenol extraction efficiency compared to conventional solvents such as distilled water, 70% ethanol, and 80% methanol. The study's results indicated that the use of DES proved to be an ideal medium for extracting polyphenols from Roselle, providing high mass transfer and remarkable anthocyanin solubility.

In addition to the importance of solvent selection, the extraction process significantly impacts anthocyanin stability. Ensuring physical and chemical stabilization during extraction is crucial to enhance stability, prolong shelf life, and improve solubility these pigments, thus enhancing technological properties and preserving sensory and nutritional characteristics (Esteve-Turrillas, Garrigues & De la Guardia 2024).

2.4.2 Processing Methods

Currently, unconventional extraction techniques such as ultrasound-assisted extraction (UAE), microwave-assisted extraction, supercritical fluid extraction (SFE), and pulsed electric field (PEF) extraction are preferred over conventional methods like Soxhlet extraction, hot water extraction, and magnetic stirring. This preference originates from the fact that conventional techniques may involve high temperatures and larger solvent quantities, leading to increased compound degradation. Moreover, conventional methods often lack control over granulometry and morphology, exhibit low precipitation rates, and offer low trapping and loading efficiency (Zielinski et al. 2021; Gull et al. 2022).

Unconventional techniques emerge with the advantage not only of preserving the 'fresh' characteristics of foods, prioritizing sensory quality, but also of enhancing the integrity of phytochemical compounds. In addition to saving time and energy, thus favoring environmental sustainability through the use of techniques considered more ecological and sustainable (Esteve-Turrillas, Garrigues & De la Guardia 2024).

Given their relevance in improving anthocyanin extraction and stability, this chapter will focus on key methodologies non-conventional.

2.4.2.1 *Ultrasound-Assisted Extraction*

One of the green methodologies employed in the extraction of anthocyanins is UAE (Esteve-Turrillas, Garrigues & De la Guardia 2024). The application of high-power ultrasound can be carried out through two types of devices: the ultrasonic bath or the ultrasound equipment using a probe. Both systems consist of a transducer as the source of ultrasound power (Chemat et al. 2017).

Ultrasound is capable of generating cavitation through high-frequency sound waves, characterized by the collapse of cells resulting in the formation of bubbles, whose vibration promotes an increase in temperature, thus facilitating the softening of the matrix and ensuring efficient diffusion of anthocyanins. However, strict temperature control during this process becomes necessary, especially the use of a probe, as it can reach very high temperatures and consequently compromise its stability (D'Alessandro et al. 2014; Silva et al. 2017; Romero-Díez et al. 2019). The UAE, being a rapid extraction method, has the advantage of reduced processing time, as well as higher yield and lower energy consumption and, easy applicability.

The type of ultrasound device used for extractions is a variable to be considered since it can influence the process of obtaining compounds. Probes have higher power due to the delivery of ultrasonic intensity through a smaller surface, limited to the tip of the probe. This more focused delivery allows the system to provide intensity in a more restricted area, giving it greater robustness (Chemat et al. 2017). Thus, high-power ultrasonic probes have been shown to be more efficient and are generally preferred for extraction applications. Additionally, the operating conditions of the equipment also need to be studied and considered as they can influence the process of obtaining anthocyanins during extraction.

2.4.2.2 *Pressurized Liquid Extraction*

Pressurized liquid extraction (PLE) utilizes high pressure and temperature, making it a rapid extraction technique that is easily automated. In this process, when the temperature exceeds the boiling point of the solvent, the pressure must be sufficient to keep it in the liquid state (above the boiling point, but below the critical point, called the subcritical state) (Zielinski et al. 2021).

In this method, the diffusion and solubility of the solvent are facilitated due to the solvation that occurs during the process, allowing for enhanced penetration of the solvent into the matrix and more efficient mass transfer, resulting in higher extraction yields in a shorter period of time, as well as requiring a reduced amount of solvent (Zielinski et al. 2021).

In PLE, it is possible to operate at high temperatures, where the solubility of the target compounds and the desorption kinetics of the matrix are increased. However, although one of the advantages of this technique is to adjust process parameters aiming at the selectivity to a specific group of compounds, the use of high temperatures can lead to the degradation of anthocyanins during the extraction process, requiring finding a balance between degradation kinetics and extraction efficiency (Silva et al. 2017).

Benvenutti et al. (2022) compared PLE methods using water, ethanol, and DES for extracting anthocyanins from jabuticaba peel. DES formulations, particularly choline chloride and propylene glycol (ChCl:Pro) and choline chloride and malic acid (ChCl:Ma), yielded significantly higher anthocyanin content under optimal conditions: 47% solvent concentration, 90°C temperature, and 5.3 mL/min flow rate, with cyanidin concentrations up to three times higher than conventional solvents.

This combination underscores the potential of PLE with aqueous DES solutions to optimize anthocyanin extraction from plant sources, offering improved efficiency and higher anthocyanin concentrations compared to traditional methods.

2.4.2.3 Supercritical Fluids Extraction

SFE is preferable for thermally unstable and easily oxidizable compounds that may be lost due to thermal degradation, such as example anthocyanins. In this process, supercritical fluids (at the critical point between liquid and vapor) are employed to extract the desired components from a matrix, whether solid or liquid (Silva et al. 2017).

Supercritical carbon dioxide (CO_2) is a nonpolar solvent and therefore can be used to remove nonpolar components, reducing interference from unwanted substances; however, its low polarity makes it unsuitable for extracting polar compounds like anthocyanins, necessitating the use of polar modifiers. Ethanol and methanol are the most commonly used modifiers for extracting bioactive compounds (Chemat et al. 2019). These modifiers are added in small quantities, and it becomes necessary to understand the thermodynamics and kinetics of the extraction process to ensure better selectivity and reduce the possibility of co-extraction of non-target compounds, such as waxes (Essien, Young & Baroutian 2020).

In this sense, the choice of solvent is crucial during the extraction of anthocyanins by this method; typically, a mixture of CO_2, ethanol, and water is used. It is essential to adjust the ratio between ethanol and water to ensure the best solubility of anthocyanins in supercritical CO_2, taking into account specific pressure and temperature conditions (Silva et al. 2017).

Among other advantages, extraction with supercritical fluids contributes to the inhibition of natural enzymes that degrade anthocyanins; and the extracts obtained by this method are recognized for complying with environmental regulations and are considered safe for human consumption (GRAS) (Silva et al. 2017).

2.4.2.4 Pulsed Electric Field

PEF technology is a non-thermal technique investigated in food processing, which involves the application of high-voltage electricity pulses to foods positioned between two electrodes for a short period of time. Higher intensity fields are more effective in microbial inactivation, while low to medium-intensity fields, with short treatment times, have been employed to enhance mass transfer in solid foods (Zhang et al. 2019; Zielinski et al. 2021), which is advantageous for extracting compounds located within cells such as anthocyanins.

Due to the electropermeabilization mechanism, PEF has proven to be a sustainable and green extraction technology, providing positive environmental impacts in food processing. One of the main advantages of this technology, besides avoiding temperature increase, is the reduction in solvent usage during extraction and shorter processing time, enabling a decrease in the consumption of natural resources and solvents (Bocker & Silva 2022).

2.4.2.5 High-Voltage Electrical Discharge

High-voltage electrical discharge (HVED) emerges as an alternative non-thermal technique, proving to be effective, ecological, and with great potential to enhance the extraction of bioactive compounds from food by-products. This technique operates based on generating discharges through a high local electric field. By applying high voltage through an aqueous solution contained in a treatment chamber between two submerged electrodes, the propagation of various electrons occurs. The result is the formation of high-pressure shock waves and cavitation phenomena leading to cell rupture and disintegration, consequently increasing mass transfer (El Kantar et al. 2023).

Díaz-Álvarez et al. (2022) showed that HVED treatment enhances mass transfer rates in conventional extraction processes. They observed modifications in blueberry cell walls treated with HVED, resulting in increased anthocyanin content by 64.66% compared to untreated samples. Additionally, HVED reduced exposure times and energy consumption while achieving positive cell permeabilization rates at high operational frequencies, highlighting its effectiveness in improving anthocyanin extraction.

A comparative analysis of various anthocyanin extraction techniques from different plant sources reveals distinct performance characteristics (Table 2.1).

In Summary, the UAE showed superior anthocyanin recovery in *H. sabdariffa* calyx (Pinela et al. 2019), while Homogenization-Assisted Extraction (HAE) was more effective for jabuticaba peel (Albuquerque et al. 2020). The PLE offered shorter extraction times and reduced solvent usage, despite fewer anthocyanins of blackberry were obtained in some cases compared to conventional methods (Machado et al. 2015). The SFE demonstrated higher selectivity and effectiveness, particularly for juçara residues Garcia-Mendoza et al. (2017). And, the PEF technology and HVED showed promise in enhancing anthocyanin extraction efficiency while minimizing environmental impact (Medina-Meza & Barbosa-Cánovas 2015).

These findings underscore the importance of selecting solvents and appropriate extraction techniques tailored to specific plant sources and desired outcomes.

2.5 Strategies to Improve the Stability of Anthocyanins

Exploring methods to make anthocyanins more stable is important for in-depth research into the functional properties of anthocyanins and their industrial application in food. Effective strategies that can reduce the degradation of anthocyanins include, for example, acylation, co-pigmentation, metal complexation, and self-association (Table 2.2).

2.5.1 Acylation

Acylation emerges as an efficient method for preserving the stability and color intensity of anthocyanins. The introduction of acyl groups shields anthocyanins against hydrophilic elements and external factors, and contributes to heightened stability. Anthocyanins modified through acylation display intricate patterns of glycosylation and

TABLE 2.1

Comparison of Anthocyanin Extraction Techniques from Different Plant Sources

Resource	Processing Method	Extract Type	Temperature	Time	Main Findings	Reference
Hibiscus sabdariffa calyx	UAE and HAE	Hydroethanolic	~30°C	26–30 min	The UAE method significantly outperformed HAE in anthocyanin recovery, yielded 51.76 mg/g of dry residue, whereas HAE yielded only 20.86 mg/g of dry residue.	Pinela et al. (2019)
Jabuticaba (*Myrciaria jaboticaba* (Vell.) O. Berg.) peel	UAE and HAE	Hydroethanolic	~47°C	21–24 min	Anthocyanin recovery from jabuticaba peel was higher in HAE, yielded 76 ± 1 mg TAC/g while UAE yielded 32 ± 1 mg TAC/g. HAE proved more effective and selective than UAE.	Albuquerque et al. (2020)
Blackberry (*Rubus fruticosus* L.) residues	PLE, Soxhlet, and maceration.	Hydroethanolic and methanolic	at room temperature – 100°C	30 min–24 h	PLE yielded fewer anthocyanins (1.02 mg/g) than Soxhlet (1.68 mg/g) and maceration (1.21 mg/g). However, it's preferred for its shorter extraction time (30 min vs. 5 and 24 h) and lower solvent use (from 40 g/g and 20 g/g to 18 g/g)	Machado et al. (2015)
Juçara (*Euterpe edulis* Mart.) residues	PLE, SFE, UAE, Soxhlet, and ABE	Ethanolic, Hydroethanolic, aqueous,	40–80°C	45 min–6 h	In pressurized liquid extraction, acidified water (pH 2.0) at 40°C yielded the most anthocyanins, with SFE (6.2 mg/g dr) outperforming PLE (1.76 mg/g dr). Both methods showed potential for juçara residue anthocyanin extraction over low-pressure methods.	Garcia-Mendoza et al. (2017)
Grapes (*Vitis vinifera* L.) peel and plums (*Prunus domestica* var. Casselman) peel	PFE, UAE, and water extraction.	Aqueous	3–70°C	58 µs–1 h	PEF with larger chamber diameter was effective for grape skin anthocyanin extraction, while 50°C ultrasound yielded higher plum skin anthocyanin yields, both surpassing conventional water extraction at 70°C.	Medina-Meza and Barbosa-Cánovas (2015).

Note: UAE: ultrasound-assisted extraction; HAE: heat-assisted extraction; PLE: pressurized liquid extraction; SFE: supercritical fluids extraction; ABE: agitated bed extraction; PFE: pulsed electric field.

acylation, which can be achieved via both chemical and enzymatic acylation. These acylated derivatives exhibit superior stability and solubility in organic solvents compared to their non-acylated counterparts. Acylation serves as a principal avenue for augmenting the polarity and molecular size of anthocyanins while altering their spatial configuration. Factors such as the active site of acylation and the types and quantities of acyl groups exert influence over the acylation process (Cai et al. 2022; Luo et al. 2022).

Notably, anthocyanins bearing multiple acyl groups demonstrate enhanced stabilization compared to those with a singular acyl group. Polyacylated derivatives, associated with peonidin and cyanidin from purple-fleshed sweet potatoes, showcase superior bioavailability and resistance to pH fluctuations, highlighting their efficacy in maintaining color the red and blue in varied pH environments. (Oliveira et al. 2019). Furthermore, hydrogen bonding and electrostatic interactions bolster the stability of anthocyanin interactions with pectin in acidic conditions, facilitating stability during gastrointestinal simulation (Koh, Xu & Wicker 2020).

The acylated anthocyanins, particularly those featuring diacylated groups and disaccharides of paeoniflorin derivatives, exhibit prolonged stability during storage, underscoring their more favorable against environmental stressors such as high temperatures, pH variations, light exposure, and oxygen presence (Quan et al. 2019).

2.5.1.1 Chemical Acylation

Chemical acylation represents a viable strategy for enhancing anthocyanin stability (Table 2.2). The selection of suitable acylating agents, catalysts, and reaction conditions significantly affects the success of chemical acylation processes. Aliphatic carbonyl acids, including acetic acid, oxalic acid, heptane acid, and lauric acid, are commonly employed as acylating agents under optimized conditions (Xu et al. 2017).

The primary mechanism in acylation is the stacking interaction (π) between the acyl group and the chromophore under acidic conditions. The cis-trans configuration plays a pivotal role in acylation, with cis-isomers offering substantial protection against hydration (Moloney et al. 2018; Sigurdson, Tang & Mónica Giusti 2018). Anthocyanin modification, particularly through increasing the carbon chain length of aliphatic carbonyls, has demonstrated enhanced resistance against high temperatures and intense light. This process contributes to color enhancement, thus expanding the applicability of anthocyanins in various industries (Xu et al. 2017).

Despite its efficacy, chemical acylation presents certain limitations such as low selectivity, environmental concerns, and uncertainty regarding the acylated structure. Challenges include the precise acylation of specific hydroxyl group positions and the potential suppression of antioxidant capacity due to indiscriminate binding to phenolic hydroxyl groups. Compared to enzymatic acylation, chemical acylation struggles more with producing specific acylation products (Cai et al. 2022). Therefore, enzymatic acylation is considered a more efficient method than chemical acylation for increasing anthocyanin stabilization in application.

2.5.1.2 Enzymatic Acylation

Enzymatic acylation represents a vital biochemical avenue dependent on activated acyl donors such as fatty acids and aromatic acids. Efficient enzymatic acylation necessitates suitable conditions including the presence of acyl-COA, acyl glucose, optimized pH, controlled temperature, and ample energy. Acylase serves as a catalyst directing the acylation

TABLE 2.2

Examples of Anthocyanin Association from Diverse Sources

Source	Acylation Type	Acylate	Acyl Donor	Catalyst	Outcome	Reference
Purple-fleshed sweet potato	Chemical acylation	Peonidin-3-(6'-hydroxybenzoyl)-sophoroside-5-glucoside; peonidin-3-(6'-hydroxybenzoyl-6"-caffeoyl)-sophoroside-5-glucoside	p-hydroxybenzoic acid, caffeic acid	None	Higher capacity in retaining the red and blue colors at acidic and basic pH values	Oliveira et al. (2019)
Blueberry	Chemical acylation	Malvidin-3-glucoside; cyanidin-3-glucoside; and delphinidin-3-glucoside	Blueberry pectin	None	Increased stability at pH 4.0, and stability of anthocyanins after simulated gastrointestinal digestion	Koh, Xu and Wicker (2020)
Purple-fleshed sweet potato	Chemical acylation	Peonidin 3-p-hydroxybenzoyl sophoroside-5-glucoside, peonidin 3-feruloyl sophoroside-5-glucoside, peonidin 3-caffeoyl sophoroside-5-glucoside, peonidin dicaffeoyl sophoroside-5-glucoside, peonidin 3-caffeoyl-p-hydroxybenzoyl sophoroside5-glucoside and peonidin caffeoyl-feruloyl sophoroside-5-glucoside.	p-hydroxybenzoic acid, ferulic acid or caffeic acid or glucosylated with glucose, or sophorose	None	The PFSPAs were more stable during storage	Quan et al. (2019)
Blueberry	Chemical acylation	Blueberry anthocyanin	acetic acid, oxalic acid, heptane acid, and lauric acid	None	Increased stability of acylated blueberry anthocyanin on temperature and light	Xu et al. (2017)
Red cabbage	Chemical acylation	Cyaniding (3,3',4',5,7-pentahydroxyflavylium)	hydroxycinnamic acids	None	Improved color stabilizing, and an improved resistance against the long-term color loss in mildly alkaline conditions	Moloney et al. (2018)
Asian eggplant and Black goji	Chemical acylation	cis and trans p-coumaroylated isomers of delphinidin-3-rutinoside-5-glucoside (Dp-3-rut-5-glu) (Pt-3-rut-5-glu) and trans p-coumaroylated derivatives of petunidin-3-rutinoside-5-glucoside (Pt-3-rut-5-glu)	Alkaline saponification	None	Cis acylation reduced hydration across pH, increasing color intensity, while trans acylation generally improved color retention over time	Sigurdson, Tang and Mónica Giusti (2018)

(Continued)

TABLE 2.2 *(Continued)*

Examples of Anthocyanin Association from Diverse Sources

Source	Acylation Type	Acylate	Acyl Donor	Catalyst	Outcome	Reference
Blackcurrant	Enzymatic acylation	Monoacylated derivatives of anthocyanins	Lauric acid	Novozym 435 lipase	The half-life extended	Yang et al. (2019)
Black Rice	Enzymatic acylation	cyanidin-3-(6''-benzoyl)- glucoside, cyanidin 3-(6''- salicyloyl)-glucoside, and cyanidin-3-(6''-cinnamoyl)-glucoside	Benzoic acid methyl ester Salicylic acid methyl ester Cinnamic acid methyl ester	Novozym 435 lipase	Improved half-life to resist thermal and light	Yan et al. (2016)
Alpine bearberry	Enzymatic acylation	Cyanidin-3-O-(6''-dodecanoyl) galactoside	Lauric acid	Novozym 435 lipase	Increased thermal stability, lipophilicity, color enhancement, inhibition of lipid peroxidation, and light resistivity of anthocyanin	Yang et al. (2018)

CO-PIGMENTATION

Source	Association Mechanism	Types of Anthocyanins	Co-Pigments	Outcome	Reference
Blueberry juice	Intermolecular co-pigmentation	Peonidin-3-glucoside	Gallic acid	Color saturation and anthocyanins stability	Zhang et al. (2020)
Purple corn	Intermolecular co-pigmentation	C-glycosyl flavones	citric acid	Better stability and significantly longer half-lives (up to 50% longer)	Chatham, Howard and Juvik (2020)
Black chokeberry	Intermolecular co-pigmentation	Cyanidin-3-O-galactoside and cyanidin-3-O-arabinoside	rosmarinic acid, syringic acid, and catechin	Hyperchromic effect most significant, magnification of color intensity and decrease of color hue, both related to a more pleasant color	Klisurova et al. (2019)
Açai fruit	Intermolecular co-pigmentation	cyanidin-3-rutinoside and cyanidin-3-glucoside	flavone-C-glycosides	Increased anthocyanin color and stability	Pacheco-Palencia (2010)
Cabernet Sauvignon grape	Intermolecular co-pigmentation	malvidin- 3-glucoside	caffeic acid	Increased stability to anthocyanin solutions under the storage conditions evaluated	Gris et al. (2007)

(Continued)

TABLE 2.2 *(Continued)*

Examples of Anthocyanin Association from Diverse Sources

CO-PIGMENTATION

Source	Association Mechanism	Types of Anthocyanins	Co-Pigments	Outcome	Reference
Grape juice	Intermolecular co-pigmentation	delphinidin 3,5-diglucoside, cyanidin 3,5-diglucoside, Petunidin 3,5-diglucoside, Peonidin 3,5-diglucoside, Malvidin 3,5-diglucoside	ascorbic acid	Increased anthocyanin retention, decreased ascorbic degradation and antioxidant capacity when compared to control juices	Pozo-Insfran et al. (2007)
Strawberry purees	Intermolecular co-pigmentation	pelargonidin-3-glucoside, pelargonidin-3-rutinoside and cyanidin-3-glucoside	sinapic acid	Improved the stability and the resistance of anthocyanins in HPP samples during storage to degradation	Terefe, Netzel and Netzel (2019)
Grape	Intramolecular co-pigmentation	delphinidin-3-O-glucoside, Cyanidin-3-O-glucoside, Petunidin-3-O-glucoside, Peonidin-3-O-glucoside, Malvidin-3-O-glucoside	ferulic acid, D-gluconic acid, caffeic acid, and vanillic acid	Increased the light and heat stability of anthocyanins and exhibit satisfactory antioxidant activity	He et al. (2018)
Purple carrot extracts	Intermolecular co-pigmentation	cyanidin-3-xylosyl(glucosyl) galactoside	L-phenylalanine; Ltyrosine; L-tryptophan; and ε-poly-L-lysine	The half-life: increased more than double	Cheng et al. (2017)
Muscadine grape (juice and wine)	Intermolecular co-pigmentation	3,5-diglucosides	formononetin, biochanin A, and prunetin	Color improvement and anthocyanin stability was observed in the wines during storage	Talcott, Peele and Brenes (2005)
Sorghum	Intermolecular co-pigmentation	3-deoxyanthocyandins, luteolinidin and apigeninidin and their methoxylated derivatives	tannic, ferulic, and O-coumaric acids and rutin	Enhanced color stability of pigments	Awika (2008)
Hydrangea sepals	Metal complexation	delphinidin	Al^{3+}	Promoted the formation of blue and stabilized	Schreiber, Swink and Godsey (2010)
Not related	Metal complexation	cyanin, deoxyanthocyanidin, and luteolinidin	Al^{3+}	Stabilized the blue ionized quinoidal base avoiding its degradation by reaction with oxygen	Moncada et al. (2003)

(Continued)

TABLE 2.2 *(Continued)*

Examples of Anthocyanin Association from Diverse Sources

CO-PIGMENTATION

Source	Association Mechanism	Types of Anthocyanins	Co-Pigments	Outcome	Reference
Red pears (D'Anjou)	Metal complexation	Not related	Stannous (Sn) ion,	Resulted in stable red pigments	Ngo and Zhao (2009)
Grape skin	Metal complexation	Not related	Montmorillonite, a synthetic nonswelling mica and silica	Enhanced stability against visible light irradiation	Kohno et al. (2009)
lderberry, black currant, red cabbage, and purple carrot	Metal complexation	Not related	ferric ions	Improvement thermal and storage stability	Buchweitz et al. (2013)
Grape	Metal complexation	Not related	Fe^{2+} and Fe^{3+}	Were shown to synergistically affect the color spectra (effect on bathochromic and hyperchromic) of grape anthocyanins at varying pHs	Celli et al. (2018)
Standard substance	Metal complexation	delphinidin-3-O-glucoside, cyanidin-3-O-glucoside, pelargonidin-3-O-glucoside	Fe^{3+}	Enhanced thermal stability	Tachibana, Kimura and Ohno (2014)
Purple potato	Metal complexation	cyanidin and peonidin	K^+, Ca^{2+}, Mg^{2+}, Zn^{2+}, Cu^{2+}, Fe^{2+}, and Al^{3+}	Hyperchromic effect and bathochromic	Li et al. (2016)
Red wine (Vitis vinifera L.)	Self-association	malvidin-3-O-glucoside (mv3glc) and malvidin-3-O-(6- coumaroylglucoside)	Coumaric acid residue	Improvement anthocyanin color stabilization	Fernandes et al. (2015)
Wine	Self-association	3-glucosides, malvidin, delphinidin,` and peonidin	Not related	Influence on the apparent hydration constant of the anthocyanins with subsequent modification in the color	González-Manzano et al. (2008)

Note: PFSPAs: purple-fleshed sweet potato anthocyanins; HPP: high-pressure processing.

reaction toward specific molecular targets, thereby enhancing catalytic efficiency under mild conditions (Cai et al. 2022).

Recent studies affirm the benefits of enzymatic acylation, highlighting improvements in thermal stability, lipophilicity, color enhancement, and inhibition of lipid peroxidation. Aromatic acids and their derivatives significantly enhance the thermal stability and light resistance of anthocyanins, particularly when employed with biocatalysts like Novozym 435 (Yan et al. 2016; Yang et al. 2018).

Investigations into enzymatic acylation of blackcurrant and black rice anthocyanins demonstrate superior thermal stability and resistance to light and heat in acylated derivatives compared to with the original anthocyanins, also known as parental anthocyanins (Yan et al. 2016; Yang et al. 2019). These findings underscore the potential of enzymatic acylation to bolster the stability and resistance of anthocyanins, thereby expanding their utility across various applications.

2.5.2 Copigmentation Effect

Copigmentation stands out as a pivotal and efficient approach aimed at fortifying anthocyanins against adverse conditions, including high temperatures, intense light exposure, and oxygen presence, thereby stabilizing these compounds (Table 2.2). This process entails the formation of molecular associations or complexes between anthocyanins and other colorless organic compounds or metal ions, resulting in color change or intensification, aimed at anthocyanin structural stability (Cavalcanti et al. 2011; Enaru et al. 2021; Cai et al. 2022; Chen et al. 2022).

For a molecule to function as a co-pigment, it must fulfill two essential criteria: (1) It must possess sufficiently extensive π-conjugated systems conducive to π-π stacking interaction. (2) It must feature hydrogen bridge donor/acceptor groups such as OH and C=O groups. Consequently, a solution containing anthocyanins in the presence of a co-pigment typically exhibits significantly more intense coloration than theoretically anticipated for the medium's pH value. It's worth noting that the phenomenon of co-pigmentation is contingent upon pH levels, with increased pH potentially leading to anthocyanin degradation and subsequent reduction in color intensity (Enaru et al. 2021).

The stability of co-pigmented complexes is susceptible to thermal degradation, wherein high temperatures may either have negligible effects on the co-pigmentation mechanism or yield counterproductive outcomes. For instance, co-pigmentation with compounds like ferulic acid or sinapic acid, or exposure to heat treatments at specific temperatures and durations, can accelerate anthocyanin degradation (Enaru et al. 2021).

The interaction between anthocyanins and co-pigments varies based on the nature and concentration of these compounds, resulting in diverse shades and intensities of color. While co-pigments may be colorless in isolation, their combination with anthocyanins induces a hyperchromic effect and batochromic shift in the absorption spectrum, leading to increased light absorption. The constituents of these pigments encompass a wide array of compounds, including alkaloids, amino acids, nucleotides, organic acids, flavonoids, metals, and even other anthocyanins (Castañeda-Ovando et al. 2009; Escribano-Bailón & Santos-Buelga 2012; Molaeafard et al. 2021; Oliveira et al. 2021).

The efficacy of co-pigmentation is profoundly influenced by the steric arrangement and size of the co-pigmented substance, with a planar structure being more conducive to stabilizing the flavylium ion, thereby facilitating a more efficient co-pigmentation effect (Escribano-Bailón & Santos-Buelga 2012).

An exemplary study involving model beverages demonstrated that the introduction of a flavone-rich extract into samples containing cyanidin- and pelargonidin-based anthocyanins resulted in an increase of up to 50% in the half-life of these anthocyanins, indicating significantly improved stability (Chatham, Howard & Juvik 2020). In addition, Klisurova et al. (2019) observed improvements in color and stability after analyzing various co-pigments for black chokeberry anthocyanins. It was found that the addition of herbal extracts induced a significant hyperchromic effect, even at much lower dosages compared to the pure co-pigments.

The interaction of anthocyanins present in wines, juices, and fruit-based products with phenolic cofactors has been associated with color enhancement and the stability of these pigments. Pacheco-Palencia and Talcott (2010) investigated the influence of the addition of polyphenolic cofactors on the phytochemical and color stability of anthocyanins in açaí fruit. The results indicated that flavone-C-glucoside acts as a color enhancer and stabilizing agent.

In this sense, Zhang et al. (2020) evaluated the influence of adding the co-pigment GA on the stability of anthocyanin and the color of blueberry juice. The findings revealed that the incorporation of GA resulted in a more pronounced crimson hue, enhanced color saturation, and improved stability of anthocyanins in the blueberry juice. Therefore, the addition of GA could serve as processing technique to produce juices with superior color quality and stable anthocyanin content.

Gris et al. (2007) conducted a study on co-pigmentation in Cabernet Sauvignon grape extracts. The spectrophotometric results revealed an interaction between the crude anthocyanin extract and caffeic acid, suggesting a co-pigmentation phenomenon. It was observed that the addition of caffeic acid significantly increased the stability of the anthocyanins. Furthermore, the introduction of partially purified co-pigments intensified the color of muscadine grape juice, improved antioxidant activity, and minimized phytochemical losses during processing and storage under high hydrostatic pressure (HHP) (Pozo-Insfran et al. 2007).

In summary, co-pigmentation has emerged as a promising strategy to enhance the stability and color of anthocyanins, with significant implications for the food and beverage industry. Anthocyanins can interact with co-pigments through diverse mechanisms, which vary based on the nature of the co-pigment involved. An association between anthocyanins and colorless co-pigments is called intermolecular co-pigmentation. Additionally, intramolecular associations occur when there is interaction between the chromophore and non-chromophore regions within the same molecule. Another mode of interaction involves anthocyanidin nuclei or their interaction with aromatic acyl groups, which is known as self-association. Consequently, both intramolecular and intermolecular co-pigmentation, metal complexation, and self-association serve as effective strategies for bolstering anthocyanin stability (Cavalcanti et al. 2011; Cao et al. 2021; Enaru et al. 2021; Chen et al. 2022).

2.5.2.1 *Intramolecular and Intermolecular Co-Pigmentation*

Intramolecular co-pigmentation is a process where the anthocyanidin's chemical structure interacts with a co-pigment that is inherently part of the anthocyanin. This interaction is facilitated by the alignment of the aromatic acyl fragments of the acylated anthocyanins with the anthocyanidins' core structure. The structural attributes of the anthocyanins play a crucial role in determining the effectiveness of this interaction. Moreover, the sugar present in the anthocyanin structure serves as a bridge between the anthocyanin framework and the acyl segment. This arrangement allows the molecule to adjust its position so that the aromatic acyl group can interact with the π system of the planar pyrrole ring, thereby shielding it from a nucleophilic attack by water. The creation of sandwich-like structures, where the anthocyanin core is nestled between two acyl segments, leads to the effective stabilization of anthocyanin pigments. This phenomenon is pronounced in anthocyanins possessing two or more aromatic acyl segments necessary for forming such complexes (Enaru et al. 2021; Cai et al. 2022).

In contrast, intermolecular co-pigmentation frequently manifests in red wines. This phenomenon arises from van der Waals interactions, hydrophobic effects, and hydrogen bonding, facilitating a non-covalent interplay between anthocyanin molecules and co-pigments. These interactions contribute to the wine's purple hues and deeper color through two mechanisms: the hyperchromic and batochromic effects. The alteration in color is attributed to conformational changes in the anthocyanin chromatophore, which prevents hydration reactions, ultimately increasing the ratio between the flavylium cation (red color) and the quinonoid base (purple color). This interaction is pivotal in safeguarding against bleaching during light and heat treatments, essential for specific product production (Malien-Aubert, Dangles & Amiot 2001).

Terefe et al. (2019) found that the co-pigmentation of anthocyanins in strawberry puree with sinapic acid led to a notable improvement in degradation resistance and antioxidant capacity under conditions of high pressure and thermal processing. Moreover, intramolecular co-pigmentation showed a beneficial effect on the stability and hyperchromicity of anthocyanins from *Vitis amurensis* Rupr (He et al. 2018). The inclusion of flavonoids, polyphenols, amino acids, peptides, and proteins to co-pigmentation can enhance the storage stability and prolong the shelf life of foods containing anthocyanins (Cheng et al. 2017).

In a distinct investigation, Talcott et al. (2005) demonstrated that isoflavonoids sourced from red clover (*Trifolium pratense*) bolster the overall color and stability of anthocyanin 3,5-diglycosides present in muscadine grape juice and wine (*Vitis rotundifolia*) via intermolecular co-pigmentation reactions. Awika (2008) delved into the impact of common phenolic co-pigments, including tannic, ferulic, and o-coumaric acids, as well as rutin, on the stability and behavior of six 3-deoxyanthocyanins. The findings unveiled that phenolic compounds offer heightened color stability, with substitutions at C-5 emerging as pivotal determinants for the general characteristics of 3-deoxyanthocyanidins. These findings underscore the importance of these compounds in maintaining color and stability in products such as grape juices and wines. Understanding these mechanisms can contribute to improving the quality and stability of food products.

2.5.2.2 Metal Complexation

Metal complexation is a widely utilized technique in the food industry to enhance the stability of anthocyanins, thereby preserving the vibrant color of food products. Through the coordination of metal cations, such as iron ions, with anthocyanins, the distribution of non-localized electrons in the pyrylium ring is modulated, resulting in profound color effects. The most intense color effects are observed when co-pigmentation occurs, with two positively charged metal ion protons or +2 or +3 valence metals being simultaneously captured by anthocyanins that have catechol or pyrogallol in the B ring. Examples of these anthocyanins include cyanidin, petunidin, delphinidin, and their derivatives. The metal ions commonly involved in this process are sodium (Na^+), iron ($Fe^{2+/3+}$), aluminum (Al^{3+}), magnesium (Mg^{2+}), and zinc (Zn^{2+}). This complexation process not only stabilizes anthocyanins but also improves their thermal and storage stability, thus contributing to prolonged shelf life and enhanced product quality (Table 2.2) (Cavalcanti et al. 2011; Enaru et al. 2021; Cai et al. 2022).

In addition, studies suggest that the blue color observed in some plants is the result of the interaction between anthocyanins and metals. For instance, it is proposed that the interaction between anthocyanins and molybdenum contributes to the stabilization of the blue coloration in Hindu cabbage tissue (Muhammad & Saad 2016). Studies indicate that the combination of o-dihydroxy anthocyanins and Fe^{3+} or Mg^{2+} ions in a solution with pH = 5 leads to the development of the blue hue in plants. This occurs under specific conditions, particularly when the stoichiometric ratio between anthocyanin and Fe^{3+} is 1:6 (Castañeda-Ovando et al. 2009; Cortez et al. 2017).

In a separate study, Schreiber et al. (2010) explored the chemical mechanism of Al^{3+} complexation with delphinidin. They proposed a two-step process: initially, the production of the blue quinoidal base anion from delphinidin occurs, which complexes with Al^{3+}. Subsequently, this complex is stabilized by the π-π stacking of a second delphinidin, acting as a flavilium cation on top of the blue complex. This process results in a bathochromic change in the spectral signature of the cation and intensifies the blue color. These findings are in line with the data of Moncada et al. (2003), who demonstrated that complexation with Al^{3+} stabilizes the ionized blue quinoidal base, preventing its degradation by reaction with oxygen. Ngo and Zhao (2009) reported that the use of an aqueous formula composed of stannous, formaldehyde, hydrochloric acid, and tannic acid, followed by steam heating at a temperature of 73°C, resulted in the anchoring and stabilization of red pigments in pear peels.

Additionally, Kohno et al. (2009) reported that incorporating natural anthocyanins into montmorillonite significantly enhanced their stability in an alkaline environment and under visible light exposure. This improved stability was ascribed to both electrostatic interactions between the intercalated dye and the montmorillonite surface and the safeguarding effect against atmospheric oxygen. Consequently, the intercalated and stabilized anthocyanin emerges as a promising candidate for eco-friendly dye applications.

In particular, iron ions (Fe^{3+} and Fe^{2+}) have been found to exert a significant stabilizing effect on anthocyanins present in various foods and beverages. Their interaction with anthocyanins leads to the formation of stable complexes, characterized by hyperchromic properties, which enhance color retention and overall product quality (Buchweitz et al. 2013; Celli and Su Ling Brooks 2017). Tachibana et al. (2014) showed that the addition of chondroitin sulfate, alginate, and low-methoxylated pectin to the anthocyanin Fe^{3+} complex can effectively contribute to the synergistic effect on the hyperchromic displacement of cyanidin-3-O-glucoside and delphinidin-3-O-glucoside. Additionally, favorable hyperchromic effects have been observed in metal complexation systems involving other cations such as K^+, Ca^{2+}, Mg^{2+}, Zn^{2+}, Cu^{2+}, Fe^{2+} and Al^3 in conjunction with anthocyanin-rich sources like purple sweet potato (Li et al. 2016).

Therefore, metal chelation represents a significant strategy for stabilizing anthocyanins present in fruit-derived products such as juices and wines. Enrichment with metals can result in an increase in added value, as well as providing a chemical stabilization mechanism, guaranteeing nutritional and sensory quality to the products obtained.

2.5.3 Self-Association

Self-association is an effective strategy for increasing the stability of anthocyanins, allowing them to form bonds with two or more other anthocyanin molecules (Table 2.2). Anthocyanin molecules have been observed to form bonds with each other, particularly when their concentration exceeds 0.1 mM. These self-association complexes, stabilized by hydrophobic interactions within their aromatic cores, exhibit a chiral structure reminiscent of left-handed screws (Cavalcanti et al. 2011; Enaru et al. 2021; Cai et al. 2022).

Various factors, including anthocyanin concentration, types, and positions within the B ring, as well as the presence of glucose at the C3 or C5 position, influence the formation and stability of these complexes. The weaker associations observed at the C3′ position and stronger interactions at the C5′ position. Furthermore, methoxyl substitutions at both the C3′ and C5′ positions significantly augment self-association and color intensity (Cai et al. 2022). Notably, acylated anthocyanins demonstrate a heightened propensity for self-association, leading to increased interaction and the formation of hydrophobic π-π complexes (Fernandes et al. 2015).

In the context of red and blackberry wine models, the chromogenic mechanism primarily involves self-association, positively impacting coloration and stabilization processes. Leveraging self-association in extract concentration and purification procedures can contribute to natural stabilization, particularly when combined with pH adjustment techniques (González-Manzano et al. 2008).

Self-association can therefore be used as an effective strategy in extract concentration and/or purification procedures. In this context, it is possible that an interaction occurs between the anthocyanins themselves. When associated with other factors, such as pH adjustment, this interaction can contribute significantly to the natural stabilization process.

2.6 Industrial Applications

Anthocyanins, renowned for their natural pigment attributes, are increasingly being incorporated into diverse food formulations. Beyond augmenting polyphenol intake, the integration of anthocyanins aims to enhance product acceptability, extend shelf life, and avert spoilage risks. This strategic utilization can elevate products to functional foods, thereby amplifying their health-promoting properties and addressing apprehensions surrounding the safety of synthetic food colors (Martins et al. 2016; Xiong et al. 2019; Echegaray et al. 2022).

In a study by Papillo et al. (2018), the addition of anthocyanin-rich extracts of Italian black rice (*Oryza sativa* L., var. Artemide) to baked cookies was examined. Employing hydroalcoholic extracts enriched with anthocyanins and stabilized through either *spray-drying* or *freeze-drying* methods, notable enhancements in both anthocyanin (between 194 and 230 µg/g cookie) content and antioxidant (482–595 vs 283 µg TE/g cookie) capacities were observed in fortified cookies compared to control samples. Similar outcomes were documented by Maner et al. (2017) following the incorporation of wine grape pomace powder into cookie formulations, resulting in enriched color (2.03–3.51 mg/g in the samples with extract vs 0.16 mg/g in the control sample), heightened antioxidant activity, and improved organoleptic attributes. In a similar vein, López et al. (2019) noted that the inclusion of an anthocyanin-rich extract of *Arbutus unedo* L. concurrently enhanced antioxidant activity and provided a more appealing color in baked wafers. These studies collectively underscore the potential of anthocyanin-rich extracts to enhance the nutritional and sensory attributes of bakery items.

Pasqualone et al. (2013) investigated the efficacy of grape pomace extract in enhancing the anthocyanin content and color intensity of cookies. Their findings revealed that incorporating a freeze-dried extract not only increased the anthocyanin content but also intensified the reddish hue of the cookies compared to the control samples. Additionally, Albuquerque et al. (2020) conducted a study focusing on enriching macarons with extracts derived from jaboticaba epicarp (*Myrciaria jaboticaba* (Vell.) Berg). Their results demonstrated that this enrichment led to a more stable color over a six-day period, surpassing the performance of the commercial colorant E136. Similarly, Sui (2017) demonstrated the enhanced resistance to lipid oxidation in cookies enriched with anthocyanins from black rice extract powder (at concentrations of 2% and 4%) compared to those containing the synthetic additive butylated hydroxyanisole (BHA).

Moreover, a study on gluten-free tortillas and cookies revealed that the integration of black bean seed husk extracts into nixtamalized corn flour enhanced the color of the end products. The results showed significant improvements in color without notable effects on texture. The authors noted that the inclusion of 3 g of extract per kg of dough maintained the dough's rheological and functional properties. However, an addition of 7 g of extract/ kg of dough, some rheological changes were observed. Notably, both scenarios retained approximately 60% of the bioactive compounds (Chávez-Santoscoy et al. 2016).

Wallace and Giusti (2008) noted that incorporation of 20 mg of Cy-3-glu from blueberry powder (*Berberis boliviana* L.) per 100 g of yogurt resulted in products exhibiting a color reminiscent of commercial blueberry yogurt. Remarkably, the color remained stable over the initial two-month period, and the products maintained high phenolic stability, irrespective of the fat matrix. In a subsequent study, Karaaslan et al. (2011) enhanced yogurts with extracts from various grape types (Cabernet Sauvignon, Chardonnay, Shyrah, and Merlot) and grape tissue callus, both of which are rich in phenolic compounds and anthocyanins (1 mL of acidified extract for every 100 mL of yogurt). The yogurts produced had superior anthocyanin content and antioxidant capacity compared to the control yogurts. Notably, yogurts fortified with white grape extracts had a reduced phenolic compound content.

In relation to a dairy product like kefir, Montibeller et al. (2018) discovered that fortifying with anthocyanins from grape skins served as a natural colorant. The enhanced kefir preserved over 50% of the total anthocyanins for the first 27 days and approximately 67% on the 16th day. However, it is crucial to acknowledge that the content of each anthocyanin gradually decreased during storage to varying extents. Particularly, peonidin-3-glucoside exhibited the highest stability, retaining 87% of its anthocyanins by day 16, while delphinidin-3-bglucoside decreased to 50% by day 12. Interestingly, kefir made with these natural colorant exhibited properties akin to kefir without additives.

Choi et al. (2017) engineered a smart film that acts as a pH indicator, where they affixed the anthocyanin extract derived from purple sweet potato to a starch agar base. This film was utilized to track the pH alterations in pork samples, with the hues fluctuating at distinct pH levels, signifying spoilage. The film's color, which transitions from red to green, can signal the spoilage of pork as the pH shifts.

Consequently, enhancing the control of factors that influence the stability of anthocyanins, coupled with a high level of selectivity and reduced production expenses, is of utmost significance. This presents a promising avenue for future investigations aimed at bolstering the stability of anthocyanins, given their association with numerous health advantages.

2.7 Conclusion

The stability of anthocyanins during food processing is a crucial issue for the food industry, as these bioactive compounds play a significant role in promoting health. The data discussed here has provided a deeper understanding of the factors that affect the stability of these pigments during the various stages of processing. Indicating that the proper selection of processing methods, temperature control, pH, and the presence of other food components can play a crucial role in preserving the beneficial properties of anthocyanins. Therefore, the identification of effective strategies to minimize the degradation of these compounds during processing opens the door to the development of healthier and more nutritious foods.

We therefore conclude that this chapter contributes to the advancement of scientific knowledge at the intersection of food processing technology and bioactive compounds, which have the potential not only to improve the quality of processed foods but also to offer additional health benefits to consumers, paving the way for the development of healthier and more sustainable food products that meet the growing demands for functional foods.

References

Albuquerque, Bianca R., José Pinela, Lillian Barros, M. Beatriz P.P. Oliveira, and Isabel C.F.R. Ferreira. 2020. "Anthocyanin-Rich Extract of Jabuticaba Epicarp as a Natural Colorant: Optimization of Heat- and Ultrasound-Assisted Extractions and Application in a Bakery Product." *Food Chemistry* 316 (June). https://doi.org/10.1016/j.foodchem.2020.126364

Aramwit, Pornanong, Nipaporn Bang, and Teerapol Srichana. 2010. "The Properties and Stability of Anthocyanins in Mulberry Fruits." *Food Research International* 43 (4): 1093–97. https://doi.org/10.1016/j.foodres.2010.01.022

Askar, Kahtan A., Zabarjad H. Alsawad, and Moayad N. Khalaf. 2015. "Evaluation of the pH and Thermal Stabilities of Rosella Anthocyanin Extracts under Solar Light." *Beni-Suef University Journal of Basic and Applied Sciences* 4 (3): 262–68. https://doi.org/10.1016/j.bjbas.2015.06.001

Awika, Joseph M. 2008. "Behavior of 3-Deoxyanthocyanidins in the Presence of Phenolic Copigments." *Food Research International* 41 (5): 532–38. https://doi.org/10.1016/j.foodres.2008.03.002

Benvenutti, Laís, Acácio Antonio Ferreira Zielinski, and Sandra Regina Salvador Ferreira. 2022. "Pressurized Aqueous Solutions of Deep Eutectic Solvent (DES): A Green Emergent Extraction of Anthocyanins from a Brazilian Berry Processing By-Product." *Food Chemistry: X* 13. https://doi.org/10.1016/j.fochx.2022.100236

Bocker, Ramon, and Eric Keven Silva. 2022. "Pulsed Electric Field Assisted Extraction of Natural Food Pigments and Colorings from Plant Matrices." *Food Chemistry: X* 15. https://doi.org/10.1016/j.fochx.2022.100398

Brouillard, Raymond, and Bernard Delaporte. 1977. "Chemistry of Anthocyanin Pigments. 2. Kinetic and Thermodynamic Study of Proton Transfer, Hydration, and Tautomeric Reactions of Malvidin 3-Glucoside." *Journal of the American Chemical Society* 26: 8461–68. https://doi.org/10.1021/ja00468a015

Brownmiller, C., L.R. Howard, and R.L. Prior. 2008. "Processing and Storage Effects on Monomeric Anthocyanins, Percent Polymeric Color, and Antioxidant Capacity of Processed Blueberry Products." *Journal of Food Science* 73: H72–79. https://doi.org/10.1111/j.1750-3841.2008.00761.x

Buchweitz, M., J. Brauch, R. Carle, and D. R. Kammerer. 2013. "Colour and Stability Assessment of Blue Ferric Anthocyanin Chelates in Liquid Pectin-Stabilised Model Systems." *Food Chemistry* 138 (2–3): 2026–35. https://doi.org/10.1016/j.foodchem.2012.10.090

Cai, Dongbao, Xusheng Li, Jiali Chen, Xinwei Jiang, Xiaoqiang Ma, Jianxia Sun, Lingmin Tian, et al. 2022. "A Comprehensive Review on Innovative and Advanced Stabilization Approaches of Anthocyanin by Modifying Structure and Controlling Environmental Factors." *Food Chemistry* 366 (January). https://doi.org/10.1016/j.foodchem.2021.130611

Cao, Qiqi, Jianwen Teng, Baoyao Wei, Li Huang, and Ning Xia. 2021. "Phenolic Compounds, Bioactivity, and Bioaccessibility of Ethanol Extracts from Passion Fruit Peel Based on Simulated Gastrointestinal Digestion." *Food Chemistry* 356 (May 2020): 129682. https://doi.org/10.1016/j.foodchem.2021.129682

Castañeda-Ovando, Araceli, Ma de Lourdes Pacheco-Hernández, Ma Elena Páez-Hernández, José A. Rodríguez, and Carlos Andrés Galán-Vidal. 2009. "Chemical Studies of Anthocyanins: A Review." *Food Chemistry.* https://doi.org/10.1016/j.foodchem.2008.09.001

Cavalcanti, Rodrigo N., Diego T. Santos, and Maria Angela A. Meireles. 2011. "Non-Thermal Stabilization Mechanisms of Anthocyanins in Model and Food Systems—An Overview." *Food Research International* 44 (2): 499–509. https://doi.org/10.1016/j.foodres.2010.12.007

Celli, Giovana B., Michael J. Selig, Chen Tan, and Alireza Abbaspourrad. 2018. "Synergistic Bathochromic and Hyperchromic Shifts of Anthocyanin Spectra Observed Following Complexation with Iron Salts and Chondroitin Sulfate." *Food and Bioprocess Technology* 11 (5): 991–1001. https://doi.org/10.1007/s11947-018-2055-z

Celli, Giovana Bonat, and Marianne Su Ling Brooks. 2017. "Impact of Extraction and Processing Conditions on Betalains and Comparison of Properties with Anthocyanins — A Current Review." *Food Research International* 100 (October): 501–9. https://doi.org/10.1016/j.foodres.2016.08.034

Chatham, Laura A., Jay E. Howard, and John A. Juvik. 2020. "A Natural Colorant System from Corn: Flavone-Anthocyanin Copigmentation for Altered Hues and Improved Shelf Life." *Food Chemistry* 310 (April). https://doi.org/10.1016/j.foodchem.2019.125734

Chávez-Santoscoy, Rocio A., Janet A. Gutiérrez-Uribe, Sergio O. Serna-Saldivar, and Esther Perez-Carrillo. 2016. "Production of Maize Tortillas and Cookies from Nixtamalized Flour Enriched with Anthocyanins, Flavonoids and Saponins Extracted from Black Bean (Phaseolus Vulgaris) Seed Coats." *Food Chemistry* 192 (July): 90–97. https://doi.org/10.1016/j.foodchem.2015.06.113

Chemat, Farid, Maryline Abert-Vian, Anne Sylvie Fabiano-Tixier, Jochen Strube, Lukas Uhlenbrock, Veronika Gunjevic, and Giancarlo Cravotto. 2019. "Green Extraction of Natural Products. Origins, Current Status, and Future Challenges." *TrAC Trends in Analytical Chemistry* 118. https://doi.org/10.1016/j.trac.2019.05.037

Chemat, Farid, Natacha Rombaut, Anne-Gaëlle Sicaire, Alice Meullemiestre, Anne-Sylvie Fabiano-Tixier, and Maryline Abert-Vian. 2017. "Ultrasound Assisted Extraction of Food and Natural Products. Mechanisms, Techniques, Combinations, Protocols and Applications. A Review." *Ultrasonics Sonochemistry* 34. https://doi.org/10.1016/j.ultsonch.2016.06.035

Chen, Yanpei, Tarun Belwal, Yanqun Xu, Quan Ma, Dong Li, Li, Hang Xiao, and Zisheng Luo. 2022. "Updated Insights into Anthocyanin Stability Behavior from Bases to Cases: Why and Why Not Anthocyanins Lose during Food Processing." *Critical Reviews in Food Science and Nutrition.* Taylor and Francis Ltd. https://doi.org/10.1080/10408398.2022.2063250

Cheng, Jing, Jian Hua Liu, Govindarajan Prasanna, and Pu Jing. 2017. "Spectrofluorimetric and Molecular Docking Studies on the Interaction of Cyanidin-3-O-Glucoside with Whey Protein, β-Lactoglobulin." *International Journal of Biological Macromolecules* 105 (December): 965–72. https://doi.org/10.1016/j.ijbiomac.2017.07.119

Choi, Inyoung, Jun Young Lee, Monique Lacroix, and Jaejoon Han. 2017. "Intelligent PH Indicator Film Composed of Agar/Potato Starch and Anthocyanin Extracts from Purple Sweet Potato." *Food Chemistry* 218 (March): 122–28. https://doi.org/10.1016/j.foodchem.2016.09.050

Cortez, Regina, Diego A. Luna-Vital, Daniel Margulis, and Elvira Gonzalez de Mejia. 2017. "Natural Pigments: Stabilization Methods of Anthocyanins for Food Applications." *Comprehensive Reviews in Food Science and Food Safety* 16 (1): 180–98. https://doi.org/10.1111/1541-4337.12244

Cunha, Robson, Pollyana Trigueiro, María del Mar Orta Cuevas, Santiago Medina-Carrasco, Thiago M. Duarte, Luzia M.de C. Honório, Dihêgo H.L. Damacena, Maria Gardennia Fonseca, Edson C. da Silva-Filho, and Josy A. Osajima. 2023. "The Stability of Anthocyanins and Their Derivatives through Clay Minerals: Revising the Current Literature." *Minerals.* MDPI. https://doi.org/10.3390/min13020268

D'Alessandro, Leandro Galván, Krasimir Dimitrov, Peggy Vauchel, and Iordan Nikov. 2014. "Kinetics of Ultrasound Assisted Extraction of Anthocyanins from Aronia melanocarpa (Black Chokeberry) Wastes." *Chemical Engineering Research and Design* 92 (10): 1818–26. https://doi.org/10.1016/j.cherd.2013.11.020

De Marchi, Laura, Laura Salemi, Maria Bellumori, Roberto Chignola, Federica Mainente, Diana Vanessa Santisteban Soto, Ilaria Fierri, Marco Ciulu, and Gianni Zoccatelli. 2024. "Thermal Degradation of Red Cabbage (*Brassica oleracea* L. var. Capitata f. rubra) Anthocyanins in a Water Model Extract under Accelerated Shelf-Life Testing." *Food Chemistry* 440. https://doi.org/10.1016/j.foodchem.2023.138272

Echegaray, Noemí, Paulo E.S. Munekata, Patricia Gullón, Christian K.O. Dzuvor, Beatriz Gullón, Francis Kubi, and José M. Lorenzo. 2022. "Recent Advances in Food Products Fortification with Anthocyanins." *Critical Reviews in Food Science and Nutrition.* Taylor and Francis Ltd. https://doi.org/10.1080/10408398.2020.1844141

Enaru, Bianca, Georgiana Drețcanu, Teodora Daria Pop, Andreea Stănilă, and Zorița Diaconeasa. 2021. "Anthocyanins: Factors Affecting Their Stability and Degradation." *Antioxidants.* MDPI. https://doi.org/10.3390/antiox10121967

Escribano-Bailón M. T., and Celestino Santos-Buelga. 2012. "Anthocyanin Copigmentation-Evaluation, Mechanisms and Implications for the Colour of Red Wines." *Current Organic Chemistry* 16 (6): 715–23.

Essien, Sinemobong O., Brent Young, and Saeid Baroutian. 2020. "Recent Advances in Subcritical Water and Supercritical Carbon Dioxide Extraction of Bioactive Compounds from Plant Materials." *Trends in Food Science and Technology* 97: 156–69. https://doi.org/10.1016/j.tifs.2020.01.014

Esteve-Turrillas, Francesc A., Salvador Garrigues, and Miguel de la Guardia 2024. "Green Extraction Techniques in Green Analytical Chemistry: A 2019–2023 Up-Date." *TrAC Trends in Analytical Chemistry* 170. https://doi.org/10.1016/j.trac.2023.117464

Fang, Zhongxiang, Min Zhang, Yunfei Sun, and Jingcai Sun. 2007. "Polyphenol Oxidase from Bayberry (Myrica Rubra Sieb. et Zucc.) and Its Role in Anthocyanin Degradation." *Food Chemistry* 103 (2): 268–73. https://doi.org/10.1016/j.foodchem.2006.07.044

Farr, Jacob E., and M. Monica Giusti. 2018. "Investigating the Interaction of Ascorbic Acid with Anthocyanins and Pyranoanthocyanins." *Molecules* 23 (4). https://doi.org/10.3390/molecules23040744

Fernandes, Ana, Natércia F. Brás, Nuno Mateus, and Victor De Freitas. 2015. "A Study of Anthocyanin Self-Association by NMR Spectroscopy." *New Journal of Chemistry* 39 (4): 2602–11. https://doi.org/10.1039/c4nj02339k

Foroutani, Zahra, Mohammad Reza Afshar Mogaddam, Zahra Ghasempour, and Nazila Ghareaghajlou. 2024. "Application of Deep Eutectic Solvents in the Extraction of Anthocyanins: Stability, Bioavailability, and Antioxidant Property." *Trends in Food Science & Technology* 104324. https://doi.org/10.1016/j.tifs.2023.104324

Garcia-Mendoza, Maria del Pilar, Faber Ariel Espinosa-Pardo, Andressa Mara Baseggio, Gerardo Fernández Barbero, Mário Roberto Maróstica Junior, Mauricio Ariel Rostagno, and Julian Martínez. 2017. "Extraction of Phenolic Compounds and Anthocyanins from Juçara (*Euterpe edulis* Mart.) Residues Using Pressurized Liquids and Supercritical Fluids." *The Journal of Supercritical Fluids* 119: 9–16. https://doi.org/10.1016/j.supflu.2016.08.014

Giusti, M. Monica, Gonzalo Miyagusuku-Cruzado, and Taylor C. Wallace. 2023. "Flavonoids as Natural Pigments." In: C. Stevens, T. Bechtold, A. Manian & T. Pham. *Handbook of Natural Colorants.* Wiley Online Library. https://doi.org/10.1002/9781119811749.ch17

González-Manzano, Susana, Celestino Santos-Buelga, Montserrat Dueñas, Julián C. Rivas-Gonzalo, and Teresa Escribano-Bailón. 2008. "Colour Implications of Self-Association Processes of Wine Anthocyanins." *European Food Research and Technology* 226 (3): 483–90. https://doi.org/10.1007/s00217-007-0560-9

Gris, E. F., E. A. Ferreira, L. D. Falcão, and M. T. Bordignon-Luiz. 2007. "Caffeic Acid Copigmentation of Anthocyanins from Cabernet Sauvignon Grape Extracts in Model Systems." *Food Chemistry* 100 (3): 1289–96. https://doi.org/10.1016/j.foodchem.2005.10.014

Gull, Amir, Mohd Aaqib Sheikh, Jasmeet Kour, Beenish Zehra, Imtyaz Ahmad Zargar, Altaf Ahmad Wani, Surekha Bhatia, and Mushtaq Ahmad Lone. 2022. "Chapter 18 – Anthocyanins." In: Kour, J. & Nayik, G.A. *Nutraceuticals and Health Care.* Academic Press, 317–29. https://doi.org/10.1016/B978-0-323-89779-2.00018-1

He, Jian, and Steven J Schwartz. 2008. "Isolation of Anthocyanin Mixtures from Fruits and Vegetables and Evaluation of Their Stability, Availability and Biotransformation in The Gastrointestinal Tract." http://rave.ohiolink.edu/etdc/view?acc_num=osu1222108733

He, Yang, Liankui Wen, Hansong Yu, Fei Zheng, Zhitong Wang, Xuanwei Xu, Hao Zhang, et al. 2018. "Effects of High Hydrostatic Pressure-Assisted Organic Acids on the Copigmentation of *Vitis amurensis* Rupr Anthocyanins." *Food Chemistry* 268 (December): 15–26. https://doi.org/10.1016/j.foodchem.2018.06.052

Hou, Zhaohua, Peiyou Qin, Yan Zhang, Songhuan Cui, and Guixing Ren. 2013. "Identification of Anthocyanins Isolated from Black Rice (*Oryza sativa* L.) and Their Degradation Kinetics." *Food Research International*: 691–97. https://doi.org/10.1016/j.foodres.2011.07.037

Karaaslan, Mehmet, Mustafa Ozden, Hasan Vardin, and Huseyin Turkoglu. 2011. "Phenolic Fortification of Yogurt Using Grape and Callus Extracts." *LWT* 44 (4): 1065–72. https://doi.org/10.1016/j.lwt.2010.12.009

Kara, Şelale, and Emine Alben Erçelebi. 2013. "Thermal Degradation Kinetics of Anthocyanins and Visual Colour of Urmu Mulberry (*Morus nigra* L.)." *Journal of Food Engineering* 116 (2): 541–47. https://doi.org/10.1016/j.jfoodeng.2012.12.030

Kim, Ah-Na, Kyo-Yeon Lee, Bo Gyeong Kim, Si Won Cha, Eun Ji Jeong, William L. Kerr, and Sung-Gil Choi. 2021. "Thermal Processing under Oxygen–Free Condition of Blueberry Puree: Effect on Anthocyanin, Ascorbic Acid, Antioxidant Activity, and Enzyme Activities." *Food Chemistry* 342. https://doi.org/10.1016/j.foodchem.2020.128345

Klisurova, Daniela, Ivalina Petrova, Manol Ognyanov, Yordan Georgiev, Maria Kratchanova, and Petko Denev. 2019. "Co-Pigmentation of Black Chokeberry (Aronia Melanocarpa) Anthocyanins with Phenolic Co-Pigments and Herbal Extracts." *Food Chemistry* 279 (May): 162–70. https://doi.org/10.1016/j.foodchem.2018.11.125

Koh, Jeewon, Zhimin Xu, and Louise Wicker. 2020. "Blueberry Pectin and Increased Anthocyanins Stability under in Vitro Digestion." *Food Chemistry* 302 (January). https://doi.org/10.1016/j.foodchem.2019.125343

Kohno, Yoshiumi, Reina Kinoshita, Shuji Ikoma, Keiko Yoda, Masashi Shibata, Ryoka Matsushima, Yasumasa Tomita, Yasuhisa Maeda, and Kenkichiro Kobayashi. 2009. "Stabilization of Natural Anthocyanin by Intercalation into Montmorillonite." *Applied Clay Science* 42 (3–4): 519–23. https://doi.org/10.1016/j.clay.2008.06.012

Li, Xiao Ding, Jie Li, Meng Wang, and Hong Jiang. 2016. "Copigmentation Effects and Thermal Degradation Kinetics of Purple Sweet Potato Anthocyanins with Metal Ions and Sugars." *Applied Biological Chemistry* 59 (1): 15–24. https://doi.org/10.1007/s13765-015-0140-9

Liu, Jinning, Yunbing Tan, Hualu Zhou, Jorge L. Muriel Mundo, and David Julian McClements. 2019. "Protection of Anthocyanin-Rich Extract from pH-Induced Color Changes Using Water-in-Oil-in-Water Emulsions." *Journal of Food Engineering* 254: 1–9. https://doi.org/10.1016/j.jfoodeng.2019.02.021

López, Cecilia Jiménez, Cristina Caleja, M. A. Prieto, Marina Sokovic, Ricardo C. Calhelha, Lillian Barros, and Isabel C.F.R. Ferreira. 2019. "Stability of a Cyanidin-3-O-Glucoside Extract Obtained from *Arbutus unedo* L. and Incorporation into Wafers for Colouring Purposes." *Food Chemistry* 275 (March): 426–38. https://doi.org/10.1016/j.foodchem.2018.09.099

Loypimai, Patiwit, Anuchita Moongngarm, and Pheeraya Chottanom. 2016. "Thermal and pH Degradation Kinetics of Anthocyanins in Natural Food Colorant Prepared from Black Rice Bran." *Journal Food Science Technology* 53: 461–70. https://doi.org/10.1007/s13197-015-2002-1

Luo, Xiu'er, Ruoyong Wang, Jinhua Wang, Ying Li, Huainan Luo, Shi Chen, Xin'an Zeng, and Zhong Han. 2022. "Acylation of Anthocyanins and Their Applications in the Food Industry: Mechanisms and Recent Research Advances." *Foods*. MDPI. https://doi.org/10.3390/foods11142166

Machado, Ana Paula Da Fonseca, José Luis Pasquel-Reátegui, Gerardo Fernández Barbero, and Julian Martínez. 2015. "Pressurized Liquid Extraction of Bioactive Compounds from Blackberry (*Rubus fruticosus* L.) Residues: A Comparison with Conventional Methods." *Food Research International* 77: 675–83. https://doi.org/10.1016/j.foodres.2014.12.042

Malien-Aubert, C., O. Dangles, and M. J. Amiot. 2001. "Color Stability of Commercial Anthocyanin-Based Extracts in Relation to the Phenolic Composition. Protective Effects by Intra- and Intermolecular Copigmentation." *Journal of Agricultural and Food Chemistry* 49 (1): 170–76. https://doi.org/10.1021/jf000791o

Maner, Salama, Ajay Kumar Sharma, and K. Banerjee. 2017. "Wheat Flour Replacement by Wine Grape Pomace Powder Positively Affects Physical, Functional and Sensory Properties of Cookies." *Proceedings of the National Academy of Sciences India Section B - Biological Sciences* 87 (1): 109–13. https://doi.org/10.1007/s40011-015-0570-5

Markakis, Pericles 1982. "Chapter 6 – Stability of Anthocyanins in Foods." In: Markakis, P. *Anthocyanins as Food Colors*. Academic Press, 163–80. https://doi.org/10.1016/B978-0-12-472550-8.50010-X

Martins, Natália, Custódio Lobo Roriz, Patricia Morales, Lillian Barros, and Isabel C.F.R. Ferreira. 2016. "Food Colorants: Challenges, Opportunities and Current Desires of Agro-Industries to Ensure Consumer Expectations and Regulatory Practices." *Trends in Food Science and Technology*. Elsevier Ltd. https://doi.org/10.1016/j.tifs.2016.03.009

Mehmet, Özkan. 2002. "Degradation of Anthocyanins in Sour Cherry and Pomegranate Juices by Hydrogen Peroxide in the Presence of Added Ascorbic Acid." *Food Chemistry*, *78*(4), 499-504. www.elsevier.com/locate/foodchem.

Mercadante, Adriana Zerlotti 2018. "Capítulo 7 – Pigmentos Naturais." In: Lajolo, F. M. and Mercadante, A. Z. *Química e Bioquímica dos Alimentos*. Rio de Janeiro: Ed. Atheneu, Vol. 2, 251–59.

Medina-Meza, Ilce Gabriela, and Gustavo V. Barbosa-Cánovas. 2015. "Assisted Extraction of Bioactive Compounds from Plum and Grape Peels by Ultrasonics and Pulsed Electric Fields." *Journal of Food Engineering*: 268–75. https://doi.org/10.1016/j.jfoodeng.2015.06.012

Molaeafard, Shahrbanu, Rashid Jamei, and Ahmad Poursattar Marjani. 2021. "Co-Pigmentation of Anthocyanins Extracted from Sour Cherry (*Prunus cerasus* L.) with Some Organic Acids: Color Intensity, Thermal Stability, and Thermodynamic Parameters." *Food Chemistry* 339 (March). https://doi.org/10.1016/j.foodchem.2020.128070

Moloney, Micheal, Rebecca J. Robbins, Tom M. Collins, Tadao Kondo, Kumi Yoshida, and Olivier Dangles. 2018. "Red Cabbage Anthocyanins: The Influence of D-Glucose Acylation by Hydroxycinnamic Acids on Their Structural Transformations in Acidic to Mildly Alkaline Conditions and on the Resulting Color." *Dyes and Pigments* 158 (November): 342–52. https://doi.org/10.1016/j.dyepig.2018.05.057

Moncada, Margarida C., Sara Moura, M. João Melo, Ana Roque, Carlos Lodeiro, and Fernando Pina. 2003. "Complexation of Aluminum(III) by Anthocyanins and Synthetic Flavylium Salts: A Source for Blue and Purple Color." *Inorganica Chimica Acta* 356 (December): 51–61. https://doi.org/10.1016/S0020-1693(03)00394-3

Montibeller, Maria Jara, Patric de Lima Monteiro, Diego Santiago Tupuna-Yerovi, Alessandro de Oliveira Rios, and Vitor Manfroi. 2018. "Stability Assessment of Anthocyanins Obtained from Skin Grape Applied in Kefir and Carbonated Water as a Natural Colorant." *Journal of Food Processing and Preservation* 42 (8). https://doi.org/10.1111/jfpp.13698

Muhammad, Muhammad Riaz, and Zia-Ul-Haq Bashar Saad. 2016. "Anthocyanins and Human Health: Biomolecular and Therapeutic Aspects." http://www.springer.com/series/10203

Muhlack, Richard A., Ravichandra Potumarthi, and David W. Jeffery. 2018. "Sustainable Wineries through Waste Valorisation: A Review of Grape Marc Utilisation for Value-Added Products." *Waste Management* 72: 99–118. https://doi.org/10.1016/j.wasman.2017.11.011

Ngo, Thao, and Yanyun Zhao. 2009. "Stabilization of Anthocyanins on Thermally Processed Red D'Anjou Pears through Complexation and Polymerization." *LWT* 42 (6): 1144–52. https://doi.org/10.1016/j.lwt.2009.02.013

Nikkhah, E, M Khayamy, R Heidari, and R Jamee. 2007. "Effect of Sugar Treatment on Stability of Anthocyanin Pigments in Berries." *Journal of Biological Sciences* 8: 1412–17.

Oancea, Simona. 2021. "A Review of the Current Knowledge of Thermal Stability of Anthocyanins and Approaches to Their Stabilization to Heat." *Antioxidants* 10 (9): 1337. https://doi.org/10.3390/antiox10091337

Oliveira, Hélder, Nuno Basílio, Fernando Pina, Iva Fernandes, Victor de Freitas, and Nuno Mateus. 2019. "Purple-Fleshed Sweet Potato Acylated Anthocyanins: Equilibrium Network and Photophysical Properties." *Food Chemistry* 288 (August): 386–94. https://doi.org/10.1016/j.foodchem.2019.02.132

Oliveira, Joana, Joana Azevedo, Natércia Teixeira, Paula Araújo, Victor De Freitas, Nuno Basílio, and Fernando Pina. 2021. "On the Limits of Anthocyanins Co-Pigmentation Models and Respective Equations." *Journal of Agricultural and Food Chemistry* 69 (4): 1359–67. https://doi.org/10.1021/acs.jafc.0c05954

Pacheco-Palencia, Lisbeth A., Palo Hawken, and Stephen T. Talcott. 2007. "Phytochemical, Antioxidant and Pigment Stability of Açai (Euterpe Oleracea Mart.) as Affected by Clarification, Ascorbic Acid Fortification and Storage." *Food Research International* 40 (5): 620–28. https://doi.org/10.1016/j.foodres.2006.11.006

Pacheco-Palencia, Lisbeth A., and Stephen T. Talcott. 2010. "Chemical Stability of Açai Fruit (Euterpe Oleracea Mart.) Anthocyanins as Influenced by Naturally Occurring and Externally Added Polyphenolic Cofactors in Model Systems." *Food Chemistry* 118 (1): 17–25. https://doi.org/10.1016/j.foodchem.2009.02.032

Papillo, Valentina Azzurra, Monica Locatelli, Fabiano Travaglia, Matteo Bordiga, Cristiano Garino, Marco Arlorio, and Jean Daniel Coïsson. 2018. "Spray-Dried Polyphenolic Extract from Italian Black Rice (*Oryza sativa* L., Var. Artemide) as New Ingredient for Bakery Products." *Food Chemistry* 269 (December): 603–9. https://doi.org/10.1016/j.foodchem.2018.07.059

Pasqualone, Antonella, Anna Maria Bianco, and Vito Michele Paradiso. 2013. "Production Trials to Improve the Nutritional Quality of Biscuits and to Enrich Them with Natural Anthocyanins." *CYTA - Journal of Food* 11 (4): 301–8. https://doi.org/10.1080/19476337.2012.753113

Patras, A., N.P. Brunton, C. O'Donnell, and B.K. Tiwari. 2010. "Effect of Thermal Processing on Anthocyanin Stability in Foods; Mechanisms and Kinetics of Degradation." *Trends in Food Science & Technology* 21 (1): 3–11. https://doi.org/10.1016/j.tifs.2009.07.004

Pinela, J., M.A. Prieto, Eliana Pereira, Inès Jabeur, Maria Filomena Barreiro, Lillian Barros, and Isabel C.F.R Ferreira. 2019. "Optimization of Heat- and Ultrasound-Assisted Extraction of Anthocyanins from *Hibiscus sabdariffa* Calyces for Natural Food Colorants." *Food Chemistry* 275: 309–21. https://doi.org/10.1016/j.foodchem.2018.09.118

Pozo-Insfran, D. Del, A. Del Follo-Martinez, S. T. Talcott, and C. H. Brenes. 2007. "Stability of Copigmented Anthocyanins and Ascorbic Acid in Muscadine Grape Juice Processed by High Hydrostatic Pressure." *Journal of Food Science* 72 (4). https://doi.org/10.1111/j.1750-3841.2007.00316.x

Quan, Wei, Wei He, Mei Lu, Bo Yuan, Maomao Zeng, Daming Gao, Fang Qin, Jie Chen, and Zhiyong He. 2019. "Anthocyanin Composition and Storage Degradation Kinetics of Anthocyanins-Based Natural Food Colourant from Purple-Fleshed Sweet Potato." *International Journal of Food Science and Technology* 54 (8): 2529–39. https://doi.org/10.1111/ijfs.14163

Rawson, A., A. Patras, B.K. Tiwari, F. Noci, T. Koutchma, and N. Brunton. 2011. "Effect of Thermal and Non Thermal Processing Technologies on the Bioactive Content of Exotic Fruits and Their Products: Review of Recent Advances." *Food Research International* 44 (7): 1875–87. https://doi.org/10.1016/j.foodres.2011.02.053

Rigolon, Thaís Caroline Buttow, Larissa Lorrane Rodrigues Borges, Amanda Lais Alves Almeida Nascimento, Janaina Gonçalves Fernandes, João Carlos Bouzas Marins, Evandro Martins, Pedro Henrique Campelo, and Paulo César Stringheta. 2024. "Study of the Stability of Hydroelectrolytic Sports Beverages Enriched With Phenolic Extract from Jaboticaba Peel or Blueberry Pulp." *Food and Humanity* 2. https://doi.org/10.1016/j.foohum.2023.100214

Rodríguez-Mena, Azucena, Luz Araceli Ochoa-Martínez, Silvia Marina González-Herrera, Olga Miriam Rutiaga-Quiñones, Rubén Francisco González-Laredo, Begoña Olmedilla-Alonso, and Sandra Vega-Maturino. 2023. "Coloring Potential of Anthocyanins from Purple Sweet Potato Paste: Ultrasound-Assisted Extraction, Enzymatic Activity, Color and Its Application in Ice Pops." *Food Chemistry Advances* 3. https://doi.org/10.1016/j.focha.2023.100358

Romero-Díez, R., M. Matos, L. Rodrigues, M.R. Bronze, S. Rodríguez-Rojo, M.J. Cocero, and A.A. Matias. 2019. "Microwave and Ultrasound Pre-Treatments to Enhance Anthocyanins Extraction from Different Wine Lees." *Food Chemistry* 272: 258–66. https://doi.org/10.1016/j.foodchem.2018.08.016

Schreiber, Henry D., Amy M. Swink, and Taylor D. Godsey. 2010. "The Chemical Mechanism for Al3+ Complexing with Delphinidin: A Model for the Bluing of Hydrangea Sepals." *Journal of Inorganic Biochemistry* 104 (7): 732–39. https://doi.org/10.1016/j.jinorgbio.2010.03.006

Sendri, Nitisha, Sarvpreet Singh, Shriya Bhat, Mahesh Gupta, and Pamita Bhandari. 2023. "Insight into the Influence of Oxygen, Sunlight and Temperature on the Stability and Color Attributes of Red Cabbage Anthocyanins and in Vitro Gastrointestinal Behaviour." *Food Chemistry Advances* 3. https://doi.org/10.1016/j.focha.2023.100359

Silva, S., E.M. Costa, C. Calhau, R.M. Morais, and M.E. Pintado. 2017. "Anthocyanin Extraction from Plant Tissues: A Review." *Critical Reviews in Food Science and Nutrition* 57 (14): 3072–83. https://doi.org/10.1080/10408398.2015.1087963

Sigurdson, Gregory T., Peipei Tang, and M. Mónica Giusti. 2018. "Cis-Trans Configuration of Coumaric Acid Acylation Affects the Spectral and Colorimetric Properties of Anthocyanins." *Molecules* 23 (3). https://doi.org/10.3390/molecules23030598

Sui, Xiaonan. 2017. *Impact of Food Processing on Anthocyanins.* Springer Theses. Singapore: Springer Singapore. https://doi.org/10.1007/978-981-10-2612-6

Tachibana, Noriko, Yukihiro Kimura, and Takashi Ohno. 2014. "Examination of Molecular Mechanism for the Enhanced Thermal Stability of Anthocyanins by Metal Cations and Polysaccharides." *Food Chemistry* 143: 452–58. https://doi.org/10.1016/j.foodchem.2013.08.017

Talcott, Stephen T., Janelle E. Peele, and Carmen H. Brenes. 2005. "Red Clover Isoflavonoids as Anthocyanin Color Enhancing Agents in Muscadine Wine and Juice." *Food Research International* 38 (10): 1205–12. https://doi.org/10.1016/j.foodres.2005.05.004

Terefe, Netsanet Shiferaw, Gabriele A. Netzel, and Michael E. Netzel. 2019. "Copigmentation with Sinapic Acid Improves the Stability of Anthocyanins in High-Pressure-Processed Strawberry Purees." *Journal of Chemistry* 2019. https://doi.org/10.1155/2019/3138608

Trouillas, Patrick, Juan C. Sancho-García, Victor De Freitas, Johannes Gierschner, Michal Otyepka, and Olivier Dangles. 2016. "Stabilizing and Modulating Color by Copigmentation: Insights from Theory and Experiment." *Chemical Reviews* 116 (9): 4937–82. ttps://doi.org/10.1021/acs.chemrev.5b00507.

Tsai, P. J., L. Delva, T. Y. Yu, Y. T. Huang, and L. Dufossé. 2005. "Effect of Sucrose on the Anthocyanin and Antioxidant Capacity of Mulberry Extract during High Temperature Heating." *Food Research International* 38: 1059–65. https://doi.org/10.1016/j.foodres.2005.03.017

Wallace, T. C., and M. M. Giusti. 2008. "Determination of Color, Pigment, and Phenolic Stability in Yogurt Systems Colored with Nonacylated Anthocyanins from *Berberis boliviana* L. as Compared to Other Natural/Synthetic Colorants." *Journal of Food Science* 73 (4). https://doi.org/10.1111/j.1750-3841.2008.00706.x

Wang, Wenjie, Jooyeoun Jung, Elizabeth Tomasino, and Yanyun Zhao. 2016. "Optimization of Solvent and Ultrasound-Assisted Extraction for Different Anthocyanin Rich Fruit and Their Effects on Anthocyanin Compositions." *LWT – Food Science and Technology* 72: 229–38. https://doi.org/10.1016/j.lwt.2016.04.041

Xiong, Yun, Pangzhen Zhang, Robyn Dorothy Warner, and Zhongxiang Fang. 2019. "3-Deoxyanthocyanidin Colorant: Nature, Health, Synthesis, and Food Applications." *Comprehensive Reviews in Food Science and Food Safety.* Blackwell Publishing Inc. https://doi.org/10.1111/1541-4337.12476

Xu, Qing, Yuanjing Zhou, Liping Luo, Zhu Huang, Fei Nie, and Guilong Gao. 2017. "Acylation of Blueberry Anthocyanins with Aliphatic Carbonyl Acids and Their Stability Analysis." In *2017 6th International Conference on Energy and Environmental Protection (ICEEP 2017)* (pp. 904–907). Atlantis Press.

Yan, Zheng, Chunyang Li, Lixia Zhang, Qin Liu, Shiyi Ou, and Xiaoxiong Zeng. 2016. "Enzymatic Acylation of Anthocyanin Isolated from Black Rice with Methyl Aromatic Acid Ester as Donor: Stability of the Acylated Derivatives." *Journal of Agricultural and Food Chemistry* 64 (5): 1137–43. https://doi.org/10.1021/acs.jafc.5b05031

Yang, Wei, Maaria Kortesniemi, Xueying Ma, Jie Zheng, and Baoru Yang. 2019. "Enzymatic Acylation of Blackcurrant (*Ribes nigrum*) Anthocyanins and Evaluation of Lipophilic Properties and Antioxidant Capacity of Derivatives." *Food Chemistry* 281 (May): 189–96. https://doi.org/10.1016/j.foodchem.2018.12.111

Yang, Wei, Maaria Kortesniemi, Baoru Yang, and Jie Zheng. 2018. "Enzymatic Acylation of Anthocyanins Isolated from Alpine Bearberry (*Arctostaphylos alpina*) and Lipophilic Properties, Thermostability, and Antioxidant Capacity of the Derivatives." *Journal of Agricultural and Food Chemistry* 66 (11): 2909–16. https://doi.org/10.1021/acs.jafc.7b05924

Yao, Liang, Jiang Xu, Liwei Zhang, Tao Zheng, Lei Liu, and Lei Zhang. 2021. "Physicochemical Stability-Increasing Effects of Anthocyanin via a Co-Assembly Approach with an Amphiphilic Peptide." *Food Chemistry* 362 (November). https://doi.org/10.1016/j.foodchem.2021.130101

Zannou, Oscar, Ilkay Koca, Turki M.S. Aldawoud, and Charis M. Galanakis. 2020. "Recovery and Stabilization of Anthocyanins and Phenolic Antioxidants of Roselle (*Hibiscus sabdariffa* L.) with Hydrophilic Deep Eutectic Solvents." *Molecules* 25 (16): 3715. https://doi.org/10.3390/molecules25163715

Zhang, Zhi-Hong, Lang-Hong Wang, Xin-An Zeng, Zhong Han, and Charles S. Brennan. 2019. "Non-Thermal Technologies and Its Current and Future Application in the Food Industry: A Review." *International Journal off Food Science and Technology* 54: 1–13. https://doi.org/10.1111/ijfs.13903

Zhang, Lingli, Wenbo Wang, Xueyang Yue, Guang Sheng Wu, Pengxiang Yue, and Xueling Gao. 2020. "Gallic Acid as a Copigment Enhance Anthocyanin Stabilities and Color Characteristics in Blueberry Juice." *Journal of Food Science and Technology* 57 (4): 1405–14. https://doi.org/10.1007/s13197-019-04175-w

Zielinski, Acácio Antonio Ferreira, Andrea del Pilar Sanchez-Camargo, Laís Benvenutti, Diego Márlon Ferro, Jônatas Lopes Dias, Sandra Regina Salvador Ferreira. 2021. "High-Pressure Fluid Technologies: Recent Approaches to the Production of Natural Pigments for Food and Pharmaceutical Applications." *Trends in Food Science & Technology* 118: 850–69. https://doi.org/10.1016/j.tifs.2021.11.008

Žilić, Slađana, Tolgahan Kocadağlı, Jelena Vančetović, and Vural Gökmen. 2016. "Effects of Baking Conditions and Dough Formulations on Phenolic Compound Stability, Antioxidant Capacity and Color of Cookies Made from Anthocyanin-Rich Corn Flour." *LWT – Food Science and Technology* 65: 597–603. https://doi.org/10.1016/j.lwt.2015.08.057

3

Encapsulation of Anthocyanins – Insights on Technological Strategies to Increase Its Stability during Food Processing and Digestion

Carolina Marques Mandaji, Matheus Ismael Gomes Rodrigues, and Renan Campos Chisté

3.1 Introduction

Anthocyanins constitute one of the groups of natural pigments responsible for the blue, purple, and red colors in the plant kingdom. They play an important role in attracting animals, promoting seed dispersal, pollination, and helping to protect plants from damages induced by ionizing radiation, such as ultraviolet (Smeriglio et al., 2016). Moreover, this group of pigments has functional properties and potential benefits to human health. The regular intake of anthocyanins, as part of healthy food habits, is associated with a reduced risk of developing chronic non-communicable diseases (Rosales et al., 2023).

Generally, the absorption of anthocyanins *in vivo* mainly depends on their stability and bioavailability in different cells and organs. Under the acidic pH gastric environment, anthocyanins are in a stable flavylium cation form, while they are rapidly converted to the unstable hemiketal, chalcone, or quinoidal base forms under neutral or alkaline pH conditions in the small intestine and colon. The gut microbiota is involved in anthocyanin metabolism and produces phenolic acids from the unstable forms of anthocyanidins. Thus, the bioavailability of anthocyanins is partially limited due to their susceptibility to chemical alterations and affinity for different cells (Shen et al., 2022).

Despite the remarkable functional properties, the incorporation of anthocyanins into food or pharmaceutical formulations by industrial sectors are challenging, as their color and stability are highly affected by the processing conditions, such as pH changes, temperature increase, light exposure, presence of copigments, self-association, presence of metallic ions, enzymatic reactions, and presence of oxygen and ascorbic acid (Cavalcanti, Santos & Meireles, 2011). Therefore, to overcome such limitations, encapsulation rises as a reliable strategy to increase the stability of anthocyanins throughout technological processing for further application by the interested industries.

Encapsulation is a technique that consists of entrapping active ingredients (solid, liquid, or gas) in polymeric coatings (encapsulating agents), allowing these ingredients to be protected from adverse environmental conditions, controlling their release or incorporating functional properties into food products. Depending on the encapsulation method and coating materials, the capsules can assume different sizes: macrocapsules (>5000 µm), microcapsules (0.2–5000 µm), and nanocapsules (<0.2 µm) (Mandaji, Pena & Chisté, 2022).

DOI: 10.1201/9781003453260-4

Different encapsulation methods are known and widely used by the food, cosmetics, and pharmaceutical industries, such as spray drying, freeze-drying, and ionic gelation (Timilsena, Haque & Adhikari, 2020), being the encapsulation of natural pigments for incorporation into food products one of the most commonly practical applications.

In addition to the molecular stability, another advantage of encapsulation is the modulation of the bioaccessibility and bioavailability of anthocyanins through their controlled release in the human body, attributed to different wall materials used to form capsules (Rosales & Fabi, 2022). Several applications for microencapsulated anthocyanins in foods are known, such as food coloring, an ingredient in formulations to confer potential functional properties to foods due to their antioxidant and antimicrobial activities and their application in "intelligent" films to inform food freshness (Shipp & Abdel-Aal, 2010; Leong et al., 2021; Santos, da Silva & Leite Neta, 2022).

Therefore, in this chapter, the most common encapsulation methods and coating materials used for the microencapsulation of anthocyanins will be approached. Additional information regarding the bioaccessibility and bioavailability of anthocyanins will also be presented as influenced by the different coating materials, followed by some technological trends for the application of microencapsulated anthocyanins.

3.2 A Quick Look on Microencapsulation Techniques and the Most Used Coating Materials to Encapsulate Anthocyanins

For contextualization purposes, a quick literature search was carried out in *Science Direct* database (Elsevier), with articles published in a one-year period (January/2023 to January/2024), accounting for 110 studies, only to illustrate the trends of the most studied methods and coating materials used for anthocyanins' microencapsulation (Figure 3.1). These trends showed that the method used for encapsulation might be a determinant factor when choosing the proper coating material.

Currently, the development of films incorporated with anthocyanins is a hot topic among the encapsulation methods, accounting for 29% of the selected studies (Figure 3.1). Many of these studies aimed to develop "intelligent" films, so they can be used as indicators of food quality since anthocyanins' molecules undergo structural modifications according to pH alteration, which results in visual color change (Figure 3.2). The main choices of coating materials to produce films with anthocyanins are not only polysaccharides, such as starch, chitosan, and sodium alginate, but also protein materials, such as gelatins and soy protein isolates are usually reported (Figure 3.1b). Films produced with anthocyanins as part of the formulation have proven to be useful in identifying the freshness of foods such as fish, pork, and milk (Francisco et al., 2024; Wu et al., 2024; Yang et al., 2024).

Spray drying is one of the most popular methods used to microencapsulate anthocyanin in the literature (Figure 3.1a), due to its ease of use on industrial scales. This method offers simplicity during execution and compatibility with a wide range of coating materials, which allows the production of encapsulated bioactive compounds with high stability during storage under different conditions, in addition to preserving their biological properties (Mandaji et al., 2022). Maltodextrin is the most common coating material used for encapsulating anthocyanins by spray drying, followed by gum arabic and proteins.

Freeze-drying is another widely used method, which is mainly used to encapsulate heat-sensitive molecules, such as anthocyanins. During freeze-drying, the aqueous mixture

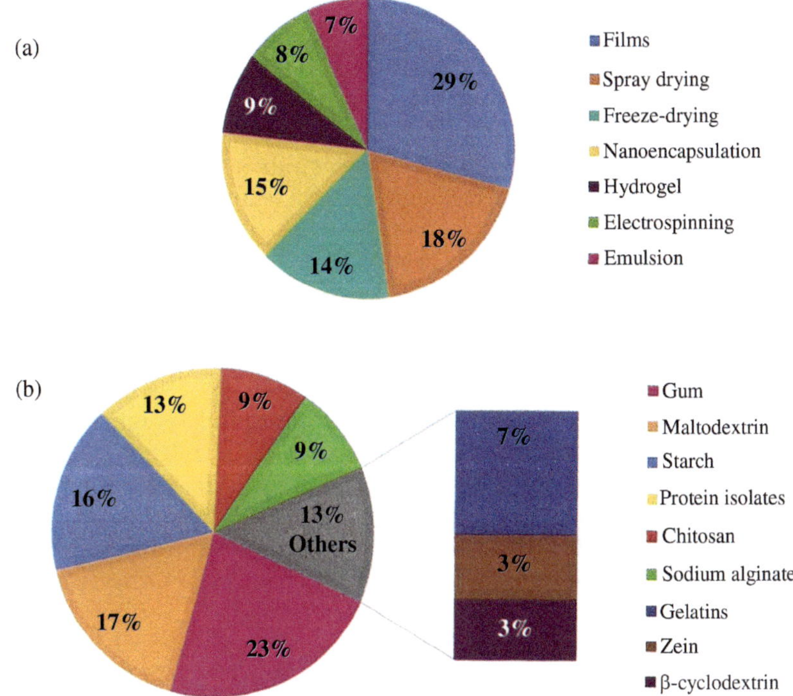

FIGURE 3.1
The main methods (a) and coating materials (b) used to encapsulate anthocyanins as published in a one-year period (*Science Direct*, jan/2023–jan/2024).

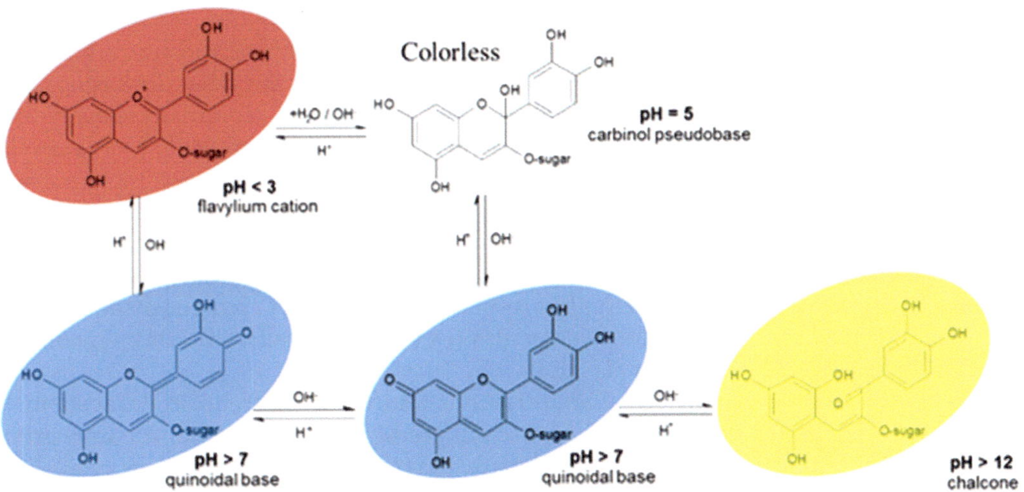

FIGURE 3.2
Structural and color changes of anthocyanins according to pH alteration.

containing the coating material, and the interested active ingredient, is kept frozen at temperatures below −40°C, followed by drying by sublimation, under low pressure, which minimizes thermal degradation during the production of microcapsules/microparticles (Mandaji et al., 2021). For comparison purposes, microcapsules/microparticles produced by freeze-drying (−50°C) allowed total monomeric anthocyanins retention of 85%, while 71% was observed for spray drying at 150°C (Yavuz-Düzgün et al., 2023). Regarding the coating materials, arabic and xanthan gums, maltodextrin, starch, and proteins are the main materials used to produce anthocyanin's microcapsules/microparticles by freeze-drying.

Other encapsulation methods are also frequently investigated (Figure 3.1a) to produce anthocyanins' microcapsules, namely hydrogel formation, electrospinning, and emulsion. Hydrogels have unique technological properties, such as high water content, high permeability, and responsiveness to the surrounding environment, which make them ideal carriers of bioactive compounds (Moriyama et al., 2023). Hydrogels are formed by stabilization processes, in which an ionic or covalent bond binds polymer chains, limiting its mobility until the liquid state is converted into a solid or gel state (Nezamdoost-Sani et al., 2024). Sodium alginate is the most widely used material to prepare hydrogels for the encapsulation of anthocyanins due to its advantageous properties, such as sensitivity to pH alteration, high solubility in water (hydrophilicity), non-toxicity, biodegradability, biocompatibility, and high economic efficiency (Hao et al., 2024). Other materials, such as chitosan, starch, and pectin are also used to a lesser extent.

Electrospinning involves an electrohydrodynamic process, in which a drop of liquid is electrified to generate a conically shaped structure known as a Taylor cone, which goes through stretching processes leading to the formation of fibers from a polymer solution (Mercante et al., 2024a). In recent decades, electrospinning has been considered an easy and efficient approach for manufacturing micro/nanofibers from polymeric solutions, which can have diameters of less than 1 μm (Mercante et al., 2024b). These ultra-thin materials have unique characteristics, such as high surface area to volume, high porosity, and high flexibility (Freitas et al., 2024). Most of the studies involving the encapsulation of anthocyanins using electrospinning technique involve the development of nanofiber blankets to monitor the freshness based on the same principle shown in Figure 3.2 (Heydarian & Shavisi, 2023; Pereira et al., 2023; Karambasti & Shavisi, 2024; Khaledian, Moshtaghi & Shahbazi, 2024; Lv et al., 2024).

Finally, an emulsion consists of two immiscible liquids, usually oil and water, with one of the liquids being dispersed as small spherical droplets into the other. Emulsions can be classified according to the relative spatial distribution of the oil and aqueous phase, as follows: oil in water (O/W), water in oil (W/O), water/oil/water (W/O/W), and oil/water/oil (O/W/O) (Lu, Kelly & Miao, 2016). Part of the studies that used the emulsion method to encapsulate anthocyanins aimed to direct the emulsion to another encapsulation technique, such as the formation of hydrogels, films, coacervation, and spray chilling. In general, the development of emulsions increased the encapsulation efficiency of anthocyanins, which may produce highly responsive films to pH changes, in addition to increase the bioaccessibility of these molecules (Devi, Das & Badwaik, 2023; Figueiredo et al., 2023; Li et al., 2023; Ozcan et al., 2023).

Importantly to note, nanoencapsulation rises as an increasing trend in encapsulation studies to comprise the possible uses of active ingredients encapsulated in nanometric-sized capsules (Singh et al., 2023). Nanoencapsulation covers a wide range of encapsulation methods, and studies on nanoencapsulation of anthocyanins have included the employment of emulsions, colloidal particles, liposomes, and electrospinning, which accounted about 15% of the selected studies published in the last year in the *Science Direct* database

(Figure 3.1a). The nanometric size of capsules has some advantages compared to microcapsules, such as increased intracellular and intestinal absorption of the active encapsulated compounds (Freixo, 2013). Further information regarding the nanoencapsulation of anthocyanins was nicely reviewed elsewhere (Rosales & Fabi, 2022).

3.3 The Effect of Encapsulation on Bioaccessibility and Bioavailability of Anthocyanins

Bioaccessibility refers to the amount of a given compound that is available for absorption in the human gastrointestinal tract, while bioavailability refers to the fraction of these compounds that reaches systemic circulation (Nascimento et al., 2023). The term "bioavailability of compounds" includes gastrointestinal digestion (release), absorption, tissue distribution, metabolism, and excretion. In turn, bioaccessibility is generally determined by a digestion procedure that includes several steps: release from the food matrix, conversion during digestion conditions, and absorption by intestinal epithelial cells (Karaś et al., 2016).

The digestive process of anthocyanins begins with chewing, where the action of microorganisms and enzymes in saliva, together with mechanical breakdown, contributes to the release and initial fragmentation of anthocyanins. In the stomach, low pH softens the food matrix and favors the release of anthocyanins, which can be absorbed in the gastric wall (Nascimento et al., 2023). Subsequently, in the small intestine, the highest absorption rates of anthocyanins occur through two mechanisms: (1) through direct absorption of the glycosidic form (intact anthocyanin) mediated by sugar transporters, and (2) through the action of enzymes, such as β-glucosidase, that promotes its conversion into aglycones (anthocyanidins), which are more easily absorbed by passive diffusion through colonocytes than anthocyanins, due its higher hydrophobicity (Rosales & Fabi, 2022; Nascimento et al., 2023). In the large intestine, the gut microbiota plays a crucial role in the breakdown of anthocyanins into metabolites that can be further absorbed or excreted (Nascimento et al., 2023). The absorption scheme for anthocyanins is summarized in Figure 3.3.

Both the bioaccessibility and bioavailability of anthocyanins in foods can be carried out through *in vitro* and *in vivo* tests. Bioaccessibility is generally assessed by *in vitro* digestion procedures, often simulating gastric and small intestinal digestion, sometimes followed by cellular uptake determination using Caco-2 cells as a model of the intestinal epithelial barrier. In terms of bioavailability, when a claim is made, it must be demonstrated that the target component was digested and absorbed efficiently, being the in vivo determination preferable (Carbonell-Capella et al., 2013; Karaś et al., 2016,). However, a number of researchers have used *in vitro* approaches, as they facilitate reproduction, standardization, and methodological repeatability (Nascimento et al., 2023).

As anthocyanins undergo structural modifications according to pH alteration (Figure 3.2), their stability limitations over the digestion steps are relevant to predict both bioaccessibility and bioavailability. As an example, the contents of anthocyanins of jussara extracts (*Euterpe edulis* Mart.) were reported to decrease significantly during the *in vitro* bioaccessibility assay by 81% after the oral phase, 48% after gastric, and 31% after the intestinal phases (Vannuchi et al., 2024). In another study, anthocyanins were isolated from blackberries, red apple skins, strawberries, red grape skins, and wine, and the stability of the anthocyanins in the gastric phase was high (100%), but low after the intestinal phase (1.6–82%), and only a small percentage of anthocyanins were intact at the end of the *in vitro* bioaccessibility

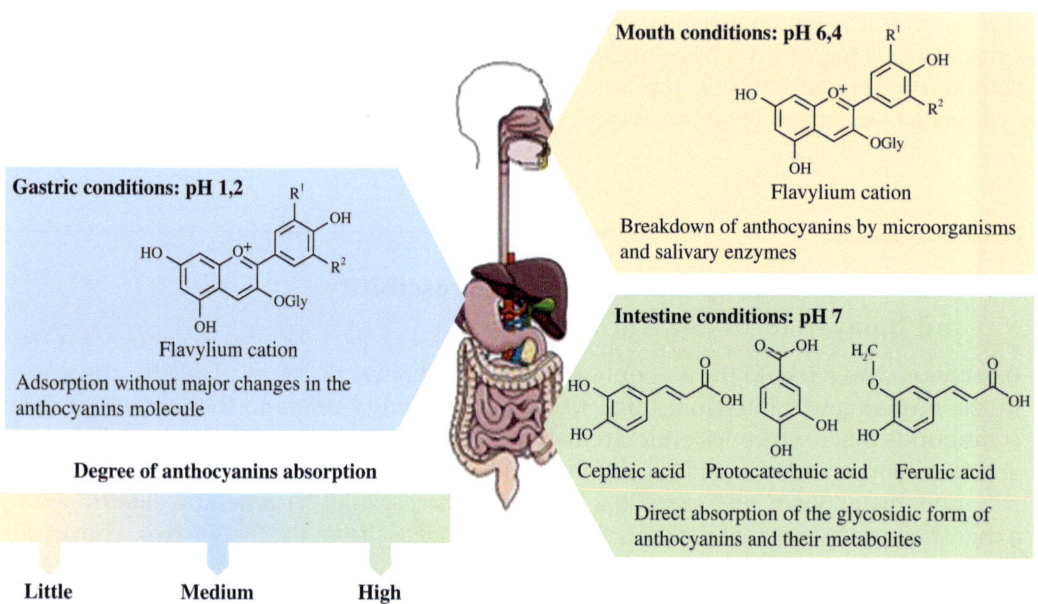

FIGURE 3.3
Scheme of anthocyanin's absorption.

assay (0.07–3.51%) (Victoria-Campos et al., 2022). Another study reported similar behavior, in which the anthocyanins extracted from blackberries presented high stability during the gastric phase, but after the intestinal phase, only 13% of the initial amount was identified (Rosales et al., 2023). These investigations corroborate the fact that anthocyanins are stable in the stomach due to the low pH values (1.5–2.0) that favor the flavylium cation forms. However, in the intestine (pH ~8.0), anthocyanin molecules are subjected to high alkaline conditions of the medium that favor the formation of less stable structures, and also the binding of anthocyanins to proteins and bile salts, followed by subsequent precipitation that decreases bioaccessibility (Victoria-Campos et al., 2022; Rosales et al., 2023).

Concerning absorption, an *in vivo* study demonstrated that the bioavailability of anthocyanins of blueberry extract in humans is very low (0.02% plasma/0.03% urine) (Mueller et al., 2017). However, these same authors showed that plasma levels of anthocyanins' metabolites (gallic acid, protocatechuic acid, 4-hydroxybenzoic acid, 4-hydroxybenzaldehyde, 3-*O*-methylgallic acid, phloroglucinol aldehyde, syringic acid, and vanillic acid) were up to 20 times higher in plasma and 38 times higher in urine than in the initial extract. In another *in vivo* study, the red radish extract was administered to mice and pelargonidin 3-glucoside (~28.9 pmol/mL) was the most bioavailable anthocyanin present in the plasma (Li et al., 2023). Additionally, these same authors also found high levels of anthocyanins' metabolites in the plasma: gallic acid (~0.07 μmol/mL), protocatechuic acid (~0.10 μmol/mL), *p*-hydroxybenzoic acid (~0.03 μmol/mL), and caffeic acid (~0.13 μmol/mL). Contrastly, in an *in vivo* study regarding the administration of red cabbage to humans, almost 14% of the total anthocyanins found in the plasma were present in the form of native compounds, and less than 4% in the form of metabolites (Wiczkowski et al., 2016). Therefore, although the estimated total bioavailability of anthocyanins increases when their metabolized products are accounted, this value remains low, with less than 2% being effectively absorbed by the gastrointestinal tract (Rosales & Fabi, 2022).

To overcome the stability issues of anthocyanins during digestion, which might directly affect their biological properties, encapsulation rises as a technological strategy to increase the bioaccessibility of anthocyanins. As an example, microencapsulation of jamelão extract (*Syzygium cumini* L.) with maltodextrin provided high stability for the anthocyanins (26%) after simulated *in vitro* digestion, when compared to the free extract (0.8%) (Paes et al., 2024). Similar results were observed for anthocyanins microencapsulated from elderberry juice (*Sambucus nigra* subsp. canadensis), with isolated soy protein, which presented ~26% of bioaccessibility, being about twice higher than for the anthocyanins from the non-encapsulated juice (12%) (Ravichandran et al., 2023).

Although systematic and deep investigations are still needed to fully understand the fate of anthocyanins during digestion and absorption, the increase in the *in vitro* bioaccessibility of anthocyanins, as influenced by the protection provided by encapsulation, might result in the increase of bioavailability. Blueberry anthocyanins were microencapsulated with α-casein, and the *in vivo* bioavailability was assessed using mice models (Lang et al., 2021). These authors observed a decrease in the excreted contents of the main anthocyanins, namely malvidin 3-*O*-galactoside, cyanidin 3-*O*-glucoside, and delphinidin 3-*O*-glucoside, followed by increase in the metabolite contents in urine (syringic acid, ferulic acid, 4-hydroxybenzoic acid, and vanillic acid), and decreased contents in feces (syringic acid, ferulic acid, and gallic acid) (Lang et al., 2021). In another study, microencapsulation of anthocyanins from blueberry extract, with pectin as coating material, increased the intestinal accessibility of these molecules during passage through the small intestine and modulated the formation of aldehyde phloroglucinol (metabolite) in human plasma (Mueller et al., 2018). In general, increasing the bioavailability of anthocyanins through encapsulation promotes a number of beneficial health effects that are linked to these molecules. Another bioavailability research carried out with anthocyanins extracted from black soybean hulls, which were microencapsulated with fructooligosaccharides (FOS) and whey protein, reported that groups of mice fed with encapsulated anthocyanins showed higher concentrations of anthocyanins in the blood stream 4 h after microcapsule intake (Zhang et al., 2020). Moreover, the results of the mentioned research also suggest that microencapsulated anthocyanins produced with dietary fibers as coating material might affect blood glucose index after postprandial meal (mice model).

3.4 Controlled Release of Encapsulated Anthocyanins

Predicting the release rates of encapsulated bioactive compounds in different media, through mathematical models, can be seen as a useful research strategy to design potential functional foods with high bioaccessibility and bioavailability. Kinetic data for the release of encapsulated active ingredients can be predicted using theoretical, empirical, or semi-empirical methods (Flores and Kong, 2017). Among the mathematical models reported in studies about the release kinetics of encapsulated compounds, four are generally used and conveniently compared: zero-order, first-order, and Higuchi and Korsmeyer-Peppas equations (Flores and Kong, 2017).

Despite the importance of predicting the release behavior of encapsulated compounds in both food and physiological systems, the literature regarding the release kinetics of microencapsulated anthocyanins is limited. However, two studies demonstrated that the Korsmeyer-Peppas (Korsmeyer et al., 1983) and Higuchi (Higuchi, 1962) models (Table 3.1) provided the best fit to predict the release of active ingredients during storage.

TABLE 3.1

Mathematical Models Used to Predict Anthocyanins' Release Kinetics

Model	Equation	Description
Korsmeyer-Peppas	$\dfrac{Mt}{M\infty} = kt^n$	Model that exponentially relates the release rate of the active ingredient to the elapsed time. Used in systems in which the release mechanisms are unknown, and analyzes when more than one type of release occurs.
Higuchi	$Mt = k\sqrt{t}$	Model that predicts the active ingredient release based on Fick's Law. If the correlation coefficient (R^2) is higher than other kinetic models, it can be inferred that the main release mechanism is controlled by diffusion.

Mt is the amount of active ingredient released as a function of time t; M∞ is the total amount of the active ingredient, Mt/M∞ is the fraction of active ingredient released in time t, k is the release constant, and n is the indicative release exponent of the transport mechanism. For spherical structures, the transport mechanisms are classified as follows: quasi-Fickian diffusion ($n < 0.43$), Fickian diffusion ($n = 0.43$), anomalous diffusion (non-Fickian) ($0.43 < n < 0.85$), case 2 – zero-order release ($n = 0.85$), and super case 2 – relaxation ($n > 0.85$). All data in Table 3.1 were based on information described by Pires (2011), Tavares (2015), and Mendes (2019).

Choosing the appropriate coating material to accomplish the purpose of the addition of microencapsulated anthocyanins may affect the release rate of these compounds during digestion, which will influence the bioaccessibility. Peanparkdee, Borompichaichartkul, and Iwamoto (2021) developed a study with different coating materials to microencapsulate anthocyanins from black rice extracts. These authors reported that two coating materials presented the most interesting results: the first one was gelatin, which, on the one hand, presented the highest release rate of anthocyanin during simulated intestinal digestion (83%). The other coating material was a combination of gelatin/gum arabic which, on the other hand, showed a low release rate of anthocyanins (9.8%), suggesting that the highest anthocyanin proportion of anthocyanins remained inside the microcapsules after digestion. In another study regarding the microencapsulation of anthocyanins extracted from red cabbage, using a combination of maltodextrin/pectin as coating material, it was seen that the free extract released ~71% of anthocyanins in the gastric digestive phase, while the encapsulated extract released 49% (Sendri et al., 2023). These studies demonstrate that encapsulation controlled the level of anthocyanins released during the gastric phase, allowing a higher amount of anthocyanins to reach the intestine, which might be available for further absorption. All these results are useful for the food and pharmaceutical industries, depending on the approach and purpose.

The anthocyanins from grape pomace extracts were microencapsulated with different coating materials, by spray drying, and subjected to the release study in simulated gastric and intestinal media (Moreno, Cocero & Rodríguez-Rojo, 2018). These authors reported that microcapsules produced with pea protein isolate showed the highest anthocyanins' release rates, being 50% released in the gastric phase and 45% in the intestinal phase, within 120 min after digestion. Among the empirical models selected to fit the release data, the Higuchi model provided the best fit to the experimental data, followed closely by the Korsmeyer-Peppas model. In all the cases, the diffusion exponent n for the Korsmeyer-Peppas model was less than 0.43, which suggested quasi-Fickian diffusion, which agrees with Higuchi's interpretation (Moreno, Cocero & Rodríguez-Rojo, 2018).

In another study, the release kinetics of anthocyanins encapsulated in alginate hydrogel spheres was investigated, being the encapsulated jamelão pulp extracts (*S. cumini*)

produced through drip extrusion (Sharma, Dash & Badwaik, 2022). After carrying out the simulated gastrointestinal experiment, the Korsmeyer-Peppas model was chosen as the most appropriate to predict the release of anthocyanins. The diffusion exponent n revealed that the jamelão extracts encapsulated inside alginate hydrogel spheres combined with guar gum ($n = 0.358$) showed release behavior described by quasi-Fickian diffusion, while the encapsulation with alginate hydrogel spheres combined with xanthan gum ($n = 0.670$), pectin ($n = 0.578$), and gum arabic ($n = 0.452$) exhibited release behavior by Non-Fickian diffusion. All the microcapsules showed high anthocyanin release rates (75–87%) in the gastric phase, within 120 min after digestion (Sharma, Dash & Badwaik, 2022). The alginate hydrogel is porous and these high-diffusion coefficients indicated that its combination with the selected coating materials was not sufficient to reduce the porosity (Carvalho et al., 2019).

3.5　Microencapsulated Anthocyanins for Food Application

The use of anthocyanins in microcapsules as food additives can be a promising strategy to prevent degradation, loss of color, and overall reduced stability of these compounds under food processing and storage conditions. The conversion of aqueous anthocyanins' extracts into dried extracts through microencapsulation results in technological products with higher physicochemical and microbiological stability than non-encapsulated extracts, enabling industrial application of the encapsulation technique for the development of food colorants, and potential functional ingredients in foods (Oliveira & Petrovick, 2010; Tarone, Cazarin & Junior, 2020; Machado et al., 2022).

The replacement of red synthetic dyes used by food industries, such as Allura Red AC (INS 129), Amaranth (INS 123), and Azorubin (E122), by natural pigments, such as anthocyanins, is challenging and represents a growing trend, which is justified by the search for healthier eating habits and concerns on the environment impacts caused by synthetic dyes production (Maronpot, Hayashi & Bastaki, 2020; Santos, da Silva & Leite Neta, 2022). Although the encapsulation of natural pigments is a promising alternative to the synthetic ones, it is noted that the ingredients and storage conditions of foods may affect the total contents or bioaccessibility/bioavailability of these compounds. Thus, the addition of microencapsulated anthocyanins in processed foods makes it possible to impart interesting colors to new food products or recover color loss due to food processing.

Shamshad et al. (2023) investigated the stability of microcapsules of anthocyanins extracted from black carrots in ice cream for 60 days, being the microcapsules prepared using maltodextrin and gum arabic by freeze-drying. Three formulations were produced containing 3% (T1), 6% (T2), and 9% (T3) of microencapsulated anthocyanins powder-enriched ice cream. As expected, the ice cream with 9% presented the highest contents of total monomeric anthocyanins (143 mg/100 g), and for all the formulations, only a slight decrease in the anthocyanin contents was observed from the 45th day until the end of the experiment (~2–3%). The authors' conclusion showed that the addition of microencapsulated anthocyanins powder in ice cream proved to stabilize black carrot anthocyanins and contributed positively to the sensory characteristics of ice cream.

In addition to contributing to the sensorial aspects of foods, anthocyanins can act as functional food ingredients due to their antioxidant properties as scavengers of free radicals, which delay lipid oxidation, both in the human body and in food systems

(Bordignon Jr et al., 2009; Shipp & Abdel-Aal, 2010). Acidic foods, such as soft drinks, ice creams, and yogurts, are preferably used in the application of anthocyanins as the mean favors the predominance of flavylium cation, as shown in Figure 3.2.

In this sense, anthocyanins of blackberry were microencapsulated with maltodextrin by spray drier and freeze-drying and with sodium alginate by ionic gelation, followed by application in natural yogurt aiming to determine the *in vitro* bioaccessibility (Santos et al., 2022). The incorporation of anthocyanin microcapsules, which were prepared by spray drier and freeze-drying, was successful, while the capsules obtained by ionic gelation partially dissolved in the yogurt and compromised the desired texture and final appearance of the product. The natural yogurt incorporated with freeze-dried microcapsules showed higher bioaccessibility of cyanidin 3-glucoside (29.9%), the major anthocyanin, than ionic gelation (21%), and spray drying (4.9%). According to Gowd et al. (2020), milk proteins and phenolic compounds are capable of forming stable complexes, which makes it possible to increase the stability of these substances during the digestive process.

Another study evaluated the antioxidant capacity, anthocyanin contents, and digestive stability of microencapsulated anthocyanins (coating materials: maltodextrin and inulin), which were extracted from red cabbage (*Brassica oleracea* L. var. capitata L. f. rubra, var. Garance), cornelian cherry (*Cornus mas* L.), and chokeberry (*Aronia melanocarpa*), and added to wheat bread formulation (Czubaszek et al., 2023). The breads made using microencapsulated anthocyanins showed higher contents of anthocyanins and antioxidant capacity than simple wheat bread, being red cabbage microcapsules the one with the highest antioxidant capacity against ABTS radical (75–76-μmol Trolox equivalents/g), which was 30 times higher than breads without microcapsules' addition. Regarding digestive stability, acylated anthocyanins in breads with red cabbage microcapsules were detected in all the digestive phases (oral, gastric, and intestinal), while breads added with microcapsules with anthocyanins of cornelian cherry and chokeberry were identified only up to the gastric phase.

Another technological feature of anthocyanin microcapsules is related to their applicability in active films for food packing. These packages incorporate active substances that improve their performance by optimizing the shelf life of products by delaying the maturation or development of microorganisms in foods, for example (Robertson, 2005). Leong et al. (2021) developed an active gelatin film incorporated with microcapsules of anthocyanins extracted from butterfly peas (*Clitoria ternatea*) to evaluate its antimicrobial activity against three Gram-positive and six Gram-negative foodborne pathogenic bacteria. Six among the nine microorganisms tested were susceptible to the anthocyanins film, including those that cause foodborne illnesses, such as *Staphylococcus aureus*, *Escherichia coli*, and *Pseudomonas aeruginosa*. The application of the active film to pack tofu as a food model system indicated the high efficiency of active films to control microbiological growth, both at room temperature and under refrigeration, providing microbiological shelf life from five to six days.

3.6 Final Considerations

Anthocyanins have widely known biological potentials that give these natural pigments a series of health-promoting effects as well as open wide avenues for technological applications. However, considering the high susceptibility of anthocyanins to degrading or altering their chemical structures during food processing or in physiological conditions, the main methods and coating materials, which were discussed in this chapter, showed that encapsulation technology rises as an excellent way to make anthocyanins more available for technological use aiming to increase bioaccessibility/bioavailability by the human body.

The technological application of encapsulated anthocyanins by the food industries is mainly focused on the replacement of synthetic food colorants, but each day more of the functional properties of anthocyanins, such as antioxidant and antimicrobial activities, have caught the attention. A number of food products prepared with the addition of microencapsulated anthocyanins have been reported in the literature, mainly the acidic ones, such as ice creams, yogurts, and soft drinks, since the low pH and temperature favor the preservation of these pigments. Another important trend focused on food science and technology is the incorporation of anthocyanins (microencapsulated or not) in biodegradable films that can be used as active packaging and as indicators of food freshness ("intelligent" films).

Importantly, regarding the encapsulation procedure, some care must be taken, especially concerning the selection of the coating materials. Depending on the material selected, anthocyanins may or may not be released efficiently throughout the gastrointestinal tract. On this matter, available literature approaching the release kinetics of encapsulated anthocyanins during digestion is still limited and represents a relevant gap to be filled to understand the mechanisms and effects of different wall materials on the bioavailability of anthocyanins.

References

Bordignon, Celso Luiz Jr; Francescatto, Vanessa; Nienow, Alexandre Augusto; Calvete, Eunice; Reginatto, Flávio Henrique Influence of the extraction solution pH on the content of anthocyanins in strawberry fruits. **Food Science and Technology**, 29(1), 183–188, 2009. https://doi. org/10.1590/S0101-20612009000100028

Carbonell-Capella, Juana M.; Buniowska, Magdalena; Barba, Francisco J.; Esteve, María J.; Frígola, Ana. Analytical methods for determining bioavailability and bioaccessibility of bioactive compounds from fruits and vegetables: A review. **Comprehensive Reviews in Food Science and Food Safety**, 13, 155–171, 2013. https://doi.org/10.1111/1541-4337.12049

Carvalho, Ana Gabriela da Silva; Machado, Mariana Teixeira da Costa; Barros, Helena Dias de Freitas Queiroz; Cazarin, Cinthia Baú Betim; Maróstica Junior, Mário Roberto; Hubinger, Miriam Dupas. Anthocyanins from jussara (*Euterpe edulis* Martius) extract carried by calcium alginate beads pre-prepared using ionic gelation. **Powder Technology**, 345, 283–291, 2019. http:// dx.doi.org/10.1016/j.powtec.2019.01.016.

Cavalcanti, R.N.; Santos, D.T.; Meireles, M.A.A. Non-thermal stabilization mechanisms of anthocyanins in model and food systems – an overview. **Food Research International**, 44, 2, 499–509, 2011. https://doi.org/10.1016/j.foodres.2010.12.007

Czubaszek, Anna; Czaja, Anna; Sokół-Łętowska, Anna; Kolniak-Ostek, Joanna; Kucharska, Alicja Z. Quality of bread enriched with microencapsulated anthocyanin extracts during in vitro simulated digestion. **Journal of Cereal Science**, 113, 103724, 2023. https://doi.org/10.1016/j. jcs.2023.103724

Devi, Lourembam Monika; Das, Amit Baran; Badwaik, Laxmikant S. Effect of gelatin and acacia gum on anthocyanin coacervated microcapsules using double emulsion and its characterization. **International Journal of Biological Macromolecules**, 235, 123896, 2023. http://dx.doi. org/10.1016/j.ijbiomac.2023.123896

Figueiredo, ayne de Abreu; Norcino, Laís Bruno; Carmo, Eloá Lourenço do; Campelo, Pedro Henrique; Botrel, Diego Alvarenga; Borges, Soraia Vilela; Souza, Sérgio Michielon de; Oliveira, Cassiano Rodrigues de. Microstructured lipid microparticles containing anthocyanins: Production, characterization, storage, and resistance to the gastrointestinal tract. **Food Research International**, 166, 112611, 2023. http://dx.doi.org/10.1016/j.foodres.2023.112611

Flores, Floirendo P.; Kong, Fanbin. In vitro release kinetics of microencapsulated materials and the effect of the food matrix. **Annual Review of Food Science and Technology**, 8, 1, 237–259, 2017. http://dx.doi.org/10.1146/annurev-food-030216-025720

Francisco, Ana Paula; Sganzerla, William Gustavo; Castro, Luiz Eduardo Nochi; Barroso, Tiago Linhares Cruz Tabosa; Silva, Aline Priscilla Gomes da; Rosa, Cleonice Gonçalves da; Nunes, Michael Ramos; Forster-Carneiro, Tânia; Rostagno, Mauricio A. Pressurized liquid extraction of bioactive compounds from grape peel and application in pH-sensing carboxymethyl cellulose films: A promising material to monitor the freshness of pork and milk. **Food Research International**, 179, 114017, 2024. http://dx.doi.org/10.1016/j.foodres.2024.114017

Freitas, Camila Fabiano de; Santos, Maria Luiza Borges; Cândido, Gabriela Braga Gomes; Colombari Neto, Jean; Oliveira, Antônia Carla de Jesus; Medeiros, Gabriela Brunosi; Pereira, Antonio Guilherme Basso; Dragunski, Douglas Cardoso; Caetano, Josiane; Corradini, Elisângela; Muniz, Edvani Curti. HIDROGÉIS CONTENDO FIBRAS ELETROFIADAS PARA LIBERAÇÃO SUSTENTADA DE SOLUTOS. In: Mercante, Luiza Amim; Corrêa, Daniel Souza. **Chapter 6 - Eletrofiação e nanofibras: Fundamentos e aplicações**. Ponta Grossa: Atena Editora, p. 138–167, 2024.

Freixo, Luís Carlos Araújo. **Micro e Nanoencapsulação como Estratégias de Estabilização de Entidades Bioativas: Proteínas, Enzimas e Bacteriófagos**. 2013. 76 f. Dissertação (Mestrado) - Curso de Ciências da Saúde, Universidade Fernando Pessoa, Porto, 2013.

Gowd, Vemana; Karim, Naymul; Xie, Lianghua; Shishir, Mohammad Rezaul Islam; Xu, Yang; Chen, Wei. In vitro study of bioaccessibility, antioxidant, and α-glucosidase inhibitory effect of pelargonidin-3-oglucoside after interacting with beta-lactoglobulin and chitosan/pectin. **International Journal of Biological Macromolecules**, 154, 380–389, 2020. https://doi.org/10.1016/j.ijbiomac.2020.03.126

Hao, Hongdan; Duan, Bingchao; Zhang, Lidan; Wang, Lan; Zhang, Lingli; Wang, Yufei; Li, Yuanyuan; Zhao, Chunmei; Jia, Guochao; Li, Yuling; Liu, Congjun; Lu, Kui. Fabrication and characterization of polyvinyl alcohol/sodium alginate loaded carvacrol/silica hollow microspheres composite hydrogel as a colourimetric freshness indicator. **Food Bioscience**, 57, 103474, 2024. http://dx.doi.org/10.1016/j.fbio.2023.103474

Heydarian, Ali; Shavisi, Nassim. Multifunctional food packaging materials: Electrospun mats based on gelatin-xanthan gum containing chitin nanofibers and black barberry anthocyanins for freshness monitoring and enhancing the shelf-life quality of pacific white shrimps. **Food Packaging and Shelf Life**, 40, 101219, 2023. http://dx.doi.org/10.1016/j.fpsl.2023.101219

Higuchi, W.I. Analysis of data on the medicament release from ointments. **Journal of Pharmaceutical Sciences**, 51, 8, 802–804, 1962. http://dx.doi.org/10.1002/jps.2600510825

Karambasti, Parya Rashidi; Shavisi, Nassim. Development of guar gum-pectin nanofiber mats containing *Papaver rhoeas* petal anthocyanins and cellulose nanocrystals for real-time visual detection of lamb meat freshness. **LWT – Food Science and Technology**, 194, 115786, 2024. http://dx.doi.org/10.1016/j.lwt.2024.115786

Karaś, Monika; Jakubczyk, Anna; Szymanowska, Urszula; Złotek, Urszula; Zielińska, Ewelina. Digestion and bioavailability of bioactive phytochemicals. **International Journal of Food Science & Technology**, 52, 2, 291–305, 2016. http://dx.doi.org/10.1111/ijfs.13323

Khaledian, Yousef; Moshtaghi, Hamdollah; Shahbazi, Yasser. Development and characterization of smart double-layer nanofiber mats based on potato starch-turnip peel anthocyanins and guar gum-cinnamaldehyde. **Food Chemistry**, 434, 137462, 2024. http://dx.doi.org/10.1016/j.foodchem.2023.137462

Korsmeyer, Richard W.; Gurny, Robert; Doelker, Eric; Buri, Pierre; Peppas, Nikolaos A. Mechanisms of solute release from porous hydrophilic polymers. **International Journal of Pharmaceutics**, 15, 1, 25–35, 1983. http://dx.doi.org/10.1016/0378-5173(83)90064-9

Lang, Yuxi; Li, Bin; Gong, Ersheng; Shu, Chi; Si, Xu; Gao, Ningxuan; Zhang, Weijia; Cui, Huijun; Meng, Xianjun. Effects of α-casein and β-casein on the stability, antioxidant activity and bioaccessibility of blueberry anthocyanins with an in vitro simulated digestion. **Food Chemistry**, 334, 127526, 2021. http://dx.doi.org/10.1016/j.foodchem.2020.127526

Leong, Chean Ring; Daud, Nurul Shahida; Tong, Woei Yenn; Cheng, See Yuan; Tan, Wen Nee; Hamin, Nurhanis Syafiqah; Pa'ee, Khairul Faizal. Gelatine film incorporated with *Clitoria ternatea*-derived anthocyanin microcapsules, a food packaging material effective against foodborne pathogens. **Food Technology & Biotechnology**, 59(4), 422, 2021. https://doi.org/10.17113/ftb.59.04.21.7069

Li, Jian; Guo, Chaofan; Cai, Shengbao; Yi, Junjie; Zhou, Linyan. Fabrication of anthocyanin–rich W1/O/W2 emulsion gels based on pectin–GDL complexes: 3d printing performance. **Food Research International**, 168, 112782, 2023. http://dx.doi.org/10.1016/j.foodres.2023.112782

Lu, Wei; Kelly, Alan L.; Miao, Song. Emulsion-based encapsulation and delivery systems for polyphenols. **Trends in Food Science & Technology**, 47, 1–9, 2016. http://dx.doi.org/10.1016/j.tifs.2015.10.015.

Lv, Haowei; Wang, Chenxi; He, Deyun; Zhao, Haibo; Zhao, Meng; Xu, Enbo; Jin, Zhengyu; Yuan, Chao; Guo, Li; Wu, Zhengzong; Liu, Pengfei; Cui, Boi. Intelligent food tag: A starch-anthocyanin-based ph-sensitive electrospun nanofiber mat for real-time food freshness monitoring. **International Journal of Biological Macromolecules**, 256, 128384, 2024. http://dx.doi.org/10.1016/j.ijbiomac.2023.128384

Machado, Michelle Heck; Da Rosa Almeida, Aline; Maciel, Matheus Vinicius de Oliveira Brisola; Vitorino, Vinicius Bittencourt; Bazzo, Giovana Carolina; Da Rosa, Cleonice Gonçalves; William Gustavo Sganzerla; Cassiana Mendes; Pedro Luiz Manique Barreto. Microencapsulation by spray drying of red cabbage anthocyanin-rich extract for the production of a natural food colorant. **Biocatalysis and Agricultural Biotechnology**, 39, 102287, 2022. https://doi.org/10.1016/j.bcab.2022.102287

Mandaji, Carolina Marques; Pena, Rosinelson da Silva; Chisté, Renan Campos. Encapsulation of bioactive compounds extracted from plants of genus Hibiscus: A review of selected techniques and applications. **Food Research International**, 151, 110820, 2022. http://dx.doi.org/10.1016/j.foodres.2021.110820

Maronpot, Robert R.; Hayashi, Shim-mo; Bastaki, Maria. Synthetic and natural food colorants. **Foods & Foods Ingredients Journal of Japan**, 225, 2, 100–110, 2020.

Mendes, Ana Paula Pais. **Cinética de liberação de clorexidina em nanocompósito de magnetita e quitosana**. 2019. 78 f. Dissertação (Mestrado) - Curso de Mestrado em Química, Universidade Federal de São Carlos, São Carlos, 2019.

Mercante, Luiza A.; Schneider, Rodrigo; Facure, Murilo H.M.; Andre, Rafaela S.; Teodoro, Kelcilene; Martins, Danilo; Migliorini, Fernanda L.; Correa, Daniel S. Chapter 2 – Eletrofiação: parâmetro e configurações de sistema, tipos de fibras e métodos de modificação. In: Mercante, Luiza Amim; Corrêa, Daniel Souza. **Eletrofiação e nanofibras: Fundamentos e aplicações**. Ponta Grossa: Atena Editora, p. 13–51, 2024a.

Mercante, Luiza Amim; Correa, Daniel Souza. Chapter 1 - Eletrofiação: histórico e fundamentos. In: Mercante, Luiza Amim; Correa, Daniel Souza. **Eletrofiação e nanofibras: Fundamentos e aplicações**. Ponta Grossa: Atena Editora, Cap. 1. p. 1–12, 2024b.

Moreno, T.; Cocero, M.J.; Rodríguez-Rojo, S. Storage stability and simulated gastrointestinal release of spray dried grape marc phenolics. **Food and Bioproducts Processing**, 112, 96–107, 2018. http://dx.doi.org/10.1016/j.fbp.2018.08.011

Moriyama, Kousuke; Inomoto, Noe; Moriuchi, Hidetoshi; Nihei, Masanobu; Sato, Miku; Miyagi; Yoshiki; Tajiri, Ayaka; Sato, Takeshi; Tanaka, Yasuhiko; Johno, Yuuki; Goto, Masahiro; Kamiya, Noriho. Characterization of enzyme-crosslinked albumin hydrogel for cell encapsulation. **Journal of Bioscience and Bioengineering**, 136, 6, 471–476, 2023. http://dx.doi.org/10.1016/j.jbiosc.2023.09.007

Mueller, Dolores; Jung, Kathrin; Winter, Manuel; Rogoll, Dorothee; Melcher, Ralph; Kulozik, Ulrich; Schwarz, Karin; Richling, Elke. Encapsulation of anthocyanins from bilberries – Effects on bioavailability and intestinal accessibility in humans. **Food Chemistry**, 248, 217–224, 2018. http://dx.doi.org/10.1016/j.foodchem.2017.12.058

Mueller, Dolores; Jung, Kathrin; Winter, Manuel; Rogoll, Dorothee; Melcher, Ralph; Richling, Elke. Human intervention study to investigate the intestinal accessibility and bioavailability of anthocyanins from bilberries. **Food Chemistry**, 231, 275–286, 2017. http://dx.doi.org/10.1016/j.foodchem.2017.03.130

Nascimento, Amanda Lais Alves Almeida; Borges, Larissa Lorrane Rodrigues; Fernandes, Janaina Gonçalves; Freitas, Valdeir Viana; Martins, Evandro; Campelo, Pedro Henrique; Stringheta, Paulo Cesar. Exploring strategies to enhance anthocyanin bioavailability and bioaccessibility in food: A literature review. **Food Bioscience**, 56, 103388, 2023. http://dx.doi.org/10.1016/j.fbio.2023.103388

Nezamdoost-Sani, Narmin; Khaledabad, Mohammad Alizadeh; Amiri, Saber; Phimolsiripol, Yuthana; Khaneghah, Amin Mousavi. A comprehensive review on the utilization of biopolymer hydrogels to encapsulate and protect probiotics in foods. **International Journal of Biological Macromolecules**, 254, 127907, 2024. http://dx.doi.org/10.1016/j.ijbiomac.2023.127907

Oliveira, Olivia Werner; Petrovick, Pedro Ros Spray drying of plant extracts: Basic remarks and application. **Revista Brasileira De Farmacognosia**, 20(4), 641–650, 2010. https://doi.org/10.1590/S0102-695X2010000400026

Ozcan, Basak Ebru; Saroglu, Oznur; Karakas, Canan Yagmur; Karadag, Ayse. Encapsulation of purple basil leaf extract by electrospraying in double emulsion (W/O/W) filled alginate-carrageenan beads to improve the bioaccessibility of anthocyanins. **International Journal of Biological Macromolecules**, 250, 126207, 2023. http://dx.doi.org/10.1016/j.ijbiomac.2023.126207

Paes, Francisco Elvino Rodrigues; Sabino, Luiz Bruno de Sousa; Silva, Larissa Morais Ribeiro da; Silva, Ivanildo José da; Ricardo, Nágila Maria Pontes Silva; Brito, Débora; Hellen Almeida de; Menezes, Fernando Lima de; Figueiredo, Raimundo Wilane de. Anthocyanins extracted from Jamelon fruits (*Syzygium cumini* L.): Effect of microencapsulation on the properties and bioaccessibility. **South African Journal of Botany**, 166, 423–431, 2024. http://dx.doi.org/10.1016/j.sajb.2024.01.051

Peanparkdee, Methavee; Borompichaichartkul, Chaleeda; Iwamoto, Satoshi. Bioaccessibility and antioxidant activity of phenolic acids, flavonoids, and anthocyanins of encapsulated Thai rice bran extracts during in vitro gastrointestinal digestion. **Food Chemistry**, 361, 130161, 2021. http://dx.doi.org/10.1016/j.foodchem.2021.130161

Pereira, Pamela F.M.; Picciani, Paulo H. de Sousa; Calado, Veronica; Tonon, Renata V. Gelatin-based films and mats as electro-sensoactive layers for monitoring volatile compounds related to meat spoilage. **Food Packaging and Shelf Life**, 36, 101049, 2023. http://dx.doi.org/10.1016/j.fpsl.2023.101049

Pires, Rafaeli Oleques. **Modelagem matemática de perfis de liberação de fármacos a partir de nano-carreadores**. 2011. 98 f. Dissertação (Mestrado) - Curso de Curso de Mestrado em Nanociências, Centro Universitário Franciscano de Santa Maria, Santa Maria, 2011.

Ravichandran, K.S.; Silva, E.S.; Moncada, M.; Perkins-Veazie, P.; Lila, M.A.; Greenlief, C.M.; Thomas, Andrew L.; Hoskin, R.T.; Krishnaswamy, K. Spray drying to produce novel phytochemical-rich ingredients from juice and pomace of American elderberry. **Food Bioscience**, 55, 102981, 2023. http://dx.doi.org/10.1016/j.fbio.2023.102981

Robertson, G. L. **Food Packaging: Principles and Practice**. CRC Press, 2005.

Rosales, Thiécla Katiane Osvaldt; Fabi, João Paulo. Nanoencapsulated anthocyanin as a functional ingredient: Technological application and future perspectives. **Colloids and Surfaces B: Biointerfaces**, 218, 112707, 2022. http://dx.doi.org/10.1016/j.colsurfb.2022.112707

Rosales, Thiécla Katiane Osvaldt; Pedrosa, Lucas de Freitas; Nascimento, Karen Rebouças; Fioroto, Alexandre Minami; Toniazzo, Taíse; Tadini, Carmen Cecília; Purgatto, Eduardo; Hassimotto, Neuza Mariko Aymoto; Fabi, João Paulo. Nanoencapsulated anthocyanins: A new technological approach to increase physical-chemical stability and bioaccessibility. **Food Hydrocolloids**, 139, 108516, 2023. http://dx.doi.org/10.1016/j.foodhyd.2023.108516

Santos, Suelen Siqueira dos; Paraíso, Carolina Moser; Romanini, Edilson Bruno; Correa, Vanessa Gesser; Peralta, Rosane Marina; Costa, Silvio Cláudio da; Santos Junior, Oscar de Oliveira; Visentainer, Jesuí Vergílio; Reis, Miria Hespanhol Miranda; Madrona, Grasiele Scaramal.

Bioavailability of blackberry pomace microcapsules by using different techniques: An approach for yogurt application. **Innovative Food Science & Emerging Technologies**, 81, 103111, 2022. https://doi.org/10.1016/j.ifset.2022.103111

Santos, N.S.; da Silva, F. L. A. T.; Leite Neta, M. T. S. Corantes naturais: importância e fontes de obtenção. **RECIMA21 - Revista Científica Multidisciplinar**, 3(3), e331165, 2022. https://doi.org/10.47820/recima21.v3i3.1165

Sendri, Nitisha; Singh, Sarvpreet; Bhatt, Shriya; Gupta, Mahesh; Bhandari, Pamita. Insight into the influence of oxygen, sunlight and temperature on the stability and color attributes of red cabbage anthocyanins and in vitro gastrointestinal behaviour. **Food Chemistry Advances**, 3, 100359, 2023. http://dx.doi.org/10.1016/j.focha.2023.100359

Shamshad, Aneela; Iahtisham-Ul-Haq; Butt, Massod Sadiq; Nayik, Gulzar Ahmad; Al Obaid, Sami; Ansari, Mohammad Javed; Karabagias, Ioannis Konstantinos; Sarwar, Nazmul; Ramniwas, Seema. Effect of storage on physicochemical attributes of ice cream enriched with microencapsulated anthocyanins from black carrot. **Food Science and Nutrition**, 11(7), 3976, 2023. http://dx.doi.org/10.1002/fsn3.3384

Sharma, Maanas; Dash, Kshirod K.; Badwaik, Laxmikant S. Physicochemical and release behaviour of phytochemical compounds based on black jamun pulp extracts-filled alginate hydrogel beads through vibration dripping extrusion. **International Journal of Biological Macromolecules**, 194, 715–725, 2022. http://dx.doi.org/10.1016/j.ijbiomac.2021.11.116

Shen, Yixiao; Zhang, Ning; Tian, Jinlong; Xin, Guang; Liu, Ling; Sun, Xiyun; Li, Bin. Advanced approaches for improving bioavailability and controlled release of anthocyanins. **Journal of Controlled Release**, 341, 285–299, 2022. http://dx.doi.org/10.1016/j.jconrel.2021.11.031

Shipp, Jaclyn; Abdel-Aal, El-Sayed M. Food applications and physiological effects of anthocyanins as functional food ingredients. **The Open Food Science Journal**, 4, 1, 2010. http://dx.doi.org/10.2174/1874256401004010007

Singh, Akhilesh Kumar; Pal, Priti; Pandey, Brijesh; Goksen, Gulden; Sahoo, Uttam Kumar; Lorenzo, Jose M.; Sarangi, Prakash Kumar. Development of "Smart Foods" for health by nanoencapsulation: Novel technologies and challenges. **Food Chemistry: X**, 20, 100910, 2023. http://dx.doi.org/10.1016/j.fochx.2023.100910

Smeriglio, Antonella; Barreca, Davide; Bellocco, Ersilia; Trombetta, Domenico Chemistry, pharmacology and health benefits of anthocyanins. **Phytotherapy Research**, 30, 1265–1286, 2016. https://doi.org/10.1002/ptr.5642

Tarone, Adriana Gadioli; Cazarin, Cinthia Baú Betim; Junior, Mario Roberto Marostica. Anthocyanins: New techniques and challenges in microencapsulation. **Food Research International**, 133, 109092, 2020. https://doi.org/10.1016/j.foodres.2020.109092

Tavares, Jucelio Kilinski **Modelagem da liberação controlada de princípios ativos (betacaroteno e lidocaína) de microcápsulas**. 2015. 212 f. Tese (Doutorado) - Curso de Pós-graduação em Engenharia Química, Universidade Federal de Santa Catarina, Florianópolis, 2015.

Timilsena, Yakindra Prasad; Haque, Md Amdadul; Adhikari, Benu. Encapsulation in the food industry: A brief historical overview to recent developments. **Food and Nutrition Sciences**, 11, 6, 481–508, 2020. https://doi.org/10.4236/fns.2020.116035

Vannuchi, Nicholas; Ramos, Sergiana dos Passos; Mazzo, Tatiana Martelli; Longo, Elson; Bonsanto, Fabiana Perrechil; Braga, Anna Rafaela Cavalcante; Rosso, Veridiana Vera de. Natural deep eutectic solvents (NADES)-extracted anthocyanins: Bioaccessibility in electrospun peo microfibers. **Food Research International**, 177, 113898, 2024. http://dx.doi.org/10.1016/j.foodres.2023.113898

Victoria-Campos, Claudia Inés; Ornelas-Paz, José de Jesús; Rocha-Guzmán, Nuria Elizabeth; Gallegos-Infante, José Alberto; Failla, Mark L.; Pérez-Martínez, Jaime David; Rios-Velasco, Claudio; Ibarra-Junquera, Vrani. Gastrointestinal metabolism and bioaccessibility of selected anthocyanins isolated from commonly consumed fruits. **Food Chemistry**, 383, 132451, 2022. http://dx.doi.org/10.1016/j.foodchem.2022.132451

Wiczkowski, Wieslaw; Szawara-Nowak, Dorota; Romaszko, Jerzy. The impact of red cabbage fermentation on bioavailability of anthocyanins and antioxidant capacity of human plasma. **Food Chemistry**, 190, 730–740, 2016. http://dx.doi.org/10.1016/j.foodchem.2015.06.021

Wu, Xiuli; Yan, Xiangxuan; Zhang, Jianwen; Wu, Xuexu; Luan, Mingran; Zhang, Qing. Preparation and characterization of pH-sensitive intelligent packaging films based on cassava starch/polyvinyl alcohol matrices containing *Aronia melanocarpa* anthocyanins. **LWT-Food Science and Technology**, 194, 115818, 2024. http://dx.doi.org/10.1016/j.lwt.2024.115818

Yang, Shuo; Ding, Qijun; Li, You; Han, Wenjia. Bacterial cellulose/gelatin-based pH-responsive functional film for food freshness monitoring. **International Journal of Biological Macromolecules**, 259, 129203, 2024. http://dx.doi.org/10.1016/j.ijbiomac.2024.129203

Yavuz-Düzgün, Merve; Kareth, Sabine; Özçelik, Beraat; Weidner, Eckhard. Black carrot extract loaded-potato protein particles by PGSS-drying: Physico-chemical properties and in vitro bioaccessability. **The Journal of Supercritical Fluids**, 203, 106065, 2023. http://dx.doi.org/10.1016/j.supflu.2023.106065

Zhang, Rui; Zhou, Lan; Li, Jia; Oliveira, Hélder; Yang, Ning; Jin, Weiping; Zhu, Zhenzhou; Li; Shuyi; He, Jingren. Microencapsulation of anthocyanins extracted from grape skin by emulsification/internal gelation followed by spray/freeze-drying techniques: Characterization, stability and bioaccessibility. **LWT-Food Science and Technology**, 123, 109097, 2020. http://dx.doi.org/10.1016/j.lwt.2020.109097

4

Anthocyanins as Food Colorants

Elivaldo Nunes Modesto Junior, Rosane Patrícia Ferreira Chaves, and Renan Campos Chisté

4.1 Introduction

Natural food colorants comprise a class of food additives widely applied by food industries, the technological functionalities of which are associated with the aggregation to develop new colored products, restoration of color loss during processing, or color enhancement of food products (Damodaran, Parkin, & Fennema, 2010). The increased global demand for natural and healthier food products makes natural pigments extracted from plants, or produced by biotechnological means, promising food colorants to attend these types of growing markets (Bocker & Silva, 2022). Thus, food manufacturers have been pressed to reformulate their products to adopt natural pigments as replacements for synthetic dyes (Chung, Rojanasasithara, Mutilangi, & McClements, 2017).

Following this trend, the expected growth of the natural dyes market is 11%, from 2018 to 2024 (Wood, 2019), and the global natural food colors market is predicted to grow at an annual growth rate of 6% until 2032 (Gamage & Choo, 2023). In this regard, modern consumers from different countries have intensified the search for natural and sustainable products that positively impact human health and support greater longevity and quality of life (Martins, Roriz, Morales, Barros, & Ferreira, 2016). Natural food colorants follow this trend because natural pigments are linked to beneficial health effects, such as the prevention of non-communicable chronic diseases (NCDs), which are diseases not transmissible directly from one person to another (Bocker & Silva, 2022). NCDs, such as heart diseases, diabetes, autoimmune diseases, chronic kidney disease, osteoporosis, Alzheimer's disease, Parkinson's disease, and cancer, are responsible for almost 70% of all deaths worldwide (Ribeiro & Veloso, 2021). Thus, it is of paramount importance the exploitation of novel and promising sources of natural pigments to accomplish the increasing global demand. Amongst the prevailing natural pigments, anthocyanins are remarkable compounds widely investigated worldwide to be used as food coloring (Gamage & Choo, 2023).

Anthocyanins are available in plant vacuoles, and depending on the pH, and chemical interactions with other compounds, they express various attractive colors in the range of red, orange, purple, and blue (Khoo, Azlan, Tang, & Lim, 2017). Despite the attractive colors and health benefits attributed to anthocyanin intake, their incorporation in food products is challenging due to their limited stability as influenced by many factors, such as the type of chemical structures, exposure to high temperature, light and oxygen, pH changes, high relative humidity conditions, presence of sulfur dioxide, sulfites, ascorbic acid, enzymes, co-pigments, and metal ions (Sharma, Gupta, Singh, Bansal, & Singh, 2016; Vidana Gamage, Lim, & Choo, 2021). Therefore, in this chapter, we will discuss promising

DOI: 10.1201/9781003453260-5

sources of anthocyanins to be used as food colorants, focusing on selected berries, along with approaches related to the limitations for the incorporation of anthocyanins by the food industries and potential technological applications.

4.2 Sources of Anthocyanins with Potential to Be Used as Food Colorants

Color is one of the most important sensory aspects to be considered in the development of food products since visual appearance is a primary influencer on consumers' perception of overall food quality (Martinez, Rando, Agante, & Abreu, 2021; Sun, Xin, & Alper, 2021). The color influence is clearly justified by the ability of colors to refer to pre-developed concepts by consumers that trigger positive or negative perceptions depending on the expected appearance, differences in tonality, brightness, and color intensity of food products (Spence, 2018; Luo, Yu, Westland, & Mahon, 2019).

Anthocyanins, one of the most frequent and widely distributed flavonoids, are water-soluble pigments in flowers, seeds, fruits, and leaves of a great number of plant species. The chemical structures of anthocyanins (C6-C3-C6) are formed from a 2-phenyl benzopyran basic structure, which is composed of two benzoyl rings connected by a heterocyclic ring, and glycosides are commonly attached to hydroxyl groups (OH) on positions C3 (most frequent), C5, and C7 (Sun, Xin, & Alper, 2021). Generally, aglycone forms of anthocyanins (anthocyanidins) are less stable and rarely frequent in plant species under natural conditions; so, they are mostly found formed by glycosidic bonds with sugars such as glucose, galactose, rhamnose, xylose, and rutinose (Zhao, Wang, Huang, & Hu, 2021). Currently, 702 different anthocyanins and 27 anthocyanidins have been identified in nature, and among the anthocyanins, cyanidin 3-*O*-glucoside is the major anthocyanin found in most plant species (Khoo, Azlan, Tang, & Lim, 2017). The different types of anthocyanins are mainly due to differences in the position of glycosidic bonds, the type and number of sugars, the number of OH groups, and the type and number of attached organic and phenolic acids (Yue et al., 2021). Anthocyanins contain unsaturated double bonds and have easily oxidized groups, which lead to their highly unstable properties. Figure 4.1 shows the six most common forms of anthocyanins found in foods, namely, cyanidin, petunidin, petonidin, pelargonidin, delphinidin, and malvidin.

There are abundant anthocyanins resources in fruits and by-products, and the high-value addition to fruit wastes is a hot topic in the food science and technology area (Tan et al., 2022). Therefore, adding value to fruit wastes targeting bioactive compounds, such as anthocyanins for food coloring purposes, not only improves the understanding of rational use of fruit residues but also stimulates the concept of circular economy during the industrial waste management strategies, which increases environmental protection by implementing sustainable approaches that result in high-value co-products with economic and social benefits.

The estimation for the anthocyanin market size is USD 355.67 million in 2024 and is expected to reach USD 457.19 million by 2029, growing at a compound annual growth rate (CAGR) of 5.15% during 2024–2029 (Mordor Intelligence, 2024). Currently, the main sources of anthocyanins, which are investigated for food coloring purposes, are fruits (mostly berries), vegetables, colored grains, and by-products of fruit and vegetable processing (Ockermann, Headley, Lizio, & Hansmann, 2021). China is the leading producer of fresh fruit worldwide with an annual output of 242 million metric tons, followed by India (105

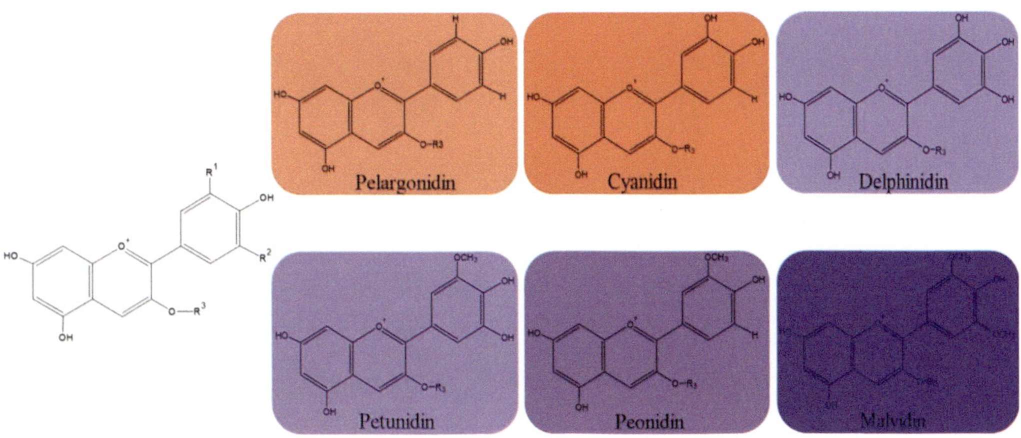

FIGURE 4.1
Chemical structures of anthocyanins, and the six most common forms found in foods. R^1 and R^2 can be replaced by H, OH, or OCH_3, and R_3 can be substituted by sugar moieties (anthocyanins) or OH (aglycone forms, anthocyanidins).

million metric tons) and Brazil (39 million metric tons) (FAOSTAT, 2024), and high production of fruit wastes is expected around the globe. About 60–70% of the flavonoids of small berries are found in their peel and seeds, which are usual by-products, being cyanidin 3-*O*-glucoside, cyanidin 3-*O*-rutinoside, and delphinidin 3-*O*-glucoside the major reported anthocyanins (Paredes-López, Cervantes-Ceja, Vigna-Pérez, & Hernández-Pérez, 2010).

Most berries are known by the different attractive colors seen in their fruit peels and pulps, which are due to the differences in the anthocyanins' composition that open great avenues to systematic investigations in different countries on the potential of anthocyanins as food colorants. Furthermore, numerous studies confirmed that anthocyanins from berries present numerous biological properties (Figure 4.2), mainly antioxidant and anti-inflammatory activities, and the frequent intake of anthocyanins has been associated with a decrease in the development of chronicle degenerative diseases, such as cardiovascular and neurodegenerative diseases, diabetes (type-2), and cancer (Kong, Chia, Goh, Chia, & Brouillard, 2003; Badescu, Badulescu, Badescu, & Ciocoiu, 2015; Koss-Mikołajczyk, Kusznierewicz, & Bartoszek, 2019; Aboonabi, Singh, & Rose'Meyer, 2020).

4.2.1 Blueberry (*Vaccinium* spp.)

Blueberry belongs to the family Ericaceae, subfamily Vacciniaceae, genus *Vaccinium* and subgenus *Cyanococcus*, being native to North America, characterized as a perennial deciduous or evergreen shrub (Yang et al., 2022), and very popular around the world for its appreciable taste (Herrera-Balandrano, Chai, Beta, Feng, & Huang, 2021). Currently, blueberry is mostly cultivated in North America (Canada and USA), China, Japan, and Europe, but also in other countries of the Southern Hemisphere, such as Chile, Argentina, New Zealand, and Australia (Routray & Orsat, 2011).

These nutritionally relevant small berries are rich in not only a variety of nutrients, such as dietary fibers, minerals, and vitamins, but also important bioactive compounds, such as hydroxycinnamic acid and resveratrol (Kalt et al., 2020; Silva et al., 2020; Wu et al., 2022; Yang et al., 2022), and highlighted by their high contents of anthocyanins (Borges, Degeneve, Mullen, & Crozier, 2010), ranging from 19 to 677-mg cyanidin 3-glucoside/100

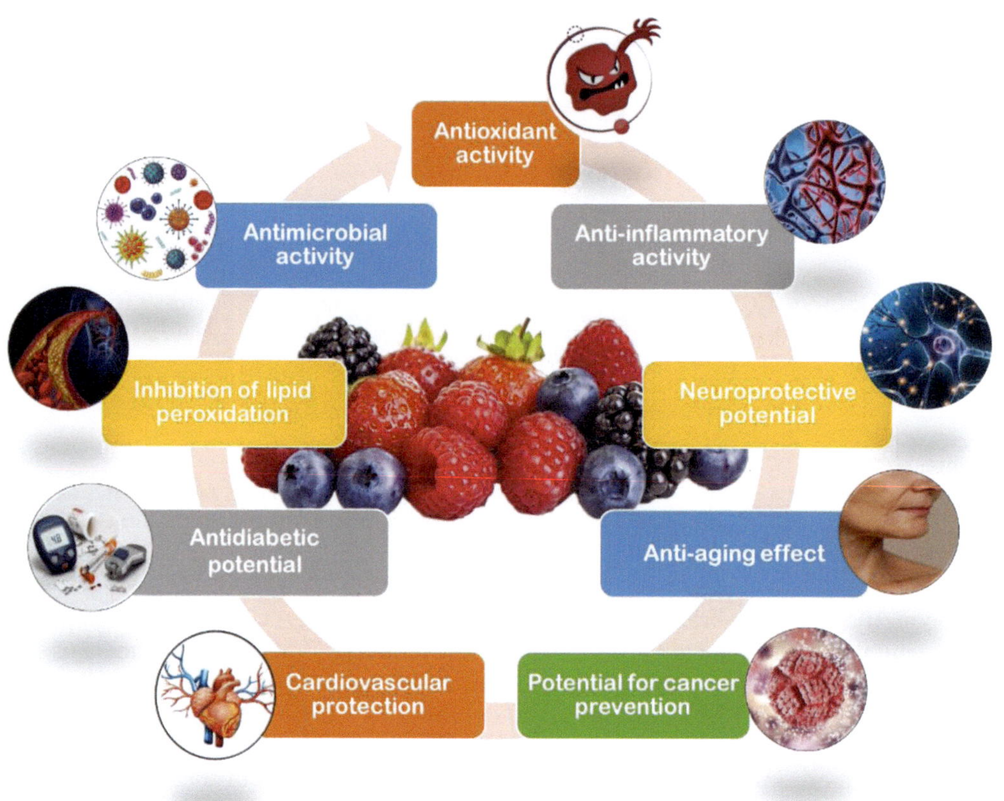

FIGURE 4.2

Biological properties associated with anthocyanins from berries.

g fresh weight, 456–1406-mg cyanidin 3,5-glucoside/100 g fresh weight, and 101–195-mg malvidin 3-glucoside/100 g fresh weigh (Koh, Xu, & Wicker, 2020).

The contents of anthocyanins in blueberry fruits are key indicators that may impact the potential functional value of blueberries. The synthesis of anthocyanins is greatly influenced by environmental factors, as well as genotypes (Wang et al., 2022); thus, expressive differences in anthocyanin levels are expected in fruits from different cultivars and origins (Wang et al., 2022; Yang et al., 2022). Wang et al. (2022) used ultra-performance liquid chromatography-mass spectrometry (UPLC-MS) to determine the anthocyanins' composition of 62 blueberry cultivars in China. Among them, "Rubel" showed the highest anthocyanin contents (534 mg/100 g fresh weight), whereas "Puru" contained the lowest contents (71 mg/100 g fresh weight). These same authors reported that among all the monomers of anthocyanins identified in the 62 cultivars, malvidin-galactoside presented the highest average content (~48 mg/100 g fresh weight).

A large number of overlapping anthocyanin components can be found among blueberry cultivars, indicating that close genetic relationships still exist among breeding cultivars (Yang et al., 2022). A recent study reported the anthocyanins' composition of 17 blueberry cultivars found in China, being delphinidin, malvidin, and petunidin glycosides as the main identified anthocyanins, accounting for more than 90% of the total anthocyanin content (Wu et al., 2023).

Anthocyanins from blueberry are diverse in composition and contents, and remarkable biological effects were reported, such as antioxidant and anti-inflammatory activities, improvement in vision quality, lowering blood pressure, and role in the prevention of cardiovascular and neurodegenerative diseases, and other effects (Routray & Orsat, 2011; Wu, Gao, Guo, Zhang, & Gong, 2018; Kalt et al., 2020; Yang et al., 2022).

4.2.2 Black Goji Berry (*Lycium ruthenicum*)

Lycium ruthenicum fruits belong to the family Solanaceae and genus *Lycium*, which is mainly distributed in China, Mongolia, Central Asia, and North Africa, and the deciduous shrub grows under the environment of special geography (high altitude) and arid climate (low temperature, less rain, and strong sunlight) (Wang et al., 2018; Liu, Liu, Wen, Tao, & Shao, 2020).

The total contents of anthocyanins of black goji berries are very high in the fresh fruits, reaching 24 mg/g, which was reported to be higher than other pigmented plants, such as blue honeysuckle (1–14 mg/g), some varieties of blueberries (4–6 mg/g), and mulberries (0.1–3 mg/g) (Luo, Yu, Westland, & Mahon, 2019). Among the 39 identified black goji anthocyanins, most are 3,5-diglycoside derivatives of petunidin (>95%), being petunidin 3-*O*-rutinoside (*trans-p*-coumaroyl)5-*O*-glucoside accounting for 80% of the total anthocyanins (Yan et al., 2022). In addition, many *Z-E* isomeric forms of anthocyanins in black goji were reported in nature, and the acyl groups in acylated anthocyanins were typically *E*-configured, with only a few *Z*-isomers (Ichiyanagi et al., 2005; Yan et al., 2022). The coexistence and abundance of both *Z*- and *E*-anthocyanin isomers in the same source can be considered another detachable feature of black goji berry fruits.

Black goji berries are popular in the food industry for preparing potential functional foods, such as black goji juices and drinks, not only due to their related biological properties but also due to the attractive black-bluish color (Zhang et al., 2019), which increased the fruit consumption and stimulated the development of functional black goji products (Wang et al., 2018; Liu, Liu, Wen, Tao, & Shao, 2020). The biological properties associated with black goji berries are derived not only from its bioactive compounds, mainly anthocyanins, but also from essential oils, organic acids, minerals, and polysaccharides (Zheng et al., 2011; Peng, Liu, Shi, & Li, 2014; Peng, Liu, Lei, & Wang, 2016; Qi et al., 2019).

4.2.3 Blackcurrant (*Ribes nigrum* L.)

Ribes nigrum L. (genus *Ribes*, family Grossulariaceae), generally called blackcurrant, is a flowering shrub that contains deep purple, bittersweet, and seed-bearing berries that can attain a diameter of about 1 cm (Gopalan et al., 2012). These edible berries can be used to prepare syrup, jelly, candy, and liqueur, among others, and are commonly used to produce dark violet pigments due to their high contents of anthocyanins (Ejaz et al., 2023). For comparison purposes, the reported total anthocyanin contents for blackcurrant (freeze-dried powder extracts) were 1741 mg/100 g (dry weight), which were higher than for blueberry (1562 mg/100 g) and blackberry (949 mg/100 g) (Lee et al., 2015).

Blackcurrants have been considered "super fruits" since they are supposed to contribute to numerous therapeutic benefits, such as the capacity to treat chronic illnesses associated with oxidative stress (Lister, Wilson, Sutton, & Morrison, 2002; Ejaz et al., 2023). Furthermore, blackcurrants are widely recognized for containing high levels of polyphenols, specifically anthocyanins and proanthocyanins, when compared with other berries (Lee et al., 2015).

Anthocyanins are the main investigated bioactive compounds in blackcurrants since they are present not only in the berries but also in its seeds and leaves (Zheng et al., 2012; Teleszko & Wojdyło, 2015; Cao, Park, Lee, & Kim, 2021). Moreover, the contents of anthocyanins were observed to show wide variation among some blackcurrant cultivars, such as "Record," "Blackdown," and "Ronix" (80–476 mg/100 mg of fresh weight) (Nour, Stampar, Veberic, & Jakopic, 2013). Concerning the anthocyanins' composition, about 97% of the anthocyanins in blackcurrants are derived from the four basic components delphinidin 3-*O*-glucoside, delphinidin 3-*O*-rutinoside, cyanidin 3-*O*-glucoside, and cyanidin 3-*O*-rutinoside (Anttonen & Karjalainen, 2006; Cao, Park, Lee, & Kim, 2021).

4.2.4 Grumixama (*Eugenia brasiliensis*)

Grumixama fruits (genus *Eugenia*, family Myrtaceae) are very similar to cherries, with 2.0-cm diameters, one to many seeds, and a thick peel that can be purple/red or yellow/white depending on the species/varieties, being purplish/reddish color the most common fruits (Da Silva, Rodrigues, Mercadante, & De Rosso, 2014; Teixeira, Bertoldi, Lajolo, & Hassimotto, 2015). The Myrtaceae family is generally considered an important family of edible fruits which comprises around 121 genera and 3800–5800 species of woody shrubs or trees, mostly evergreen and distributed mainly in the tropics and subtropics (Stefanello, Pascoal, & Salvador, 2011). The fruits from this family have significant economic potential due to their nutritional values, especially as promising sources of bioactive compounds with antioxidant properties of wide interest to the food, cosmetics, and pharmaceutical industries (Sardi et al., 2017).

Grumixama fruits have purple/red peel due to the presence of anthocyanins, especially cyanidin 3-*O*-glucoside and delfinidin 3-*O*-glucoside (Fischer et al., 2005; Flores et al., 2012; Da Silva, Rodrigues, Mercadante, & De Rosso, 2014), and the first accounts for ~94% of the total monomeric anthocyanins (Da Silva, Rodrigues, Mercadante, & De Rosso, 2014). The total contents of anthocyanins were reported in the range of 239–643 mg/100 g (fresh weight) (Araújo, Neri-Numa, de Paulo Farias, da Cunha, & Pastore, 2019; Modesto Junior, Martins, Pereira, Chisté, & Pena, 2023a).

The grumixama fruits are commonly used to prepare homemade foods, such as frozen fruits, frozen pulps, and jellies. The anthocyanins recovered from the edible part of purple grumixama were reported to have great potential to be used as a natural colorant in the food industry (Modesto Junior, Martins, Pereira, Chisté, & Pena, 2023a). Modesto Junior et al. (2023b) prepared grumixama juice to stimulate the development of healthy options of food products with high contents of antioxidant compounds, especially anthocyanins, and reported delphinidin 3-glucoside (10.68 mg/L) and cyanidin 3-glucoside (13.67 mg/L) as the predominant compounds.

Furthermore, systematic investigations concerning the biological effects of grumixama fruits are needed to understand its full technological potential, since only a few reports are available (Araújo, Neri-Numa, de Paulo Farias, da Cunha, & Pastore, 2019), such the anti-inflammatory capacity of fruit pulp extracts to reduce NF-κB activation and release of TNF-α (Lazarini et al., 2018), and the antiproliferative activity against human breast cancer cells reported for grumixama juice (Teixeira et al., 2017).

4.2.5 Mulberry (*Morus* spp.)

Mulberry from the genus *Morus* is a perennial woody plant in the Moraceae family (Erden, 2021), which is a fast-growing deciduous plant found in a wide range of climatic, topographical, and soil conditions, and it is widely distributed, from temperate to subtropical regions of the Northern hemisphere to the tropics of the Southern hemisphere (Wen et al., 2019).

Since ancient times, the Chinese Materia Medica (CMM) describes many medicinal benefits associated with mulberry, and these beneficial effects along with the promising chemical composition have attracted increasing research interest throughout the world (China, Japan, Korea, and some European countries), mainly regarding the bioactive compounds, such as flavonoids (including anthocyanins), other non-flavonoids, alkaloids, and polysaccharides (Hunyadi, Martins, Hsieh, Seres, & Zupkó, 2012; Wen et al., 2019). In most mulberry-growing countries, the fruits are eaten fresh, dried, or processed into wines, fruit juices, and jams, because of their highly appreciated taste, low-calorie contents, and appealing color due to the presence of high contents of anthocyanins (Huang, Zhou, Meng, Wu, & He, 2017; Aybastier, 2021).

As anthocyanin contents and profiles may vary with cultivars, environmental conditions, agricultural practices, processing, and storage, Kim and Lee (2020) investigated the composition of 12 cultivars of mulberry from Korea, by LC-MS, and 16 anthocyanins were identified. These authors reported that cyanidin 3-*O*-glucoside was the major anthocyanin in mulberries (33–1951 mg/100 g, dry weight), accounting for ~67% of the total contents, followed by cyanidin 3-*O*-rutinoside (18–865 mg/100 g, dry weight) and pelargonidin 3-glucoside (0–51 mg/100 g). In this same study, several anthocyanins were identified for the first time in mulberry fruits and included malvidin hexoside, cyanidin malonyl hexose hexoside, cyanidin pentoside, cyanidin malonyl hexoside, petunidin deoxyhexose hexoside, and cyanidin deoxyhexoside (Kim & Lee, 2020).

In addition to imparting attractive colors, anthocyanins from mulberry have known pharmacological activities and health-enhancing potential, such as antioxidant, and anti-inflammatory, and are associated with the decrease in the development of chronic degenerative diseases (Wen et al., 2019).

4.3　Limitations for the Application of Anthocyanins by the Food Industries

The chemical instability of anthocyanins probably results in their low bioavailability (0.26–1.8%) (Cladis, Debelo, Lachcik, Ferruzzi, & Weaver, 2020) and seriously limits their application in food, cosmetic, and pharmaceutical industries. These compounds are hydrophilic and highly susceptible to degradation and significantly lose color due to several environmental factors, mainly pH variation, high temperature, and light and oxygen exposure (Herrera-Balandrano, Chai, Beta, Feng, & Huang, 2021; Oancea, 2021; Rashwan et al., 2021). Thus, the application of anthocyanins or its natural extracts in food products that are subjected to food processing is challenging, especially on an industrial scale. Therefore, understanding the main factors that limit the stability of anthocyanins and allow exploring technological alternatives to increase the stabilization of anthocyanins, in both food and physiological systems, have high industrial relevance worldwide.

Anthocyanin molecules undergo reversible chemical alterations in their structures as influenced by pH changes, which will be one of the most important decision factors to take into consideration before using anthocyanins as food colorant (Enaru, Drețcanu, Pop, Stănilă, & Diaconeasa, 2021).

In general, in an acidic environment, at pH values from about 1 to 3, anthocyanins are found as flavylium cation (red color), and this structural form is responsible for the production of intense reddish and/or purplish colors (Khoo, Azlan, Tang, & Lim, 2017). When the pH increases between around 4 and 6, colorless carbinol pseudobases may take place (Kang, Ko, & Chung, 2021). Finally, at a pH higher than 7–8, quinoidal blueish structures are formed, followed by yellowish chalcones formation (pH > 8), due to C-ring opening. Different hues and color intensities may be observed as dependent on the anthocyanin's composition in the medium.

Therefore, considering the high susceptibility of anthocyanins to pH variation, the addition of anthocyanins in food formulations is more appropriate for food products with acidic characteristics. Another very interesting application of anthocyanins as a food colorant is about its incorporation in the formulation of "intelligent" films, which have proven to be reliable in indicating the freshness of foods such as fish, pork, and milk, based on their pH alterations due to the loss of quality during storage (Francisco et al., 2024; Wu et al., 2024; Yang, Ding, Qijun, Li, & Han, 2024). Figure 4.3 illustrates the reversible color changes of anthocyanins as influenced by pH alteration.

Another important limiting factor to be considered before the application of anthocyanins as food colorant is their susceptibility to loss of color and degradation during food processing when high temperatures are used. Thermal degradation of anthocyanins, in particular at pH above 4.0 (colorless carbinol pseudobase), follows the pathway of intermediate chalcone and final derivatives of aldehyde and phenolic acids, structurally differentiated based on the corresponding anthocyanin (Sinela et al., 2017; Wang et al., 2019),

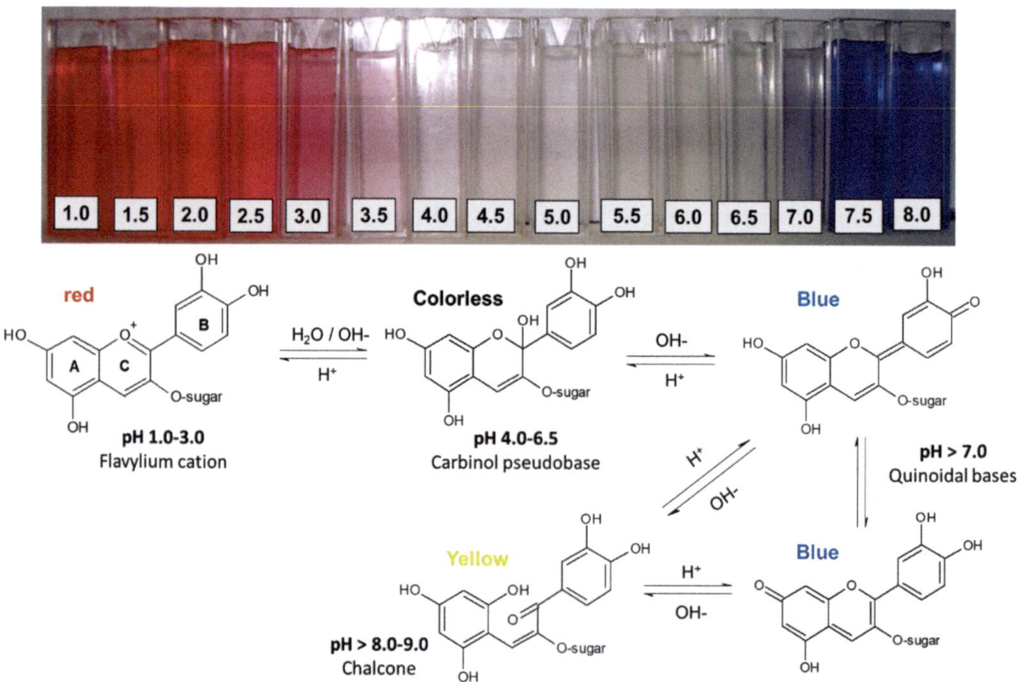

FIGURE 4.3
Structural modifications of anthocyanins molecules and associated color changes as induced by pH alteration.

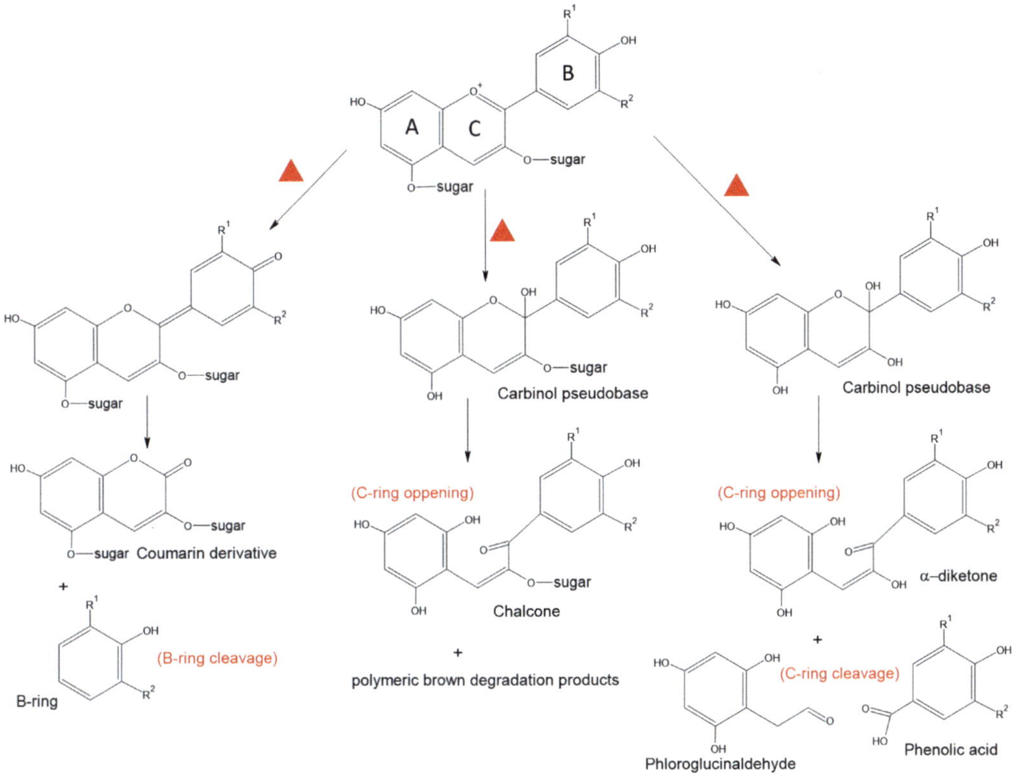

FIGURE 4.4
Possible mechanisms of thermal degradation of anthocyanins. R^1 and R^2 can be OH, H, and/or OCH_3 to indicate the other structures of anthocyanins. (Adapted from Damodaran, Parkin, & Fennema, 2010.)

as illustrated in Figure 4.4. Combined hydrolytic and auto-oxidative reactions are probably involved (Dangles & Fenger, 2018).

Obviously, thermal degradation of anthocyanins occurs to different extents, depending on the raw material, pH, structural characteristics, and presence of co-pigments. Importantly, the stability of anthocyanins increases as methylation increases or acylation is found in the structure and decreases with an increasing number of OH groups in the B-ring (Fei et al., 2020). Thus, extracts composed of highly acylated anthocyanins tend to be more stable than the non-acylated alternatives, and this fact calls our attention to the importance of investigations concerning the elucidation of anthocyanin's composition in food matrices. Furthermore, deglycosylation is the main thermal degradation pathway in glycosylated anthocyanins, as in the case of cyanidin 3-glucoside, where the sugar moiety attached to C3 is removed, inducing the cleavage of B- or A-ring that generates phloroglucinolaldehyde or 4-hydroxybenzoic acid (Patras, Brunton, O'Donnell, & Tiwari, 2010).

In addition to pH changes and temperature increases, the light exposure and molecular oxygen presence are also important frequent conditions that limit the industrial use of anthocyanins as food colorants. In nature, plants are exposed to light, which stimulates anthocyanin synthesis and accumulation; hence, light is another important parameter in the stability of these compounds (Sikorski, 2018). However, anthocyanins are affected by light exposure in two ways: light is required for anthocyanins production, but it also speeds up their breakdown (Cavalcanti, Santos, & Meireles, 2011). In addition, the amount of molecular oxygen in the medium increases the rate of light-induced breakdown; thus, it should be mentioned that intense color fade is expected when anthocyanins absorb light after a period of sensitization (Remini et al., 2018). Studies carried out to understand the impact of light on the stability of anthocyanin extracts from various biological sources can be found in the literature. As an example, Modesto Junior et al. (2023b) exposed extracts of anthocyanins of *E. brasiliensis* to different light sources, namely, fluorescent, incandescent, and ultraviolet. These authors observed that UV light, which has higher energy than visible light, induced lower half-life time ($t_{1/2}$ = 0.18 h) to anthocyanins than incandescent ($t_{1/2}$ = 59.63 h) and fluorescent ($t_{1/2}$ = 45.61) lights.

4.4 Considerations on the Application of Anthocyanins in Food Formulations

The demand for foods and natural extracts with high contents of anthocyanins has increased the attention of the scientific community due to the claimed health benefits they may offer (Giusti & Wrolstad, 2003). Consumers and food manufacturers are searching for natural food colorants over synthetic dyes (Oplatowska-Stachowiak & Elliott, 2017). According to a 2007 trial conducted on children aged three and between eight and nine years, mixtures of artificial colorants such as Tartrazine (E102), Sunset Yellow (E110), Carmoisine (E122), Ponceau 4R (E124), Allura Red AC (E129), and Quinoline Yellow (E104), when combined with sodium benzoate (E211), statistically seem to be associated with considerable rise of hyperactivity in children (McCann et al., 2007).

Anthocyanins impart distinct color hues in the range of red, orange, purple, and blue, as demonstrated in this chapter, and they have been used as natural, water-soluble, Codex Alimentarius-approved food colorant (E163) for the application in a range of food and beverage products (Gençdağ, Özdemir, Demirci, Görgüç, & Yılmaz, 2022; Silva, Reboredo, & Lidon, 2022). As an example, anthocyanins hold great importance in wine industries as they play a major role in affecting wine properties, such as wine color, mouthfeel, aging, stability, and overall quality (Ruta & Farcasanu, 2019). Moreover, adding extracts rich in anthocyanins to baked goods such as cookies, biscuits, and macarons can attribute attractive visual colors to characterize new products, in addition to potentially increase its antioxidant properties (Papillo et al., 2018; Albuquerque, Pinela, Barros, Oliveira, & Ferreira, 2020), increasing the market appeal and overall food's acceptability. Furthermore, anthocyanins have proved to exhibit high stability during storage in acidic fermented products, such as kefir, yogurt, and various beverages, and these products might be thought of as ideal foods for anthocyanin addition as a food coloring (Wallace & Giusti, 2008; Karaaslan, Ozden, Vardin, & Turkoglu, 2011). Due to the higher stability of acylated anthocyanins during food processing conditions, researchers around the world are frequently investigating the technological potential of promising sources of acylated structures, such as red cabbage, radish, and black carrots, which are known to contain high amounts of these type of anthocyanins (B'kowska-Barczak, 2005).

Therefore, anthocyanins have high technological potential for the application in the food development sectors to stimulate the design of a number of functional foods and beverages, which may contribute to beneficial effects to human health.

4.5 Final Considerations

Over the years, anthocyanins (isolated or extracts with high contents) have been investigated as natural colorants in a number of food applications mainly due to their attractive colors, in addition to the known biological properties that can add value in the development of potential functional foods. The high instability of anthocyanins over a number of food processing and storage conditions, mainly pH change, temperature increase, light, and oxygen exposure, which result in intense color fade, is the main limiting factor to consolidating these natural pigments in the food industry as the first choice to impart reddish-purplish colors to food products.

Therefore, systematic studies to optimize different strategies to protect anthocyanins during food processing are still needed, such as encapsulation techniques, as well as further scientific investigations to undercover new plant sources with high contents of selected and more stable anthocyanins to make them feasible to be exploited by sustainable approaches.

References

Aboonabi, A., Singh, I., & Rose'Meyer, R. (2020). Cytoprotective effects of berry anthocyanins against induced oxidative stress and inflammation in primary human diabetic aortic endothelial cells. *Chemico-Biological Interactions*, 317, 108940. https://doi.org/10.1016/j.cbi.2020.108940

Albuquerque, B. R., Pinela, J., Barros, L., Oliveira, M. B. P. P., & Ferreira, I. C. F. R. (2020). Anthocyanin-rich extract of jabuticaba epicarp as a natural colorant: Optimization of heat-and ultrasound-assisted extractions and application in a bakery product. *Food Chemistry*, 316, 126364. https://doi.org/10.1016/j.foodchem.2020.126364

Anttonen, M. J., & Karjalainen, R. O. (2006). High-performance liquid chromatography analysis of black currant (*Ribes nigrum* L.) fruit phenolics grown either conventionally or organically. *Journal of Agricultural and Food Chemistry*, 54, 20, 7530–7538. https://doi.org/10.1021/jf0615350

Araújo, F. F., Neri-Numa, I. A., de Paulo Farias, D., da Cunha, G. R. M. C., & Pastore, G. M. (2019). Wild Brazilian species of Eugenia genera (Myrtaceae) as an innovation hotspot for food and pharmacological purposes. *Food Research International*, 121, 57–72. https://doi.org/10.1016/j.foodres.2019.03.018

Aybastier, Ö. (2021). Isolation and determination of antioxidant compounds from black mulberry (*Morus nigra* L.) by chromatographic methods. *Food*, 46, 1, 32–41. https://doi.org/10.15237/gida.GD20114

Badescu, M., Badulescu, O., Badescu, L., & Ciocoiu, M. (2015). Effects of *Sambucus nigra* and *Aronia melanocarpa* extracts on immune system disorders within diabetes mellitus. *Pharmaceutical Biology*, 53, 4, 533–539. https://doi.org/10.3109/13880209.2014.931441

Bqkowska-Barczak, A. (2005). Acylated anthocyanins as stable, natural food colorants – A review. *Polish Journal of Food and Nutrition Sciences*, 14, 55.

Bocker, R., & Silva, E. K. (2022). Pulsed electric field assisted extraction of natural food pigments and colorings from plant matrices. *Food Chemistry: X*, 15, 30, 100398. https://doi.org/10.1016/j.fochx.2022.100398

Borges, G., Degeneve, A., Mullen, W., & Crozier, A. (2010). Identification of flavonoid and phenolic antioxidants in black currants, blueberries, raspberries, red currants, and cranberries. *Journal of Agricultural and Food Chemistry*, 58, 7, 3901–3909. https://doi.org/10.1021/jf902263n

Cao, L., Park, Y., Lee, S., & Kim, D.-O. (2021). Extraction, identification, and health benefits of anthocyanins in blackcurrants (*Ribes nigrum* L.). *Applied Sciences*, 11, 1863. https://doi.org/10.3390/app11041863

Cavalcanti, R. N., Santos, D. T., & Meireles, M. A. A. (2011). Non-thermal stabilization mechanisms of anthocyanins in model and food systems—An overview. Food research international, 44(2), 499–509.

Cavalcanti, R. N., Santos, D. T., & Meireles, M. A. A. (2021). Non-thermal stabilization mechanisms of anthocyanins in model and food systems—An overview. *Food Research International*, 44, 2 499–509. https://doi.org/10.1016/j.foodres.2010.12.007

Chung, C., Rojanasasithara, T., Mutilangi, W., & McClements, D. J. (2017). Stability improvement of natural food colors: Impact of amino acid and peptide addition on anthocyanin stability in model beverages. *Food Chemistry*, 218, 277–284. https://doi.org/10.1016/j.foodchem.2016.09.087

Cladis, D. P., Debelo, H., Lachcik, P. J., Ferruzzi, M. G., & Weaver, C. M. (2020). Increasing doses of blueberry polyphenols alters colonic metabolism and calcium absorption in ovariectomized rats. *Molecular Nutrition in Food Research*, 64, 12, 2000031. https://doi.org/10.1002/mnfr.202000031

Da Silva, N. A., Rodrigues, E., Mercadante, A. Z., & De Rosso, V. V. (2014). Phenolic compounds and carotenoids from four fruits native from the Brazilian Atlantic forest. *Journal of Agricultural and Food Chemistry*, 62, 22, 5072–5084. https://doi.org/10.1021/jf501211p

Damodaran, S., Parkin, K. I., & Fennema, O. R. (2010). *Fennema's Food Chemistry*, 4th Ed., 900 pp. Portuguese version authorized from English language edition published by CRC Press, part of Taylor & Francis Group LLC, Porto Alegre.

Dangles, O., & Fenger, J. A. (2018). The chemical reactivity of anthocyanins and its consequences in food science and nutrition. *Molecules*, 23, 8, 1970. https://doi.org/10.3390/molecules23081970

Ejaz, A., Waliat, S., Afzaal, M., Saeed, F., Ahmad, A., Din, A., Ateeq, H., Asghar, A., Shah, Y. A., Rafi, A., & Khan, M. R. (2023). Biological activities, therapeutic potential, and pharmacological aspects of blackcurrants (*Ribes nigrum* L): A comprehensive review. *Food Science & Nutrition*, 11, 10, 5799–5817. https://doi.org/10.1002/fsn3.3592

Enaru, B., Drețcanu, G., Pop, T. D., Stănilă, A., & Diaconeasa, Z. (2021). Anthocyanins: Factors affecting their stability and degradation. *Antioxidants*, 10, 12. https://doi.org/10.3390/antiox10121967

Erden, Y. (2021). Sour black mulberry (*Morus nigra* L.) causes cell death by decreasing mutant p53 expression in HT-29 human colon cancer cells. *Food Bioscience*, 42, 101113. https://doi.org/10.1016/j.fbio.2021.101113

FAOSTAT (2024). Food and Agriculture Data. Food and Agriculture Organization Corporate Statistical Database. Available online: https://www.fao.org/faostat/en/#home. Accessed on: 04/27/2024.

Fei, P., Zeng, F., Zheng, S., Chen, Q., Hu, Y., & Cai, J. (2020). Acylation of blueberry anthocyanins with maleic acid: Improvement of the stability and its application potential in intelligent color indicator packing materials. *Dyes and Pigments*, 184, 108852. https://doi.org/10.1016/j.dyepig.2020.108852

Fischer, D. C., Limberger, R. P., Henriques, A. T., & Moreno, P. R. (2005). Essential oils from leaves of two *Eugenia brasiliensis* specimens from southeastern Brazil. *Journal of Essential Oil Research*, 17, 499–500. https://doi.org/10.1080/10412905.2005.9698977

Francisco, A. P., Sganzerla, W. G., Castro, L. E. N., Barroso, T. L. C. T., Silva, A. P. G., Rosa, C. G., Nunes, M. R., Forster-Carneiro, T., & Rostagno, M. A. (2024). Pressurized liquid extraction of bioactive compounds from grape peel and application in pH-sensing carboxymethyl cellulose films: A promising material to monitor the freshness of pork and milk. *Food Research International*, 179, 114017. http://dx.doi.org/10.1016/j.foodres.2024.114017

Flores, G., Dastmalchi, K., Paulino, S., Whalen, K., Dabo, A. J., Reynertson, K. A., Foronjy, R. F., D'Armiento, J. M., & Kennelly, E. J. (2012). Anthocyanins from *Eugenia brasiliensis* edible fruits as potential therapeutics for COPD treatment. *Food Chemistry*, 134, 3, 1256–1262. https://doi.org/10.1016/j.foodchem.2012.01.086

Gamage, G. V., & Choo, W. S. (2023). Effect of hot water, ultrasound, microwave, and pectinase-assisted extraction of anthocyanins from black goji berry for food application. *Heliyon*, 9, e14426. https://doi.org/10.1016/j.heliyon.2023.e14426

Gençdağ, E., Özdemir, E. E., Demirci, K., Görgüç, A., & Yılmaz, F. M. (2022). Copigmentation and stabilization of anthocyanins using organic molecules and encapsulation techniques. *Current Plant Biology*, 29, 100238. https://doi.org/10.1016/j.cpb.2022.100238

Giusti, M. M., & Wrolstad, R. E. (2003). Acylated anthocyanins from edible sources and their applications in food systems. *Biochemical Engineering Journal*, 14, 3, 217–225. https://doi.org/10.1016/S1369-703X(02)00221-8

Gopalan, A., Reuben, S. C., Ahmed, S., Darvesh, A. S., Hohmann, J., & Bishayee, A. (2012). The health benefits of blackcurrants. *Food & Function*, 3, 8, 795–809. https://doi.org/10.1039/C2fo30058c

Herrera-Balandrano, D. D., Chai, Z., Beta, T., Feng, J., & Huang, W. (2021). Blueberry anthocyanins: Na updated review on approaches to enhancing their bioavailability. *Trends in Food Science & Technology*, 118, Part B, 808–821. https://doi.org/10.1016/j.tifs.2021.11.006

Huang, L., Zhou, Y., Meng, L., Wu, D., & He, Y. (2017). Comparison of different CCD detectors and chemometrics for predicting total anthocyanin content and antioxidant activity of mulberry fruit using visible and near infrared hyperspectral imaging technique. *Food Chemistry*, 224, 1–10. https://doi.org/10.1016/j.foodchem.2016.12.037

Hunyadi, A., Martins, A., Hsieh, T. J., Seres, A., & Zupkó, I. (2012). Chlorogenic acid and rutin play a major role in the in vivo anti-diabetic activity of *Morus alba* leaf extract on type II diabetic rats. *PLoS One*, 7, 11, e50619. https://doi.org/10.1371/journal.pone.0050619

Ichiyanagi, T., Kashiwada, Y., Shida, Y., Ikeshiro, Y., Kaneyuki, T., & Konishi, T. (2005). Nasunin from eggplant consists of cis-trans isomers of delphinidin 3-[4-(ρ-coumaroyl)-l-rhamnosul (1→6) glucopyranoside]-5-glucopyranoside. *Journal of Agricultural and Food Chemistry*, 53, 24, 9472–9477. https://doi.org/10.1021/jf051841y

Kalt, W., Cassidy, A., Howard, L., Krikorian, R., Stull, A., Tremblay, F., & Zamora-Ros, R. (2020). Recent research on the health benefits of blueberries and their anthocyanins. *Advances in Nutrition*, 11, 2, 224–236. https://doi.org/10.1093/advances/nmz065

Kang, H.-J., Ko, M.-J., & Chung, M.-S. (2021). Anthocyanin structure and pH dependent extraction characteristics from blueberries (*Vaccinium corymbosum*) and chokeberries (*Aronia melanocarpa*) in subcritical water state. *Foods*, 10, 3, 527. https://doi.org/10.3390/foods10030527

Karaaslan, M., Ozden, M., Vardin, H., & Turkoglu, H. (2011). Phenolic fortification of yogurt using grape and callus extracts. *LWT – Food Science and Technology*, 44, 4, 1065–1072. https://doi.org/10.1016/j.lwt.2010.12.009

Khoo, H. E., Azlan, A., Tang, S. T., & Lim, S. M. (2017). Anthocyanidins and anthocyanins: Colored pigments as food, pharmaceutical ingredients, and the potential health benefits. *Food & Nutrition Research*, 61, 1, 1361779. https://doi.org/10.1080/16546628.2017.1361779

Kim, I., & Lee, J. (2020). Variations in anthocyanin profiles and antioxidant activity of 12 genotypes of mulberry (*Morus* spp.) fruits and their changes during processing. *Antioxidants*, 9, 242. https://doi.org/doi:10.3390/antiox9030242

Koh, J., Xu, Z., & Wicker, L. (2020). Blueberry pectin and increased anthocyanins stability under in vitro digestion. *Food Chemistry*, 302, 125343–125350. https://doi.org/10.1016/j.foodchem.2019.125343

Koss-Mikołajczyk, I., Kusznierewicz, B., & Bartoszek, A. (2019). The relationship between phytochemical composition and biological activities of differently pigmented varieties of berry fruits; comparison between embedded in food matrix and isolated anthocyanins. *Foods*, 8, 12, 646. https://doi.org/10.3390/foods8120646

Kong, J. M., Chia, L. S., Goh, N. K., Chia, T. F., & Brouillard, R. (2003). Analysis and biological activities of anthocyanins. *Phytochemistry*, 64, 5, 923–933. https://doi.org/10.1016/S0031-9422(03)00438-2

Lazarini, J. G., Sardi, J. D. C. O., Franchin, M., Nani, B. D., Freires, I. A., Infante, J., ... & Rosalen, P. L. (2018). Bioprospection of Eugenia brasiliensis, a Brazilian native fruit, as a source of anti-inflammatory and antibiofilm compounds. *Biomedicine & Pharmacotherapy*, 102, 132–139.

Lee, S. G., Vance, T. M., Nam, T. G., Kim, D. O., Koo, S. I., & Chun, O. K. (2015). Contribution of anthocyanin composition to total antioxidant capacity of berries. *Plant Foods for Human Nutrition*, 70, 4, 427–432. https://doi.org/10.1007/s11130-015-0514-5

Lister, C. E., Wilson, P. E., Sutton, K. H., & Morrison, S. C. (2002). Understanding the health benefits of blackcurrants. *Acta Horticulturae*, 585, 443–449.

Liu, Z., Liu, B., Wen, H., Tao, Y., & Shao, Y. (2020). Phytochemical profiles, nutritional constituents and antioxidant activity of black wolfberry (*Lycium ruthenicum* Murr.). *Industrial Crops and Products*, 154, 112692. https://doi.org/10.1016/j.indcrop.2020.112692

Luo, D., Yu, L., Westland, S., & Mahon, N. (2019). The influence of colour and image on consumer purchase intentions of convenience food. *Journal of the International Colour Association*, 24, 11–23. http://www.aic-colour.org/journal.htm

Martinez, L. M., Rando, B., Agante, L., & Abreu, A. M. (2021). True colors: Consumers' packaging choices depend on the color of retail environment. *Journal of Retailing and Consumer Services*, 59, 102372. https://doi.org/10.1016/j.jretconser.2020.102372

Martins, N., Roriz, C. L., Morales, P., Barros, L., & Ferreira, I. C. F. R. (2016). Food colorants: Challenges, opportunities and current desires of agro-industries to ensure consumer expectations and regulatory practices. *Trends in Food Science & Technology*, 52, 1–15. https://doi.org/10.1016/j.tifs.2016.03.009

McCann, D., Barrett, A., Cooper, A., Crumpler, D., Dalen, L., Grimshaw, K., Kitchin, E., Lok, K., Porteous, L., Prince, E., Sonuga-Barke, E., Warner, J. O., & Stevenson, J. (2007). Food additives and hyperactive behaviour in 3-year-old and 8/9-year-old children in the community: A randomised, double-blinded, placebo-controlled trial. *The Lancet*, 370, 9598, 1560–1567. https://doi.org/10.1016/S0140-6736(07)61306-3

Modesto Junior, E. N., Martins, M. G., Pereira, G. A., Chisté, R. C., & Pena, R. D. S. (2023a). Stability kinetics of anthocyanins of grumixama berries (*Eugenia brasiliensis* Lam.) during thermal and light treatments. *Foods*, 12, 3, 565. https://doi.org/10.3390/foods12030565

Modesto Junior, E. N., Chaves, R. P. F., Martins, M. G., Pereira, G. A., Chisté, R. C., & Pena, R. D. S. (2023b). Phenolic compounds stability of grumixama (*Eugenia brasiliensis*) juice during processing and storage. *Beverages*, 9, 4, 91. https://doi.org/10.3390/beverages9040091

Mordor Intelligence. Anthocyanin Market Size & Share Analysis – Growth Trends & Forecasts (2024–2029). Available online: https://www.mordorintelligence.com/industry-reports/anthocyanin-market. Accessed on 04/27/2024.

Nour, V., Stampar, F., Veberic, R., & Jakopic, J. (2013). Anthocyanins pro-file, total phenolics and antioxidant activity of black currant ethan-olic extracts as influenced by genotype and ethanol concentration. *Food Chemistry*, 141, 2, 961–966. https://doi.org/10.1016/j.foodchem.2013.03.105

Oancea, S. (2021). A review of the current knowledge of thermal stability of anthocyanins and approaches to their stabilization to heat. *Antioxidants*, 10, 9, 1337. https://doi.org/10.3390/antiox10091337

Ockermann, P., Headley, L., Lizio, R., & Hansmann, J. (2021). A review of the properties of anthocyanins and their influence on factors affecting cardiometabolic and congnitive health. *Nutrients*, 13, 8, 2831. https://doi.org/10.3390/nu13082831

Oplatowska-Stachowiak, M., & Elliott, C. T. (2017). Food colors: Existing and emerging food safety concerns. *Critical Reviews in Food Science and Nutrition*, 57, 3, 524–548. https://doi.org/10.1080/10408398.2014.889652

Papillo, V. A., Locatelli, M., Travaglia, F., Bordiga, M., Garino, C., Arlorio, M., & Coïsson, J. D. (2018). Spray-dried polyphenolic extract from Italian black rice (*Oryza sativa* L., var. Artemide) as new ingredient for bakery products. *Food Chemistry*, 269, 603–609. https://doi.org/10.1016/j.foodchem.2018.07.059

Paredes-López, O., Cervantes-Ceja, M. L., Vigna-Pérez, M., & Hernández-Pérez, T. (2010). Berries: Improving human healthy aging, and promoting quality life – A review. *Plant Foods for Human Nutrition*, 65, 299–308. https://doi.org/10.1007/s11130-010-0177-1

Patras, A., Brunton, N. P., O'Donnell, C., & Tiwari, B. K. (2010). Effect of thermal processing on anthocyanin stability in foods; mechanisms and kinetics of degradation. *Trends in Food Science and Technology*, 21, 1, 3–11. https://doi.org/10.1016/j.tifs.2009.07.004

Peng, Q., Liu, H., Shi, S., & Li, M. (2014). *Lycium ruthenicum* polysaccharide attenuates inflammation through inhibiting TLR4/NF-κB signaling pathway. *International Journal of Biological Macromolecules*, 67, 330–335. https://doi.org/10.1016/j.ijbiomac.2014.03.023

Peng, Q., Liu, H., Lei, H., & Wang, X. (2016). Relationship between structure and immunological activity of an arabinogalactan from *Lycium ruthenicum*. *Food Chemistry*, 194, 595–600. https://doi.org/10.1016/j.foodchem.2015.08.087

Qi, Y., Zhu, C., Chen, J., Liu, G., Yang, Z., & Chen, W. (2019). Comparative analysis of the quality and health-promoting compounds of two-shaped fruits of wild *Lycium ruthenicum* Murr. from the Qinghai–Tibet Plateau. *Acta Physiologiae Plantarum*, 41, 1–10. https://doi.org/10.1007/s11738-019-2888-8

Rashwan, A. K., Karim, N., Xu, Y., Xie, J., Cui, H., Mozafari, M. R., & Chen, W. (2021). Potential micro-/nano-encapsulation systems for improving stability and bioavailability of anthocyanins: An updated review. *Critical Reviews in Food Science and Nutrition*, 63, 19, 3362–3385. https://doi.org/10.1080/10408398.2021.1987858

Remini, H., Dahmoune, F., Sahraoui, Y., Madani, K., Kapranov, V., & Kiselev, E. (2018). Recent advances on stability of anthocyanins. *RUND Journal of Agronomy and Animal Industries*, 13, 4, 257–286. https://doi.org/10.22363/2312-797X-2018-13-4-257-286

Ribeiro, J. S., & Veloso, C. M. (2021). Microencapsulation of natural dyes with biopolymers for application in food: A review. *Food Hydrocolloids*, 112, 106374. https://doi.org/10.1016/j.foodhyd.2020.106374

Routray, W., & Orsat, W. (2011). Blueberries and their anthocyanins: Factors affecting biosynthesis and properties. *Comprehensive Reviews in Food Science and Food Safety*, 10, 303–320. https://doi.org/10.1111/j.1541-4337.2011.00164.x

Ruta, L. L., & Farcasanu, I. C. (2019). Anthocyanins and anthocyanin-derived products in yeast-fermented beverages. *Antioxidants*, 8, 6, 182. https://doi.org/10.3390/antiox8060182

Sardi, J. d. C. O., Freires, I. A., Lazarini, J. G., Infante, J., de Alencar, S. M., & Rosalen, P. L. (2017). Unexplored endemic fruit species from Brazil: Antibiofilm properties, insights into mode of action, and systemic toxicity of four *Eugenia* spp. *Microbial Pathogenesis*, 105, 280–287. https://doi.org/10.1016/J.MICPATH.2017.02.044

Sharma, R. J., Gupta, R. C., Singh, S., Bansal, A. K., & Singh, I. P. (2016). Stability of anrhocyanins-and anthocyanidins-enriched extracts, and formulations of fruit pulp of *Eugenia jambolana* ('jamun'). *Food Chemistry*, 190, 808–817. https://doi.org/10.1016/j.foodchem.2015.06.029

Sikorski, Z. E. (2018). Fennema's food chemistry (fifth edition). Edited by Srinivasan Damodaran and Kirk L. Parkin. *Journal of Food Biochemistry*, 42, 2, e12483. https://doi.org/10.1111/jfbc.12483

Silva, S., Costa, E. M., Veiga, M., Morais, R. M., Calhau, C., & Pintado, M. (2020). Health promoting properties of blueberries: A review. *Critical Reviews in Food Science and Nutrition*, 60, 181–200. https://doi.org/10.1080/10408398.2018.1518895

Silva, M. M., Reboredo, F. H., & Lidon, F. C. (2022). Food colour additives: A synoptical overview on their chemical properties, applications in food products, and health side effects. *Foods*, 11, 3, 379. https://doi.org/10.3390/foods11030379

Sinela, A., Rawat, N., Mertz, C., Achir, N., Fulcrand, H., & Dornier, M. (2017). Anthocyanins degradation during storage of *Hibiscus sabdariffa* extract and evolution of its degradation products. *Food Chemistry*, 214, 234–241. https://doi.org/10.1016/j.foodchem.2016.07.071

Spence, C. (2018). Background colour & its impact on food perception & behaviour. *Food Quality and Preference*, 68, 156–166. https://doi.org/10.1016/j.foodqual.2018.02.012

Stefanello, M. E. A., Pascoal, A. C. R. F., & Salvador, M. J. (2011). Essential oils from neotropical Myrtaceae: Chemical diversity and biological properties. *Chemistry & Biodiversity*, 8, 1, 73–94. https://doi.org/10.1002/cbdv.201000098

Sun, L., Xin, F., & Alper, H. S. (2021). Bio-synthesis of food additives and colorants – A growing trend in future food. *Biotechnology Advances*, 47, 107694. https://doi.org/10.1016/j.biotechadv.2020.107694

Tan, J., Han, Y., Han, B., Qi, X., Cai, X., Ge, S., & Xue, H. (2022). Extraction and purification of anthocyanins: A review. *Journal of Agriculture and Food Research*, 8, 100306. https://doi.org/10.1016/j.jafr.2022.100306

Teixeira, L. d. L., Bertoldi, F. C., Lajolo, F. M., & Hassimotto, N. M. A. (2015). Identification of ellagitannins and flavonoids from *Eugenia brasiliensIs* Lam. (Grumixama) by HPLC-ESI-MS/MS. *Journal of Agricultural and Food Chemistry, 63*, 22, 5417–5427. https://doi.org/10.1021/acs.jafc.5b01195

Teixeira, L. L., Costa, G. R., Dörr, F. A., Ong, T. P., Pinto, E., Lajolo, F. M., & Hassimotto, N. M. A. (2017). Potential antiproliferative activity of polyphenol metabolites against human breast cancer cells and their urine excretion pattern in healthy subjects following acute intake of a polyphenol-rich juice of grumixama (*Eugenia brasiliensis* Lam.). *Food & Function, 8*, 6, 2266–2274. https://doi.org/10.1039/C7FO00076F

Teleszko, M., & Wojdyło, A. (2015). Comparison of phenolic compounds and antioxidant potential between selected edible fruits and their leaves. *Journal of Functional Foods, 14*, 736–746. https://doi.org/10.1016/j.jff.2015.02.041

Vidana Gamage, G. C., Lim, Y. Y., & Choo, W. S. (2021). Sources and relative stabilities of acylated and nonacylated anthocyanins in beverage systems. *Journal of Food Science and Technology, 59*, 3, 831–845. https://doi.org/10.1007/s13197-021-05054-z

Wallace, T. C., & Giusti, M. M. (2008). Determination of color, pigment, and phenolic stability in yogurt systems colored with nonacylated anthocyanins from *Berberis boliviana* L. as compared to other natural/synthetic colorants. *Journal of Food Science, 73*, 4, C241–C248. https://doi.org/10.1111/J.1750-3841.2008.00706.x

Wang, H., Li, J., Tao, W., Zhang, X., Gao, X., Yong, J., Zhao, J., Zhang, L., Li, Y., & Duan, J. A. (2018). *Lycium ruthenicum* studies: Molecular biology, phytochemistry and pharmacology. *Food Chemistry, 240*, 759–766. https://doi.org/10.1016/j.foodchem.2017.08.026

Wang, F., Li, H., Qin, Y., Mao, Y., Zhang, B., & Deng, Z. (2019). Effects of heat, ultrasound, and microwave processing on the stability and antioxidant activity of delphinidin and petunidin. *Journal of Food Biochemistry, 43*, 5, e12818. https://doi.org/10.1111/jfbc.12818

Wang, S., Wang, B., Dong, K., Li, J., Li, Y., & Sun, H. (2022). Identification and quantification of anthocyanins of 62 blueberry cultivars via UPLC-MS. *Biotechnology & Biotechnological Equipment, 36*, 587–597. https://doi.org/10.1080/13102818.2022.2090857

Wen, P., Hu, T. G., Linhardt, R. J., Liao, S. T., Wu, H., & Zou, Y. X. (2019). Mulberry: A review of bioactive compounds and advanced processing technology. *Trends in Food Science & Technology, 83*, 138–158. https://doi.org/10.1016/j.tifs.2018.11.017

Wood, L. (2019). $5 Billion Natural Dyes Market – Global Outlook and Forecasts 2019–2024. Retrieved from: https://www.prnewswire.com/news-releases/5-billion-natural-dyes-market—global-outlook-and-forecasts-2019-2024-300797306.html Accessed 10/02/2023.

Wu, T., Gao, Y., Guo, X., Zhang, M., & Gong, L. (2018). Blackberry and blueberry anthocyanin supplementation counteract high-fat-diet-induced obesity by alleviating oxidative stress and inflammation and accelerating energy expenditure. *Oxidative Medicine and Cellular Longevity*, 4051232. https://doi.org/10.1155/2018/4051232

Wu, Y., Yang, H., Yang, H., Zhang, C., Lyu, L., Li, W., & Wu, W. (2022). A physiological and metabolomic analysis reveals the effect of shading intensity on blueberry fruit quality. *Food Chemistry: X, 15*, 100367. https://doi.org/10.1016/j.fochx.2022.100367

Wu, Y., Han, T., Yang, H., Lyu, L., Li, W., & Wu, W. (2023). Known and potential health benefits and mechanisms of blueberry anthocyanins: A review. *Food Bioscience, 55*, 103050. https://doi.org/10.1016/j.fbio.2023.103050

Wu, X., Yan, X., Zhang, J., Wu, X., Luan, M., & Zhang, Q. (2024). Preparation and characterization of pH-sensitive intelligent packaging films based on cassava starch/polyvinyl alcohol matrices containing *Aronia melanocarpa* anthocyanins. *LWT – Food Science and Technology, 194*, 115818. http://dx.doi.org/10.1016/j.lwt.2024.115818

Yan, Y., Nisar, T., Fang, Z., Wang, L., Wang, Z., Gu, H., Wang, H., & Wang, W. (2022). Current developments on chemical compositions, biosynthesis, color properties and health benefits of black goji anthocyanins: An updated review. *Horticulturae, 8*, 1033. https://doi.org/10.3390/horticulturae8111033

Yang, S., Ding, Q., Li, Y., & Han, W. (2024). Bacterial cellulose/gelatin-based pH-responsive functional film for food freshness monitoring. *International Journal of Biological Macromolecules, 259*, 129203. http://dx.doi.org/10.1016/j.ijbiomac.2024.129203

Yang, W., Guo, Y., Liu, M., Chen, X., Xiao, X., Wang, S., Gong, P., Ma, Y., & Chen, F. (2022). Structure and function of blueberry anthocyanins: A review of recent advances. *Journal of Functional Foods*, 88, 104864. https://doi.org/10.1016/j.jff.2021.104864

Yue, E., Yu, Y., Wang, X., Liu, B., Bai, Y., & Yang, B. (2021). Anthocyanin protects cardiac function and cardiac fibroblasts from high-glucose induced inflammation and myocardial fibrosis by inhibiting IL-17. *Frontiers in Pharmacology*, 11, 593633. https://doi.org/10.3389/fphar.2020.593633

Zhang, J., Sun, L., Dong, Y., Fang, Z., Nisar, T., Zhao, T., & Guo, Y. (2019). Chemical compositions and α-glucosidase inhibitory effects of anthocyanidins from blueberry, blackcurrant and blue honeysuckle fruits. *Food Chemistry*, 299, 125102. https://doi.org/10.1016/j.foodchem.2019.125102

Zhao, Y. W., Wang, C. K., Huang, X. Y., & Hu, D. G. (2021). Anthocyanin stability and degradation in plants. *Plant Signaling & Behavior*, 16, 12, 1987767. https://doi.org/10.1080/15592324.2021.1987767

Zheng, J., Ding, C., Wang, L., Li, G., Shi, J., Li, H., & Suo, Y. (2011). Anthocyanins composition and antioxidant activity of wild *Lycium ruthenicum* Murr. from Qinghai-Tibet Plateau. *Food Chemistry*, 126, 3, 859–865. https://doi.org/10.1016/j.foodchem.2010.11.052

Zheng, J., Yang, B., Ruusunen, V., Laaksonen, O., Tahvonen, R., Hellsten, J., & Kallio, H. (2012). Compositional differences of phenolic compounds between black currant (*Ribes nigrum* L.) cultivars and their response to latitude and weather conditions. *Journal of Agricultural and Food Chemistry*, 60, 26, 6581–6593. https://doi.org/10.1021/jf3012739

5

Antioxidant Properties of Anthocyanins

Shiva Serwan Hasso and Javed Ahamad

5.1 Introduction

Anthocyanins (ACNs) are part of the flavonoid family which is a subclass of polyphenols and are found widely in plants (Byamukama et al., 2006). Till now more than 700 ACNs have been reported, and these different ACNs are found in 27 families and 72 genera of plants (Zhang et al., 2019). ACNs give attractive colors ranging from red to magenta, purple, and blue in flowers, fruits, and vegetables (de Pascual-Teresa and Sanchez-Ballesta, 2008; Wallace and Giusti, 2015). ACNs have been widely studied for their potential health benefits, including antioxidant, anti-inflammatory, and anticancer activity as well as other health benefits (de Pascual-Teresa and Sanchez-Ballesta, 2008). The most common ACNs found in fruits and vegetables are pelargonidin, peonidin, petunidin, cyanidin, delphinidin, and malvidin (Castañeda-Ovando et al., 2009).

ACNs are common in red and blue fruits such as berries, pomegranates, apples, and grapes as well as vegetables (red onion, red cabbage, and red lettuce) but their content varies between the various types of species, according to cultivar or diversity, growing area, climate, farming technics, harvest time, maturity, seasonal variability, processing and storing, temperature, and light exposure.

Berries such as strawberries, blueberries, blackberries, blackcurrants, and raspberries contain large numbers of ACNs, ranging from about 100–700 mg/100g of the fresh product, but the largest amount of ACN is found in elderberries and chokeberries which contain about 1.4–1.8g/100g of product (de Pascual-Teresa and Sanchez-Ballesta, 2008; Wallace and Giusti, 2015). The most common plants that contain ACNs include blackberry (*Rubus fruticosus*), blueberry (*Vaccinium oxycoccos*), bilberry (*Vaccinium myrtillus*), elderberries (*Sambucus canadensis*), blackcurrant (*Ribes nigrum*), strawberry (*Fragaria ananassa*), pomegranate (*Punica granatum*), red cabbage (*Brassica oleracea*), red grape (*Vitis vinifera*), chokeberries (*Aronia melanocarpa*), pistachio (*Pistacia vera*), red onion (*Allium cepa*), apple (*Malus domestica*), red oak leaf (*Quercus rubra*), red lollo lattuce (*Lactuca sativa*), and raspberries (*Rubus idaeus*) (Ogawa et al., 2008; Pojer et al., 2013; Ramos et al., 2014).

ACNs have been reported as powerful antioxidants in many *in-vitro* and *in-vivo* studies (Kähkönen and Heinonen, 2003; Einbond et al., 2004; Zafra-Stone et al., 2007). Due to their antioxidant potential, ACNs are used for the prevention and treatment of several human illnesses like diabetes, cardiovascular diseases (CVDs), cancer, liver disorders, neurodegenerative diseases, etc. (Wang and Stoner, 2008; Wallace, 2011; Gowd et al., 2017; Mehmood et al., 2021). The present book chapter comprises the recent studies on ACNs and discusses important progress made in ACNs' role as antioxidants and further understanding its role

DOI: 10.1201/9781003453260-6

in the prevention and treatment of various diseases like cancer, metabolic syndrome, liver, and neurodegenerative disorders.

5.2 Chemistry, Biogenesis, and Sar of Anthocyanin

5.2.1 Chemistry of Anthocyanins

ACNs belong to a large class of polyphenolic compounds known as flavonoids. Flavonoids are categorized into various subclasses, such as flavanols, flavonols, anthocyanidins, flavanones, flavones, and isoflavones. They possess a similar 15-carbon atom-based skeleton known as the 2-phenyl-1-benzopyrylium or flavylium cation backbone, which is composed of a C-6 (A ring) – C-3 (C ring) – C-6 (B ring) structure (Harborne, 1998). The variations between the subclasses are due to the presence of double bonds and carbonyl groups in the C ring. Chemically, ACNs are the glycosylated forms of anthocyanidins (aglycones). Anthocyanidins are formed by a flavylium cation backbone hydroxylated in different positions, typically on carbons C3, C5, C6, C7, and C3′, C4′, C5′, resulting in various anthocyanidins (Castañeda-Ovando et al., 2009). Currently, approximately 700 different ACNs and 25 distinct anthocyanidins (aglycone forms of ACNs) have been identified in nature. However, only six anthocyanidins are commonly found such as cyanidin (Cy), delphinidin (Dp), malvidin (Mv), pelargonidin (Pg), peonidin (Pn), and petunidin (Pt), accounting for more than 90% of all identified ACNs (Feng et al., 2023). The variations in ACN structure arise from differences in the quantity and placements of OH-groups and the extent of their methylation, the type, number, and position of sugars attached to the aglycone, as well as the type and number of aliphatic and aromatic acids bonded to these sugars (Feng et al., 2023; Liu et al., 2023). The OH-groups on the aglycone can be attached to sugar molecules in which ACN is formed, which can further connect with other sugars through glycosidic bonds. However, the most common attached sugars are glucose (glu) and rhamnose (rha), but galactose (gal), xylose (xyl), arabinose (ara), rutinose (rut), and other sugars are also frequently observed. Although, the glycosyl moieties of ACNs can also be acylated with aromatic or aliphatic acids (such as cinnamic, malonic, and acetic acids) through ester bonds. Both glycosylation and acylation impact the physical and chemical properties of ACNs by altering the size and polarity of the molecule. Glycosylation increases water solubility while acylation decreases it. Since the aglycone form of ACNs is rarely found in nature due to its instability, glycosylation enhances stability by creating an intramolecular hydrogen bonding network within the molecule. Therefore, the presence of different sugar moieties and various acyl groups (including aliphatic and cinnamic acids) contributes to the diverse chemical structures of ACNs found in nature (He and Giusti, 2010). The chemical structures of the most important ACNs are presented in Figures 5.1 and 5.2.

Anthocyanidin	R1	R2	R3
Pelargonidin (Pg)	–H	–OH	–H
Cyanidin (Cy)	–OH	–OH	–H
Peonidin (Pn)	–OCH$_3$	–OH	–H
Delphinidin (Dp)	–OH	–OH	–OH
Petunidin (Pt)	–OH	–OH	–OCH$_3$
Malvidin (Mv)	–OCH$_3$	–OH	–OCH$_3$

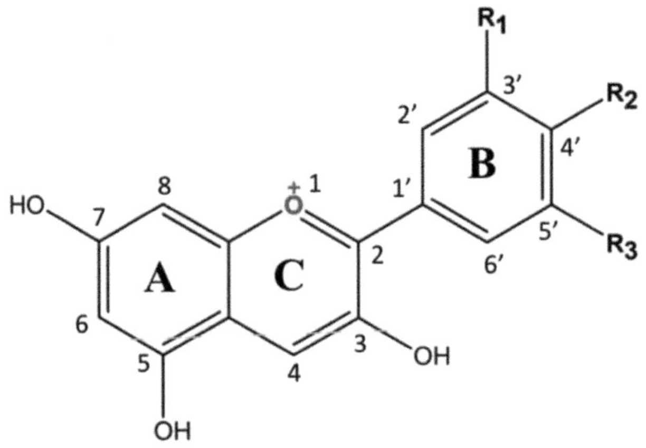

FIGURE 5.1
Chemical structures of the basic skeleton of anthocyanin.

5.2.2 Biogenesis of Anthocyanins

ACNs are derived from phenylalanine or tyrosine which is derived from the shikimate pathway, and further, it is transformed into 4-coumaroyl-CoA through the phenylpropanoid pathway or the combined action of TAL and 4CL enzymes. From that point on, the ACNs are synthesized *via* the general flavonoid pathway, whereby one molecule of 4-coumaroyl-CoA combines with three molecules of malonyl-CoA to produce naringenin chalcone, which is then converted to naringenin by CHI enzyme. Naringenin serves as a major intermediate compound and undergoes various hydroxylation reactions to form different anthocyanidins. These anthocyanidins are further modified through glycosylation and other processes to produce ACNs (Figure 5.3) (Zha and Koffas, 2017).

5.2.3 Structure-Activity Relationships (SAR) of Anthocyanins

The antioxidant properties of ACNs depend on their specific chemical structure. Whereby, factors such as the positions, numbers, and types of chemical groups (such as hydroxyl, methyl, acyl, and glycosyl) attached to the aromatic rings of ACNs are important for their structure-activity relationship (Apak et al., 2007; Gulcin, 2020). For flavonoids, three structural requirements: the catechol group in the B-ring, the conjugated double bond with the 4-oxo group in the C-ring, and hydroxyl groups in positions 3 and 5 contribute to their strong antioxidant activity. However, for ACNs, which lack of double bond at positions C2–C3 and a 4-oxo group, the substituents in the B-ring greatly influence their antioxidant capacity. The presence of OH-groups is more effective than $-OCH_3$ groups or hydrogen (H) alone. Although, the configuration of a B-ring with 3′,4′-dihydroxyl groups is particularly important for determining antioxidant activity in ACNs (Rossetto et al., 2007). The arrangement, substitution, and total number of hydroxyl groups have a significant impact on the antioxidant activity of ACNs. Specifically, the presence of OH-groups in positions C3′ and C4′ of ring B is crucial for the free radical scavenging abilities of ACNs (Kongpichitchoke et al., 2015). Although, previous studies have shown that delphinidin (Dp) exhibits stronger radical scavenging activity than cyanidin (Cy) (Noda et al., 2002; Kähkönen and Heinonen, 2003; Azevedo et al., 2010), due to the presence of three hydroxyl on the B-ring of delphinidin (Dp). However, the last findings have been reported by Wang

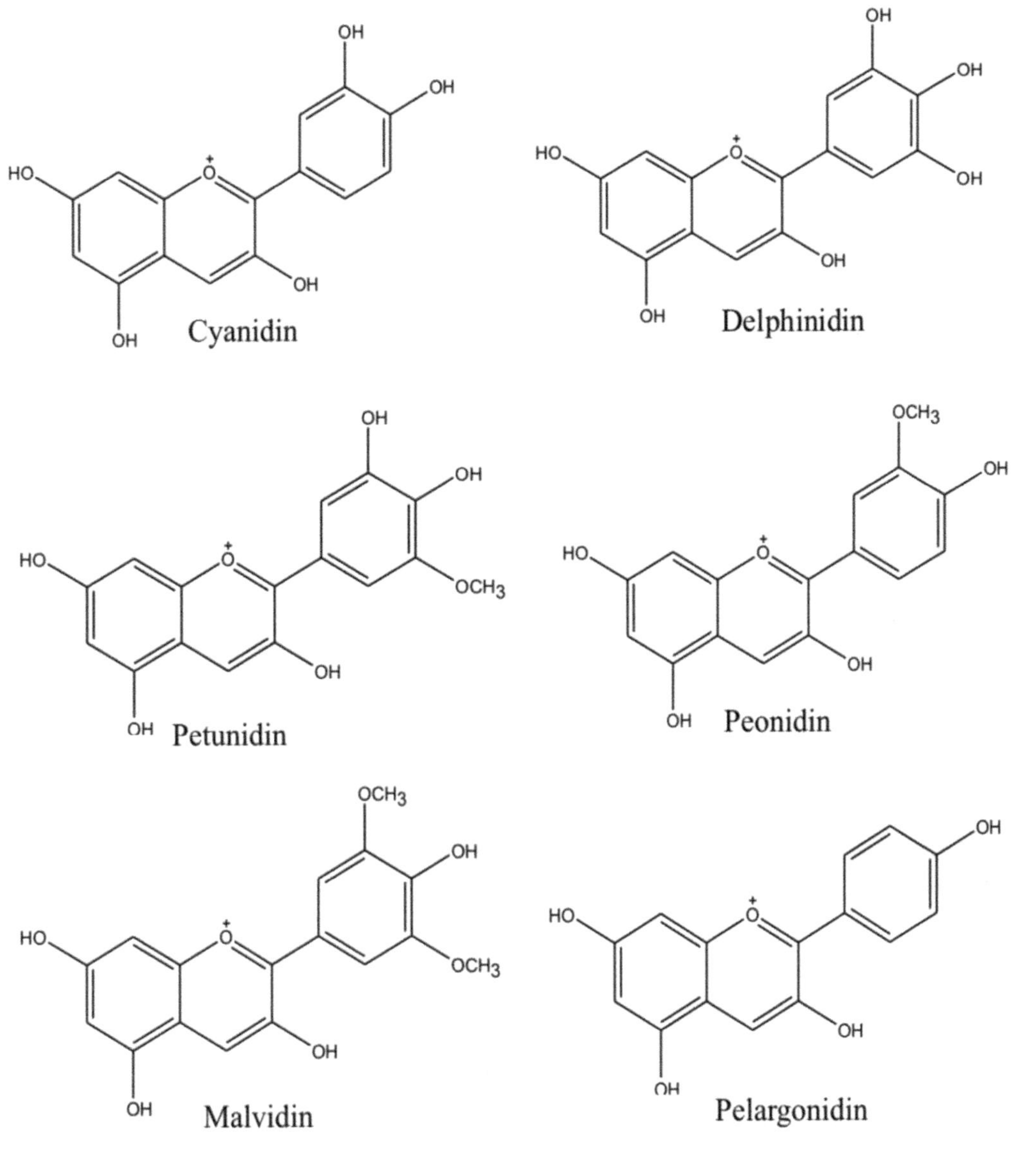

FIGURE 5.2
Chemical structures of major anthocyanins.

et al., where purified delphinidin (Dp) with three hydroxyl groups on the B-ring had lower oxygen radical absorbance capacity (ORAC) activity compared to cyanidin (Cy), pelargonidin (Pg), malvidin (Mv), and peonidin (Pn) compounds having one or two hydroxyl groups (Wang et al., 1997). These studies suggest that while hydroxylation at positions 3′ and 4′ in the B-ring is important for radical scavenging potential in ACNs, additional hydroxylation at position 5′ may decrease ORAC activity when combined with 3′ and 4′-OH. Compared to Cy and Dp, Pg, Mv, and Pn compounds which have only one hydroxyl group in the B-ring, exhibit lower activities in terms of ORAC, ABTS, and DPPH assays (Wang et al., 1997; Kähkönen and Heinonen, 2003; Chen et al., 2019). These findings highlight the

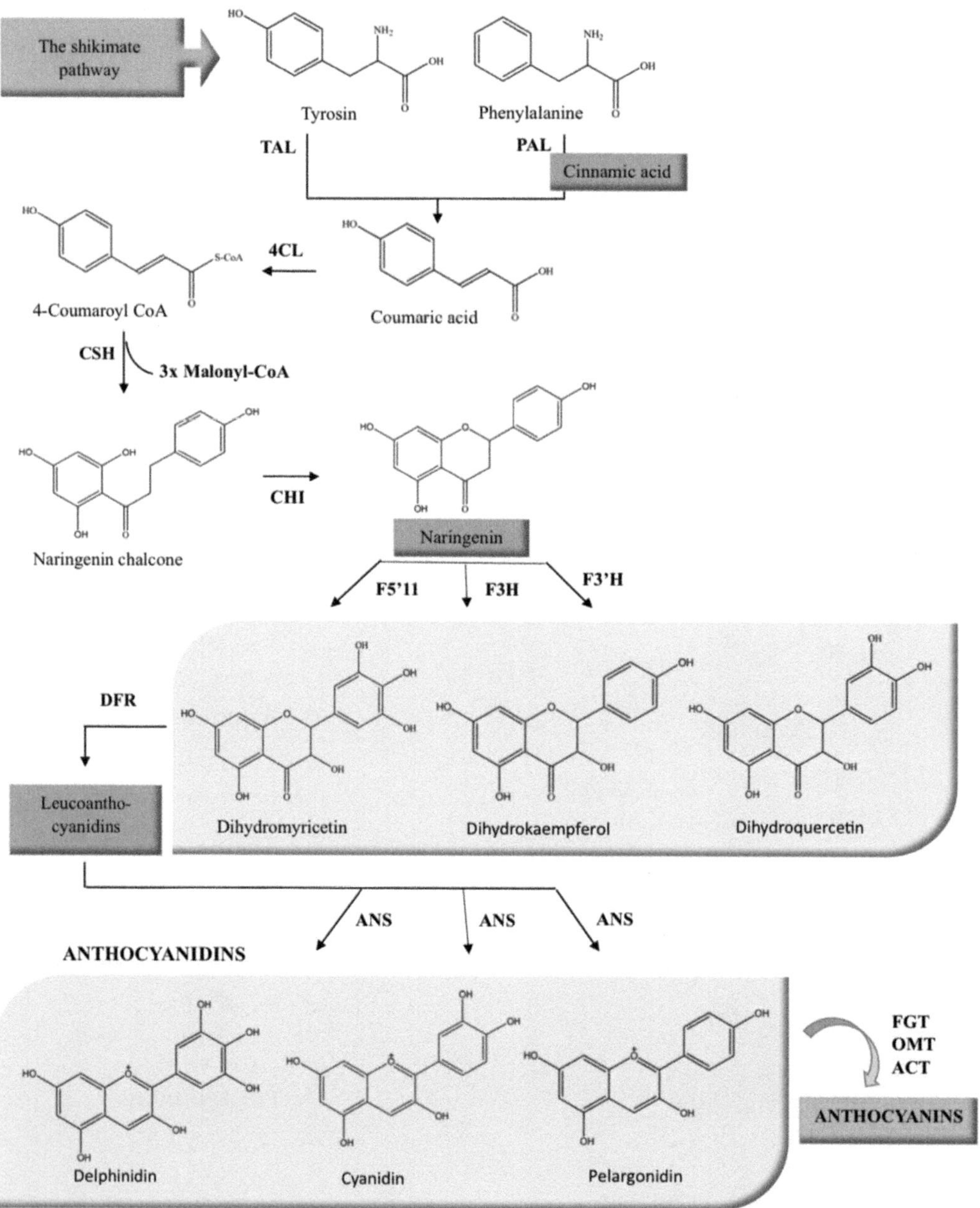

FIGURE 5.3

Biosynthetic pathway of major anthocyanins.

Abbreviations: TAL, tyrosine ammonia lyase; PAL, phenylalanine ammonia-lyase; C4H, cinnamate 4-hydroxylase; CPR, cytochrome P450 reductase; 4CL, 4-coumaroyl-CoA ligase; CHI, chalcone isomerase; CHS, chalcone synthase; F3H, flavanone 3-hydroxylase; F30 H, flavonoid 30-hydroxylase; F30 50 H, flavonoid 30, 50 -hydroxylase; DFR, dihydroflavonol reductase; ANS, anthocyanidin synthase; FGT, flavonoid glucosyltransferase; ACT, ACN acyltransferase, OMT, O-methyltransferase.

importance of the 3′–4′ catechol (dihydroxy benzene) structure in determining scavenger activity (Atmani et al., 2009). In addition to the hydroxyl groups, different arrangements of sugars can also impact the antioxidant effects of ACNs (Kähkönen and Heinonen, 2003). The position, degree, as well as sugar groups of glycosylation and anthocyanidins, play a significant role. Usually, ACNs tend to reduce the antioxidant power compared to their aglycone parts (Wang and Stoner, 2008). However, the impact differs depending on a few factors. For instance, glucosylation can either decrease (Mv3G < Mv), increase (Cy3G > Cy), or not significantly affect (Pg = Pg3G) the ORAC activity of aglycones. In some cases, the presence and quantity of sugar groups affect antioxidant activity. Glucose and rhamnose-glucose can increase ORAC activity for Cy while decreasing it when galactose is present (Wang et al., 1997). Additionally, ACNs with single sugars like glucose and galactose tend to show higher antioxidant capabilities compared to those with double sugars (Kähkönen and Heinonen, 2003; Reis et al., 2016). Other changes in the structure, like adding methyl or acyl groups to the hydroxyl groups of ACNs, significantly influence their antioxidant effects. Generally, when methyl groups are present at C3′, C4′, and C3 positions, they tend to lessen the antioxidant effects of ACNs (Reis et al., 2016). For instance, Mv and Mv3G, which contain two –OCH3 (methoxy) substituents, showed lower antiradical and reducing capabilities compared to Cy and Dp, Cy3G, and Dp3G, which lack –OCH3 groups (Azevedo et al., 2010). However, some studies found that Pn, with a methoxy group in the 3′-position exhibited more activity than Pg with a lack of methoxy group (Kähkönen and Heinonen, 2003). Moreover, a study by Sintzing et al. demonstrated that 5-glycosylation reduced ORAC value, while acylation increased the antioxidant potency (Stintzing et al., 2002).

5.3 Antioxidant Mechanism of Anthocyanins

In a typical and healthy state of the human body, there is an equilibrium between free radicals and the internal defense mechanisms of antioxidants. Certain free radicals, including superoxide radical anion, nitric oxide (NO), and associated reactive oxygen species (ROS) and/or reactive nitrogen species (RNS), play vital roles in maintaining cellular signaling pathways, the immune system, and various other physiological processes (Dröge, 2002; Jing, 2006). However, an overproduction of free radicals disrupts the body's oxidative equilibrium, leading to oxidative stress, that can harm essential cellular components such as proteins, DNA, and membrane lipids (Figure 5.4) (Tandon et al., 2005). Acute inflammation serves as the primary response to injury and pathogens, typically followed by

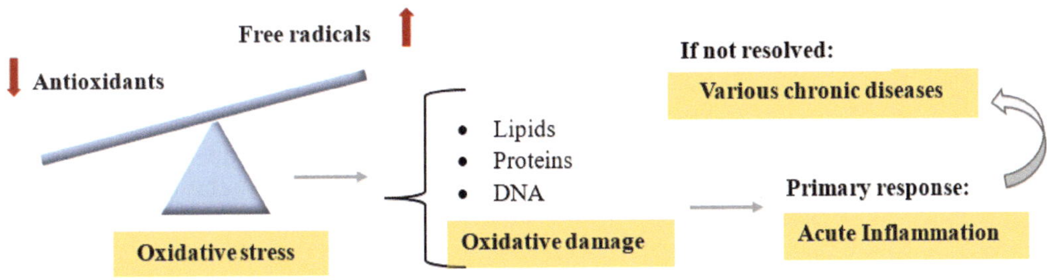

FIGURE 5.4
Pathophysiology and oxidative equilibrium in the human body.

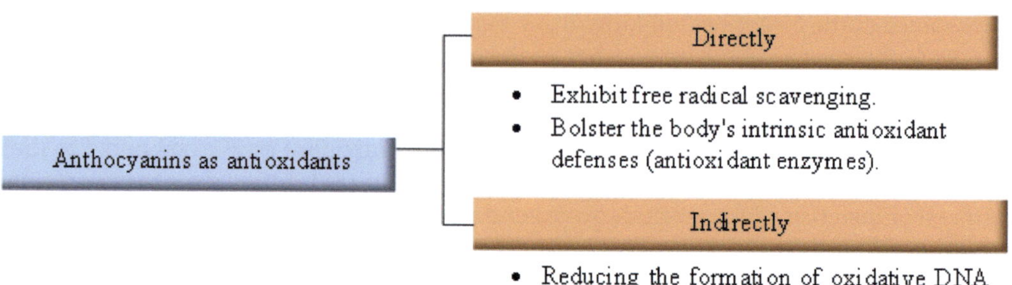

FIGURE 5.5
Antioxidant mechanism of anthocyanins.

inflammation resolution (Salehi et al., 2019b). Persistent inflammation due to the resolution failure results in the progression of numerous chronic diseases, including CVDs, diabetes mellitus, neurodegenerative disorders, and cancers (Arulselvan et al., 2016; Salehi et al., 2018g).

ACNs are the most safe, potent, and naturally occurring water-soluble scavengers of free radicals discovered to date. Their effectiveness as antioxidants is primarily related to their molecular structures (Fernandes et al., 2014). ACNs can show their antioxidant effects through both direct and indirect mechanisms. Directly, ACNs exhibit free radical scavenging capabilities due to their ability to donate hydrogen (electrons), or they may bolster the body's intrinsic antioxidant defenses, either by increasing or restoring the activities of key antioxidant enzymes such as superoxide dismutase (SOD) and glutathione peroxidase (GPx), as reported by Toufektsian et al. (2008), or by directly acting on the genes responsible for encoding these enzymes (Shih et al., 2010). Moreover, these compounds exert their antioxidant effects indirectly by reducing the formation of oxidative DNA chemical alterations and endogenous reactive oxygen species (ROS) by inhibiting enzymes like xanthine oxidase, and NADPH oxidase or by modulating arachidonic metabolism and mitochondrial respiration, as observed by Pojer et al. (2013). The mechanism of action of ACNs is presented in Figure 5.5.

5.4 Antioxidant Activity of Anthocyanins

5.4.1 Cardiovascular Protection

Cardiovascular dysfunction arises due to the production of free radicals from various sources such as the mitochondrial electron transport chain, inducible NO synthase, the Angiotensin II (ANG-II) mediated NADPH oxidase system, and xanthine oxidase (Munzel et al., 2017). Free radicals and ROS which are the leading cause of oxidation can result in cardiomyocyte loss and endothelial cell destruction in the blood vessel leading to cellular signaling for hypertrophic responses (Alam et al., 2021). The oxidation of low-density lipoprotein (LDL) initiates the accumulation of macrophage white blood cells in the artery wall, contributes to plaque formation, and eventually causes atherosclerosis and CVDs (Aviram, 2000; Aviram et al., 2005).

Previous studies show that antioxidant therapy may alleviate oxidative stress and cardiac remodeling (Ulla et al., 2017; Alam, 2019). Endothelium-derived NO deficiency correlates closely with hypertension (Jousilahti et al., 1999). To date, various antioxidants and scavengers of oxygen free radicals have been demonstrated to activate NO synthase (NOS), thereby reducing blood pressure (Kojšová et al., 2006). Although, redox imbalance is a significant risk factor for stroke that contributes to endothelial dysfunction. ACNs possess the potential to prevent harmful factors affecting endothelial cells (Manolescu et al., 2019). ACNs derived from purple sweet potatoes have been studied to prevent ischemic stroke by decreasing levels of apoptosis-inducing factors, enhancing brain-derived neurotrophic factor signaling crucial for stroke recovery, and activating other antioxidant mechanisms (Alvarez-Suarez et al., 2014; Adnyana et al., 2019). Therefore, ACNs and their vascular endothelium-related metabolites play a role in preventing CVDs, including stroke. Furthermore, oxidative stress can induce cardiomyocyte apoptosis and impair cellular function. Lingonberry ACNs protect cardiac cells from apoptotic cell death caused by oxidative stress (Isaak et al., 2017). The cardioprotective mechanisms of ACNs on mitochondria involve preventing apoptosis by reducing cytosolic cytochrome-C expression and promoting oxidative phosphorylation in ischemia-damaged mitochondria, thereby maintaining electron transfer between cytochrome-C and NADH dehydrogenase (Liobikas et al., 2016). Dietary antioxidants, including ACNs, have been observed to enhance total antioxidant capacity, antioxidant defense enzymes, and high-density lipoprotein (HDL) antioxidant properties, as well as protection against LDL oxidation and CVDs.

5.4.2 Neuroprotective Effects of Anthocyanins

ACNs have demonstrated an extraordinary antioxidant capacity that could be effective in several models of neurodegenerative diseases (Traustadottir et al., 2009; Shih et al., 2011; Miller et al., 2018). Their high oxygen radical absorbance capacity (ORAC) value accounts for a large part of this neuroprotective role (Zafra-Stone et al., 2007; Zhu et al., 2010). Additionally, their activities correlate with the antioxidant features as ACNs are effective free radical trappers, which prevent the formation of ROS in the affected cells. For instance, the generation of ROS is inhibited by ACNs in *in-vitro* models of α-β-peptide-induced toxicity (Shih et al., 2011) as well as in hydrogen peroxide injury (Hwang et al., 2012). The previous studies in animals showed the neuroprotection effects of black soybean (*Glycine max*) ACNs against ethanol- and glutamate-induced oxidative stress *via* inhibition of neuronal apoptosis, glutamate-related neurotransmission as well as protection against ROS production, glutamate-induced AMPK induction, neuroinflammation, and neurodegeneration (Shah et al., 2015; Shah et al., 2016). Moreover, the oral administration of ACN-rich extract of *V. myrtillus* (100 mg/kg) has been shown to suppress the psychological stress-induced cerebral oxidative stress in distressed mice (Rahman et al., 2008). The memory impairment reduction was also observed in rats that received 80 mg/kg ACN-rich extract of *Lycium ruthenicum* in comparison to other rat models with Alzheimer's disease. Chen et al. (2019) also confirmed that *L. ruthenicum* ACNs can show neuroprotective effects in *D*-galactose treated rats by reducing the level of receptors for glycation end products, suppressing oxidative stress, and reducing levels of inflammation markers. Sustained antioxidant enzyme levels, a decreased expression of aging-associated monoamine oxidase-B, and a decreased concentration of the lipid peroxidation product malondialdehyde (a biomarker for oxidized lipids) were also demonstrated in *D*-galactose treated mice after the administration 30 and 60 mg/kg of black rice ACNs (Lu et al., 2014). ACNs from black chokeberry (*A. melanocarpa*) with 30 mg/kg retained the total SOD and GPx levels in the brain tissue of *D*-gal-treated mice (Wei et al., 2017). Furthermore, a study recently used a model of streptozotocin-induced dementia of sporadic Alzheimer disease and showed that pretreatment with an ACN extract from grape skins by 200 mg/kg for 25 days can prevent behavioral alterations and protect against the alterations in levels of ROS and antioxidant enzyme (SOD, catalase, and GPx) (Pacheco et al., 2018).

5.4.3 Hepatoprotective Effects of Anthocyanins

The liver plays an important role in metabolizing drugs and harmful substances that can impair its function and cause damage. Arjinajarn et al. (2017) observed that rice bran extract, with high amounts of cyanidin-3-glucoside (Cy3glc) and peonidin-3-glucoside (Pn3glc) effectively protected the liver from damage induced by gentamicin. The extract significantly increased the activity of the antioxidant enzyme SOD, reduced hepatic levels of malondialdehyde, and prevented the elevation of liver injury. These protective effects play a role in the suppression of oxidative and inflammatory pathways. Furthermore, Hou et al. (2013) discovered that black rice bran extracts abundant in ACNs like Cy3glc and Pn3glc protected mice livers from carbon tetrachloride (CCl_4) toxicity (Ziberna et al., 2010). Mice treated with 200–800 mg/kg of black rice exhibited increased activities of GPx and SOD along with elevated levels of reduced glutathione compared to the model group with CCl4-intoxication (Hou et al., 2013). The same results about the protective effects of blueberry ACN extract were shown by (Chen et al., 2012) on CCl_4-induced liver injury *in vivo*. However, it was detected that blueberry ACNs reduce ROS generation and tissue oxidative damage, decrease inflammation, and suppress the activity of hepatic stellate cells (Sun et al., 2018). Blackberry fruit extract was also shown to reduce lipid peroxidation and to recover the antioxidant enzyme activity in rats treated with CCl_4 (Cho et al., 2011).

5.4.4 Nephroprotective Effects of Anthocyanins

ACNs in food are believed to play a significant role in preventing kidney diseases. For instance, recently a study explored the effect of a bilberry extract rich in ACNs (200 mg/kg daily) on animals with CCl_4-induced toxicity. The extract pretreatment notably decreased levels of pro-oxidative markers (such as H_2O_2, xanthine oxidase, and oxidized glutathione) and pro-inflammatory markers (including myeloperoxidase, TNF-α, and NO), while elevating levels of antioxidant enzymes (including SOD, catalase, S-transferase, and GPx). The ACNs were found to reduce damage to both proximal and distal tubules in the kidney cortex (Popović et al., 2019). Similar effects were observed in another study, which showed that *Malva sylvestris* extract rich in ACNs alleviates renal toxicity induced by gentamicin, resulting in improved kidney function, reduced oxidative stress, pro-inflammatory markers, and diminished tissue injuries (Yarijani et al., 2019). Furthermore, a study using a rat model of gentamicin-induced nephrotoxicity revealed the beneficial effects of a bilberry's daily diet (100 mg/kg), showing the correlation between the diet's high ACN levels, antioxidant activity, and improvements in serum catalase, malondialdehyde, and advanced oxidation protein products (Veljković et al., 2017). Additionally, studies by Qi et al. (2017) and Lee and Bae (2019) on ACN-rich fruits of *Panax ginseng* have shown promise in reducing renal tissue injury, plasma nitrite and nitrate production, and malondialdehyde levels, while restoring total glutathione content and antioxidant enzyme activity in kidney tissues of mice.

5.4.5 Antidiabetic Activity of Anthocyanins

The biggest challenge in treating diabetes, regardless of its type, is in preserving and restoring functional pancreatic β-cells. Free radicals and oxidative stress play significant roles in the autoimmune destruction of β-cells in type 1 diabetes (T1D) and the malfunction and damage of β-cells due to insulin resistance and glucotoxicity in type 2 diabetes (T2D) (Dinić et al., 2022). Current diabetes treatments include hypoglycemic drugs, insulin, and dietary adjustments. These treatments aim to stimulate insulin secretion, reduce liver gluconeogenesis, and increase peripheral glucose absorption, and delay intestinal glucose absorption (Đorđević et al., 2022). However, less than half of diabetic patients achieve glycemic control with the mentioned therapies, and the outcomes are far from expectations (Edelman and Polonsky, 2017). Patients often develop resistance to treatments and encounter side effects or toxicity from drugs, especially, life-threatening hypoglycemia due to insulin therapy (American Diabetes Association, 2019). For that reason, various strategies have emerged, including islet transplantation, stimulating β-cells proliferation, and transplantation of *in vitro*-generated β-cells. However, these strategies mostly stay ineffective or inaccessible. While, other crucial strategies, such as reducing the ROS or inducing antioxidant enzymes to alleviate oxidative stress in isolated islets before transplantation have been developed (Burke et al., 2015; Remedi and Emfinger, 2016). Additionally, the use of antioxidants, such as ACNs, shows positive results in protecting β-cells from glucose-induced oxidative stress. Studies have demonstrated that ACNs from boysenberries can prevent oxidative stress in diabetic rats induced by streptozotocin and prevent the development of diabetes-related oxidation (Bottino et al., 2004; Padmasekar et al., 2013).

5.4.6 Anticancer Activity of Anthocyanins

Oxidative stress is primarily responsible for inducing mutations in DNA of normal cells that lead to cellular proliferation and the formation of tumors. Additionally, the generation of free radicals can change cellular signaling cascades that contribute to cellular differentiation and inflammation within the tumor environment (Aggarwal et al., 2019). ACNs and related natural compounds may prevent cancer and tumor development by several mechanisms, including scavenging ROS directly, stimulating the expression of Phase II detoxification enzymes, enhancing the cells' oxygen-radical absorbing capacity as well as, reducing the formation of oxidative DNA adducts. ACNs also inhibit cellular differentiation, proliferation, and angiogenesis (Wang and Stoner, 2008; Lin et al., 2017). They have been studied to induce Phase II antioxidant and detoxifying enzymes in cultured cells, contributing to their anti-carcinogenic properties (Shih et al., 2005). Both *in-vitro* and *in-vivo* trials have shown that ACNs significantly reduce cancer cell proliferation and inhibit the formation of tumors (Kang et al., 2003; Medic et al., 2019). Blueberries, known for their antioxidant capacity, are thought to show protective effects against DNA oxidation (Pepe et al., 2013). The duration of blueberry consumption appears to play an important role in preventing mutagenesis. Del Bo' et al. (2015) observed a reduction in H_2O_2-induced DNA damage in mice not after four weeks but after eight weeks of blueberry consumption, for that reason, continuous ingestion of blueberries is necessary for DNA damage prevention. In breast cancer, ACNs inhibit key modulators promoting progression and development by directly affecting DNA fragmentation and inducing death in MCF-7 cancer cells (Hui et al., 2010). In lung cancer, cyanidin-3-glucoside (Cy-3-glu) and cyanidin 3-rutinoside (Cy-3-rut), isolated from mulberries, inhibit the migration and invasion of A549 cells (adenocarcinomic human alveolar basal epithelial cells) and decrease

MMP-2 and uPA while enhancing PAI and TIMP-2 (Wang et al., 2012). ACNs also prevent the growth of carcinogenic cells causing colon cancer, modulate macrophages in the immune response, and induce apoptosis (Mauray et al., 2010). In oral and cervical cancer, treatment with peonidin-3-glucoside and cyanidin-3-glucoside diminishes the invasion of HeLa cells and SCC-4 cells (Sankhari et al., 2012). The structural properties of ACNs significantly affect their biological activities, including the type of sugars, aglycones, acylated acids, and the position and degree of acylation and glycosylation, which seems to be the most important factors affecting their anticancer properties. The anticancer properties of various ACN-rich extracts were compared on the human colon cancer HT29 cell line, reporting various degrees of growth inhibitory activity rates among different sources of ACNs. The highest rate was observed in purple corn and the lowest rate in both radish and elderberry (Jiang et al., 2017).

5.5 Conclusion

ACNs are important colored secondary plant metabolites found in fruits, berries, and vegetables. The most common sources of ACNs include strawberries, blueberries, blackberries, blackcurrants, raspberries, pomegranates, apples, grapes, red onion, red cabbage, and red lettuce. In recent decades, due to a surge in metabolic disorders, and cancers, natural antioxidants such as ACNs play an important role in the prevention and treatment of such illness conditions. The present review compiles data from various *in-vitro*, preclinical, and animal studies showing the beneficial activities of ACNs through antioxidant mechanisms in the prevention and treatment of various types of cancers, diabetes, CVDs, neurodegenerative disorders, kidney, and liver diseases.

References

Adnyana IMO, Sudewi R, Samatra P, Suprapta S. Balinese cultivar of purple sweet potato improved neurological score and BDNF and reduced caspase-independent apoptosis among Wistar rats with ischemic stroke. Open Access Macedonian Journal of Medical Sciences 2019;7:38–44.

Aggarwal V, Tuli HS, Varol A, Thakral F, Yerer MB, Sak K, Varol M, Jain A, Khan MA, Sethi G. Role of reactive oxygen species in cancer progression: Molecular mechanisms and recent advancements. Biomolecules. 2019;9:735.

Alam MA. Anti-hypertensive effect of cereal antioxidant ferulic acid and its mechanism of action. Frontiers in Nutrition. 2019;6:121.

Alam MA, Islam P, Subhan N, Rahman MM, Khan F, Burrows GE, Nahar L, Sarker SD. Potential health benefits of anthocyanins in oxidative stress related disorders. Phytochemistry Reviews. 2021;20(4):705–749.

Alvarez-Suarez JM, Giampieri F, Tulipani S, Casoli T, Di Stefano G, González-Paramás AM, Santos-Buelga C, Busco F, Quiles JL, Cordero MD; et al. One-month strawberry-rich anthocyanin supplementation ameliorates cardiovascular risk, oxidative stress markers and platelet activation in humans. Journal of Nutritional Biochemistry. 2014, 25, 289–294.

American Diabetes Association. Pharmacologic approaches to glycemic treatment: Standards of medical care in diabetes. Diabetes Care. 2019;42(Suppl. 1):S90–S102.

Apak R, Güçlü K, Demirata B, Ozyürek M, Ozyurt D. Comparative evaluation of various total anti-oxidant capacity assays applied to phenolic compounds with the CUPRAC assay. Molecules. 2007;12(7):1496–1547.

Arjinajarn P, Chueakula N, Pongchaidecha A, Jaikumkao K, Chatsudthipong V, Mahatheeranont S, Norkaew O, Chattipakorn N, Lungkaphin A. Anthocyanin-rich Riceberry bran extract attenuates gentamicin-induced hepatotoxicity by reducing oxidative stress, inflammation and apoptosis in rats. Biomedicine & Pharmacotherapy. 2017;92:412–420.

Arulselvan P, Fard MT, Tan WS, Gothai S, Fakurazi S, Norhaizan ME, Kumar SS. Role of anti-oxidants and natural products in inflammation. Oxidative Medicine and Cellular Longevity. 2016 Oct 10;2016.

Atmani D, Chaher N, Atmani D, Berboucha M, Debbache N, Boudaoud H. Flavonoids in human health: From structure to biological activity. Current Nutrition & Food Science. 2009;5(4):225–237.

Aviram M. Review of human studies on oxidative damage and antioxidant protection related to cardiovascular diseases. Free Radical Research. 2000;33:S85–S97.

Aviram M, Kaplan M, Rosenblat M, Fuhrman B. Dietary antioxidants and paraoxonases against LDL oxidation and atherosclerosis development. Atherosclerosis: Diet and Drugs. 2005;263–300.

Azevedo J, Fernandes I, Faria A, Oliveira J, Fernandes A, Freitas VD, Mateus N. Antioxidant proper-ties of anthocyanidins, anthocyanidin-3-glucosides and respective portisins. Food Chemistry. 2010;119(2):518–523.

Bottino R, Balamurugan AN, Tse H, Thirunavukkarasu C, Ge X, Profozich J, et al. Response of human islets to isolation stress and the effect of antioxidant treatment. Diabetes. 2004;53:2559–2568.

Burke GW, Vendrame F, Virdi SK, Ciancio G, Chen L, Ruiz P, et al. Lessons from pancreas trans-plantation in type 1 diabetes: Recurrence of islet autoimmunity. Current Diabetes Reports. 2015;15:1–9.

Byamukama R, Jordheim M, Kiremire B, Namukobe J, Andersen Ø M. Anthocyanins from flowers of Hippeastrum cultivars. Scientia Horticulturae. 2006;109(3):262–266.

Castañeda-Ovando A, de Lourdes Pacheco-Hernández M, Páez-Hernández ME, Rodríguez JA, Galán-Vidal CA. Chemical studies of anthocyanins: A review. Food Chemistry. 2009;113(4):859–871.

Chen B, Ma Y, Li H, Chen X, Zhang C, Wang H, Deng Z. The antioxidant activity and active sites of delphinidin and petunidin measured by DFT, in vitro chemical-based and cell-based assays. Journal of Food Biochemistry. 2019;43(9):12968.

Chen J, Sun H, Sun A, Lin Q, Wang Y, Tao X. Studies of the protective effect and antioxidant mech-anism of blueberry anthocyanins in a CC14-induced liver injury model in mice. Food and Agricultural Immunology. 2012;23:352–362.

Chen S, Zhou H, Zhang G, et al. Anthocyanins from *Lycium ruthenicum* Murr. Ameliorated d-galactose-induced memory impairment, oxidative stress, and neuroinflammation in adult rats. Journal of Agricultural and Food Chemistry. 2019;67:3140–3149.

Cho BO, Ryu HW, Jin CH, Choi DS, Kang SY, Kim DS, Byun M-W, Jeong IY. Blackberry extract atten-uates oxidative stress through up-regulation of Nrf2-dependent antioxidant enzymes in carbon tetrachloride-treated rats. Journal of Agricultural and Food Chemistry. 2011;59:11442–11448.

de Pascual-Teresa S, Sanchez-Ballesta MT. Anthocyanins: From plant to health. Phytochemistry Reviews. 2008;7:281–299.

Del Bo' C, Fracassetti D, Lanti C, Porrini M, Riso P. Comparison of DNA damage by the comet assay in fresh versus cryopreserved peripheral blood mononuclear cells obtained following dietary intervention. Mutagenesis. 2015;30(1):29–35.

Dinić S, Arambašić JJ, Uskoković A, Mihailović M, Grdović N, Tolić A, Rajić J, Đorđević M, Vidaković M. Oxidative stress-mediated beta cell death and dysfunction as a target for diabetes manage-ment. Frontiers in Endocrinology. 2022;13:1006376.

Đorđević MM, Tolić A, Rajić J, Mihailović M, Arambasic JJ, Uskokovic A, et al. Centaurium ery-thraea methanol extract improves the functionality of diabetic liver and kidney by mitigating hyperglycemia-induced oxidative stress. Journal of Functional Foods. 2022;90:104975.

Dröge W. Free radicals in the physiological control of cell function. Physiological Reviews. 2002;82:47–95.

Edelman SV, Polonsky WH. Type 2 diabetes in the real world: The elusive nature of glycemic control. Diabetes Care. 2017;40(11):1425.

Einbond LS, Reynertson KA, Luo XD, Basile MJ, Kennelly EJ. Anthocyanin antioxidants from edible fruits. Food Chemistry. 2004;84(1):23–28.

Feng Y, Tian X, Liang W, Nan X, Zhang A, Li W, Ma Z. Genome-wide identification of grape ANS gene family and expression analysis at different fruit coloration stages. BMC Plant Biology. 2023;23(1):632.

Fernandes I, Faria A, Calhau C, de Freitas V, Mateus N. Bioavailability of anthocyanins and derivatives. Journal of Functional Foods. 2014;7:54–66.

Gowd V, Jia Z, Chen W. Anthocyanins as promising molecules and dietary bioactive components against diabetes–A review of recent advances. Trends in Food Science & Technology. 2017;68:1–3.

Gulcin I. Antioxidants and antioxidant methods: An updated overview. Archives of Toxicology. 2020;94(2):651–715.

Harborne AJ. Phytochemical Methods a Guide to Modern Techniques of Plant Analysis. Springer Science & Business Media; 1998.

He J, Giusti MM. Anthocyanins: Natural colorants with health-promoting properties. Annual Review of Food Science and Technology. 2010;1:163–87.

Hou F, Zhang R, Zhang M, Su D, Wei Z, Deng Y, Zhang Y, Chi J, Tang X. Hepatoprotective and antioxidant activity of anthocyanins in black rice bran on carbon tetrachloride-induced liver injury in mice. Journal of Functional Foods. 2013;5:1705–1713.

Hui C, Bin Y, Xiaoping Y, Long Y, Chunye C, Mantian M, Wenhua L. Anticancer activities of an anthocyanin-rich extract from black rice against breast cancer cells in vitro and in vivo. Nutrition and Cancer. 2010;62(8):1128–1136.

Hwang J-W, Kim E-K, Lee S-J, et al. Antioxidant activity and protective effect of anthocyanin oligomers on H2O2-triggered G2/M arrest in retinal cells. Journal of Agricultural and Food Chemistry. 2012;60:4282–4288.

Isaak CK, Petkau JC, Blewett H, Karmin, Siow YL. Lingonberry anthocyanins protect cardiac cells from oxidative stress-induced apoptosis. Canadian Journal of Physiology and Pharmacology. 2017;904–910.

Jiang Y, Dai M, Nie W-J, Yang X-R, Zeng X-C. Effects of the ethanol extract of black mulberry (*Morus nigra* L.) fruit on experimental atherosclerosis in rats. Journal of Ethnopharmacology. 2017;200:228–235.

Jing P. Purple Corn Anthocyanins: Chemical Structure, Chemopreventive Activity and structure/function Relationships [dissertation]. The Ohio State University; 2006. p: 5–90.

Jousilahti P, Vartiainen E, Tuomilehto J, Puska P. Sex, age, cardiovascular risk factors, and coronary heart disease: A pro spective follow-up study of 14,786 middle-aged men and women in Finland. Circulation. 1999;99:1165–1172.

Kähkönen MP, Heinonen M. Antioxidant activity of anthocyanins and their aglycons. Journal of Agricultural and Food Chemistry. 2003;51(3):628–633.

Kang S-Y, Seeram NP, Nair MG, Bourquin LD. Tart cherry anthocyanins inhibit tumor development in Apc (Min) mice and reduce proliferation of human colon cancer cells. Cancer Letters. 2003;194:13–19.

Kojšová S, Jendeková L, Zicha J, Kuneš J, Andriantsitohaina R, Pecháňová O. The effect of different antioxidants on nitric oxide production in hypertensive rats. Physiological Research. 2006;55.

Kongpichitchoke T, Hsu JL, Huang TC. Number of hydroxyl groups on the B-ring of flavonoids affects their antioxidant activity and interaction with phorbol ester binding site of PKCδ C1B domain: In vitro and in silico studies. Journal of Agricultural and Food Chemistry. 2015;63(18):4580–4586.

Lee I-C, Bae J-S. Pelargonidin protects against renal injury in a mouse model of sepsis. Journal of Medicinal Food. 2019;22:57–61.

Li P, Feng D, Yang D, Li X, Sun J, Wang G, Tian L, Jiang X, Bai W. Protective effects of anthocyanins on neurodegenerative diseases. Trends in Food Science & Technology. 2021;117:205–17.

Lin B-W, Gong C-C, Song H-F, Cui Y-Y. Effects of anthocyanins on the prevention and treatment of cancer. British Journal of Pharmacology. 2017;174:1226–1243.

Liobikas J, Skemiene K, Trumbeckaite S, Borutaite V. Anthocyanins in cardioprotection: A path through mitochondria. Pharmacological Research. 2016;113:808–815.

Liu S, Lou Y, Li Y, Zhao Y, Feng X, Capozzi V, Laaksonen O, Yang B, Li P, Gu Q. Comparison of anthocyanin and volatile organic compounds in juices and fruit wines made from blood oranges (*Citrus sinensis* L. Osbeck) at different maturity stages. Food Bioscience. 2023;56:103194.

Lu X, Zhou Y, Wu T, Hao L. Ameliorative effect of black rice anthocyanin on senescent mice induced by D-galactose. Food & Function. 2014;5:2892–2897.

Manolescu BN, Oprea E, Mititelu M, Ruta LL, Farcasanu IC. Dietary anthocyanins and stroke: A review of pharmacokinetic and pharmacodynamic studies. Nutrients. 2019;11:1479.

Mauray A, Felgines C, Morand C, Mazur A, Scalbert A, Milenkovic D. Nutrigenomic analysis of the protective effects of bilberry anthocyanin-rich extract in apo E-deficient mice. Genes & Nutrition. 2010;5:343–353.

Medic N, Tramer F, Passamonti S. Anthocyanins in colorectal cancer prevention. A Systematic Review of the Literature in Search of Molecular Oncotargets. Front Pharmacol. 2019;10:675.

Mehmood A, Zhao L, Wang Y, Pan F, Hao S, Zhang H, Iftikhar A, Usman M. Dietary anthocyanins as potential natural modulators for the prevention and treatment of non-alcoholic fatty liver disease: A comprehensive review. Food Research International. 2021;142:110180.

Miller MG, Hamilton DA, Joseph JA, Shukitt-Hale B. Dietary blueberry improves cognition among older adults in a randomized, double-blind, placebo-controlled trial. European Journal of Nutrition. 2018;57:1169–1180.

Munzel T, Camici GG, Maack C, Bonetti NR, Fuster V, Kovacic JC. Impact of oxidative stress on the heart and vasculature: part 2 of a 3-part series. Journal of the American College of Cardiology. 2017;70:212–229.

Noda Y, Kaneyuki T, Mori A, Packer L. Antioxidant activities of pomegranate fruit extract and its anthocyanidins: Delphinidin, cyanidin, and pelargonidin. Journal of Agricultural and Food Chemistry. 2002;50(1):166–171.

Ogawa K, Sakakibara H, Iwata R, Ishii T, Sato T, Goda T, Shimoi K, Kumazawa S. Anthocyanin composition and antioxidant activity of the crowberry (*Empetrum nigrum*) and other berries. Journal of Agricultural and Food Chemistry. 2008;56(12):4457–4462.

Pacheco SM, Soares MSP, Gutierres JM, et al. Anthocyanins as a potential pharmacological agent to manage memory deficit, oxidative stress and alterations in ion pump activity induced by experimental sporadic dementia of Alzheimer's type. Journal of Nutritional Biochemistry. 2018;56:193–204.

Padmasekar M, Lingwal N, Samikannu B, Chen C, Sauer H, Linn T. Exendin-4 protects hypoxic islets from oxidative stress and improves islet transplantation outcome. Endocrinology. 2013;154:1424–1433.

Pepe G, Grossi MR, Berni A, Filippi S, Shanmugakani RK, Papeschi C, Mosesso P, Natarajan AT, Palitti F. Effect of blueberries (BB) on micronuclei induced by N-methyl-N'-nitro-nitrosoguanidine (MNNG) and 7,12-dimethylbenz(a)anthracene (DMBA) in mammalian cells, assessed in in vitro and in vivo assays. Mutation Research. 2013;758(1-2):6–11.

Pojer E, Mattivi F, Johnson D, Stockley CS. The case for anthocyanin consumption to promote human health: A review. Comprehensive Reviews in Food Science and Food Safety. 2013;12:483–508.

Popović D, Kocić G, Katić V, Jović Z, Zarubica A, Janković Veličković L, Nikolić V, Jović A, Kundalić B, Rakić V, et al. Protective effects of anthocyanins from bilberry extract in rats exposed to nephrotoxic effects of carbon tetrachloride. Chemico-Biological Interactions. 2019;304:61–72.

Qi Z-L, Wang Z, Li W, Hou J-G, Liu Y, Li X-D, Li H-P, Wang Y-P. Nephroprotective effects of anthocyanin from the fruits of *Panax ginseng* (GFA) on cisplatin-induced acute kidney injury in mice. Phytotherapy Research. 2017;31:1400–1409.

Rahman MM, Ichiyanagi T, Komiyama T, Sato S, Konishi T. Effects of anthocyanins on psychological stress-induced oxidative stress and neurotransmitter status. Journal of Agricultural and Food Chemistry. 2008;56:7545–7550.

Ramos P, Herrera R, Moya-León MA. Anthocyanins: Food sources and benefits to consumer's health. Handbook of Anthocyanins: Food Sources, Chemical Applications and Health Benefits. 2014: 363–384.

Reis JF, Monteiro S, Gomes, R de S do Carmo MM, da Costa GV, Ribera PC, et al. Action mechanism and cardiovascular effect of anthocyanins: A systematic review of animal and human studies. Journal of Translational Medicine. 2016;14:315.

Remedi MS, Emfinger C. Pancreatic beta-cell identity in diabetes. Diabetes, Obesity and Metabolism. 2016;18(Suppl 1):110–116.

Rossetto M, Vanzani P, Lunelli M, Scarpa M, Mattivi F, Rigo A. Peroxyl radical trapping activity of anthocyanins and generation of free radical intermediates. Free Radical Research. 2007;41(8):854–859.

Salehi B, Lopez-Jornet P, Pons-Fuster López E, Calina D, Sharifi-Rad M, Ramírez-Alarcón K, et al. Plant-derived bioactives in oral mucosal lesions: A key emphasis to curcumin, lycopene, chamomile, aloe vera, green tea and coffee properties. Biomolecules. 2019; 9:106. doi: 10.3390/biom9030106.

Salehi B, Martorell M, Arbiser JL, Sureda A, Martins N, Maurya PK, et al. Antioxidants: Positive or negative actors? Biomolecules. 2018; 8. doi: 10.3390/biom8040124.

Sankhari JM, Thounaojam MC, Jadeja RN, Devkar RV, Ramachandran AV. Anthocyanin-rich red cabbage (*Brassica oleracea* L.) extract attenuates cardiac and hepatic oxidative stress in rats fed an atherogenic diet. Journal of the Science of Food and Agriculture. 2012;92:1688–1693.

Shah SA, Amin FU, Khan M, et al. Anthocyanins abrogate glutamate-induced AMPK activation, oxidative stress, neuroinflammation, and neurodegeneration in postnatal rat brain. Journal of Neuroinflammation. 2016;13:286.

Shah SA, Yoon GH, Kim MO. Protection of the developing brain with anthocyanins against ethanol-induced oxidative stress and neurodegeneration. Molecular Neurobiology. 2015;51:1278–1291.

Shih PH, Chan YC, Liao JW, Wang MF, Yen GC. Antioxidant and cognitive promotion effects of anthocyanin-rich mulberry (*Morus atropurpurea* l.) On senescence-accelerated mice and prevention of Alzheimer's disease. Journal of Nutritional Biochemistry. 2010;21:598–605.

Shih P-H, Wu C-H, Yeh C-T, Yen G-C. Protective effects of anthocyanins against amyloid b-peptide-induced damage in neuro-2A cells. Journal of Agricultural and Food Chemistry. 2011;59: 1683–1689.

Shih P-H, Yeh C-T, Yen G-C. Effects of anthocyanidin on the inhibition of proliferation and induction of apoptosis in human gastric adenocarcinoma cells. Food and Chemical Toxicology. 2005;43:1557–1566.

Stintzing FC, Stintzing AS, Carle R, Frei B, Wrolstad RE. Color and antioxidant properties of cyanidin-based anthocyanin pigments. Journal of Agricultural and Food Chemistry. 2002;50(21):6172–6181.

Sun J, Wu Y, Long C, He P, Gu J, Yang L, Liang Y, Wang Y. Anthocyanins isolated from blueberry ameliorates CCl4 induced liver fibrosis by modulation of oxidative stress, inflammation and stellate cell activation in mice. Food and Chemical Toxicology. 2018;120:491–499.

Tandon VR, Verma S, Singh J, Mahajan A. Antioxidants and cardiovascular health. Journal of Medical Education Research. 2005;7(2):115–118.

Toufektsian MC, De Lorgeril M, Nagy N, Salen P, Donati MB, Giordano L, et al. Chronic dietary intake of plant-derived anthocyanins protects the rat heart against ischemia-reperfusion injury. Journal of Nutritional. 2008; 138: 747–752.

Traustadottir T, Davies SS, Stock AA, Su Y, Heward CB, Roberts LJ 2nd, Harman SM. Tart cherry juice decreases oxidative stress in healthy older men and women. Journal of Nutritional. 2009;139:1896–1900.

Ulla A, Mohamed MK, Sikder B, Rahman AT, Sumi FA, Hossain M, Reza HM, Rahman GMS, Alam MA. Coenzyme Q10 prevents oxidative stress and fibrosis in isoprenaline induced cardiac remodeling in aged rats. BMC Pharmacology & Toxicology. 2017;18:1–10.

Veljković M, Pavlović DR, Stojiljković N, Ilić S, Jovanović I, Poklar Ulrih N, Rakić V, Veličković L, Sokolović D. Bilberry: Chemical profiling, in vitro and in vivo antioxidant activity and nephroprotective effect against gentamicin toxicity in rats. Phytotherapy Research. 2017;31:115–123.

Wallace TC, Giusti MM. Anthocyanins. Advances in Nutrition. 2015;6(5):620–622.

Wallace TC. Anthocyanins in cardiovascular disease. Advances in Nutrition. 2011;2(1):1–7.

Wang D, Xia M, Gao S, Li D, Zhang Y, Jin T, Ling W. Cyanidin-3-O-β-glucoside upregulates hepatic cholesterol 7α-hydroxylase expression and reduces hypercholesterolemia in mice. Molecular Nutrition & Food Research. 2012;56:610–621.

Wang D, Xia M, Yan X, Li D, Wang L, Xu Y, Jin T, Ling W. Gut microbiota metabolism of anthocyanin promotes reverse cholesterol transport in mice via repressing miRNA-10b. Circulation Research. 2012;111:967–981.

Wang H, Cao G, Prior RL. Oxygen radical absorbing capacity of anthocyanins. Journal of Agricultural and Food Chemistry. 1997;45(2):304–309.

Wang L-S, Stoner GD. Anthocyanins and their role in cancer prevention. Cancer Letters. 2008;269:281–290.

Wei J, Zhang G, Zhang X, et al. Anthocyanins from black chokeberry (*Aronia melanocarpa* Elliot) delayed aging-related degenerative changes of brain. Journal of Agricultural and Food Chemistry. 2017;65:5973–5984.

Yarijani ZM, Najafi H, Shackebaei D, Madani SH, Modarresi M, Jassemi SV. Amelioration of renal and hepatic function, oxidative stress, inflammation and histopathologic damages by malva sylvestris extract in gentamicin induced renal toxicity. Biomedicine & Pharmacotherapy. 2019;112:108635.

Zafra-Stone S, Yasmin T, Bagchi M, Chatterjee A, Vinson JA, Bagchi D. Berry anthocyanins as novel antioxidants in human health and disease prevention. Molecular Nutrition & Food Research. 2007;51:675–683.

Zha J, Koffas MA. Production of anthocyanins in metabolically engineered microorganisms: Current status and perspectives. Synthetic and Systems Biotechnology. 2017;2(4):259–266.

Zhang J, Celli GB, Brooks MS. Natural sources of anthocyanins. Anthocyanins from Natural Sources: Exploiting Targeted Delivery for Improved Health. 2019;1–33.

Zhu F, Cai Y-Z, Yang X, Ke J, Corke H. Anthocyanins, hydroxycinnamic acid derivatives, and antioxidant activity in roots of different Chinese purple-fleshed sweetpotato genotypes. Journal of Agricultural and Food Chemistry. 2010;58:7588–7596.

Ziberna L, Lunder M, Moze S, et al. Acute cardioprotective and cardiotoxic effects of bilberry anthocyanins in ischemia-reperfusion injury: Beyond concentration dependent antioxidant activity. Cardiovascular Toxicology. 2010;10:283–294.

6

Health Effects of Anthocyanins

Semih Ötleş and Emine Nakilcioglu

6.1 Introduction

There are numerous plant species on Earth that are categorized on the basis of various structures such as primary and secondary metabolites. Primary metabolites encompass chemicals necessary for the growth and development of plants, while secondary metabolites serve signaling functions in the defense and pollination processes of plants (de la Rosa et al., 2019).

Secondary metabolites are divided into different groups based on their structural characteristics. In plant tissues, phenolic compounds are commonly found as secondary metabolites. Phenolic compounds, which play a major role in determining the color, aroma, and taste of plants, are highly beneficial due to their various biological effects on human health, like antioxidative and anti-inflammatory properties. Foods obtained from plants are rich sources of phenolic compounds that can have a range of positive influences, including preventing heart diseases, reducing the risk of cancer and diabetes, and decreasing mutagenesis rates in human cells (Khoddami et al., 2013). The protection provided by the consumption of plant products like legumes, vegetables, and fruits is often colleagued with the presence of phenolic compounds. Additionally, phenolic compounds also have an impact on the quality characteristics of foods. They can enhance the shelf life of vegetables and fruits by activating antioxidant defense mechanisms. As a result, the acceptability of fruits and vegetables by consumers is improved.

Phenolic compounds including at least one hydroxyl group and one aromatic ring in their structures are divided into two groups in plants depending upon the phenol unit numbers in the molecule like simple phenols and polyphenols. The most common simple phenols found in plants are flavonoids and phenolic acids (Khoddami et al., 2013; de la Rosa et al., 2019). Additionally, plants can contain lignin, lignan, and coumarin, as well as condensed and hydrolyzable tannins (Khoddami et al., 2013).

Thousands of flavonoids have been identified in different conditions, tissues, and species, and these flavonoids serve various ecophysiological functions, such as attracting animals for seed dispersal or pollinators (Ray et al., 2017).

Flavonoids with high bioactivity are one of the main products of plant metabolism and are the most common phenolic compounds found in vegetables and fruits. Flavonoids contain a heterocyclic pyrrole ring and are classified into groups based on variations in this pyrrole ring. In plants, pigment compounds in the flavonoid structure are divided into two categories based on their fundamental properties: fat-soluble and water-soluble pigments (Li et al., 2022). Anthocyanins are the only ionic flavonoid polyphenols soluble in water (Wrolstad, 2004).

DOI: 10.1201/9781003453260-7

6.1.1 Anthocyanins

Anthocyanins are glycoside forms and play a major role in the coloration of many plants. Anthocyanins, as an important subgroup of flavonoids, are responsible for the different colors of several vegetables, fruits, and flowers depending on pH levels (Wrolstad, 2004). Acting as natural pH indicators, anthocyanins appear reddish pink at low pH levels, blue-purple at neutral conditions, and greenish yellow at high pH levels (de la Rosa et al., 2019).

In the structure of anthocyanins, sugar moieties are located in the third position of the carbon ring or in the fifth to seventh position of the A ring can be linked to the position ( ) (Prior and Wu, 2006). Glucose (glc), arabinose (arab), galactose (gal), xylose (xyl), and rhamnose (rham), in mono-, di-, or tri-sugar forms are the common sugars that bind to anthocyanins. While approximately 17 anthocyanins exist in nature, only six of them, pelargonidin (Pg), peonidin (Pn), petunidin (Pt), delphinidin (Dp), cyanidin (Cy), and malvidin (Mv), are commonly encountered. The chemical structures of these six anthocyanins differ from each other because of the radicals at three, four, and five of the B ring (Figure 6.1).

Cyanidin 3-*O*-galactoside (Cy3Gal) is the most common and dominant pigment responsible for the red and pink coloration of fruits, along with pelargonidins. Cy3Gal is particularly found in fruits with red skin or flesh, such as red-skinned apples, blueberries, and cranberries (Liang et al., 2021).

The chemical stability of Cy3Gal is influenced by such various factors as acidity, exposure to light, temperature, humidity, metal ions presence, gas composition, and other external factors. The content of Cy3Gal changes depending on the plant species. Aronia, blueberries, and blackberries are considered ideal sources. Similar to numerous other anthocyanins, Cy3Gal also demonstrates a noteworthy antioxidative capability.

Other anthocyanins are less commonly found. Delphinidin, petunidin, and malvidin are responsible for the purple and blue hues in plants. Additionally, factors like pH levels and copigmentation can influence the perceived color tones. Copigmentation enhances the vibrancy of anthocyanin colors, increasing their hue, brightness, and resistance to color changes. It adds to the steadfastness of anthocyanins.

6.1.2 History of Anthocyanin

The presence of plant pigments was first noticed in the mid-19th century. By the mid-19th century, chemical analysis methods were developed to study these pigments in more detail.

Anthocyanidin	R₁	R₂	R₃
Delphinidin (Dp)	OH	OH	OH
Pelargonidin (Pg)	H	OH	H
Cyanidin (Cy)	OH	OH	H
Malvidin (Mv)	OMe	OH	OMe
Peonidin (Pn)	OMe	OH	H
Petunidin (Pt)	OMe	OH	OH

FIGURE 6.1
Chemical structures of six anthocyanins (Prior and Wu, 2006).

In the 20th century, anthocyanins were discovered by Mendel who examined flower colors in peas (Holton and Cornish, 1995). From this period onward, research on the genetic and biochemical nature of pigment production in plants gained momentum. Scientists used spectroscopic methods to determine the chemical composition and structures of anthocyanins. Through these studies, it became clear that anthocyanins are responsible for color production in plants and enable the formation of different colors in various plant species.

6.1.3 Chemical Properties and Color of Anthocyanins

Numerous factors hinder the practical application of anthocyanin colorants, including their vulnerability to heat and light, as well as their proneness to degradation reactions. Specifically, ascorbic acid can expedite the deterioration processes of anthocyanins. Anthocyanins predominantly exist as resilient flavylium cations only at approximately pH 2. This distinctive chemical configuration is among the fundamental elements affecting the metabolism, absorption, and biological reactions to anthocyanins. Most anthocyanin colorants exhibit an effective pH range constrained to acidic foods, so color alterations and instability arise above pH 4. Numerous commercially used anthocyanin colorants may exhibit low color stability and could potentially harbor undesirable flavors.

Anthocyanins undergo structural transformations that can be reversed by pH changes, which have a significant impact on color. Different anthocyanins can vary considerably in terms of stability and color properties. Differences in anthocyanin structures can influence the stability of pigments, while the composition of the food matrix and pigment concentration can also exert an impact on pigment stability. Anthocyanins exhibit increased stability at low water activity, making them appropriate for use in intermediate moisture and dried foods (Wrolstad, 2004).

Nature has documented the existence of over 600 distinct anthocyanins, and their chemical structures vary based on the factors below:

1. The quantity and arrangement of hydroxyl and methoxy groups in the fundamental anthocyanidin structure.

2. The specific type, quantity, and locations of sugars attached to the anthocyanidin structure.

3. The nature of the acylating agent used and the extent of sugar acylation.

These factors contribute to the wide array of anthocyanin compounds discovered in different plant species and play an important role in determining their distinct properties and functions.

The chemical configurations of anthocyanins can exhibit notable differences contingent upon the extent of glycosylation (the bonding of sugars) and acylation (the bonding of acyl groups). These chemical modifications can impact the color, stability, and biological activity of anthocyanins.

In the case of most anthocyanins, their molecular weights usually fall within the range of 400–1200. Unlike other flavonoid subgroups sharing the same C6-C3-C6 skeleton, anthocyanins bear a positive charge under acidic pH conditions. In a solution, anthocyanins are actually in equilibrium among four molecular forms: quinoidal base, the flavylium cation, chalcone, and hemiacetal base. The relative ratios of these four states fluctuate depending on both the pH level and the specific structure of the anthocyanin molecules.

Anthocyanins in nature are found in the glycosylated form of anthocyanidins (Liang et al., 2021). This is due to the inherent instability of anthocyanidins in nature, and their stability is significantly influenced by environmental factors like temperature and pH. The stability of anthocyanidins is enhanced through various mechanisms by the addition of glucose. This enables them to establish hydrogen bonds, elevate van der Waals interactions, and enhance electron distribution, consequently improving stability. The stability of anthocyanins is intricately connected to various common physicochemical factors utilized in food processing, including pH and temperature. Factors such as fluctuations in pH, changes in temperature, the presence of ascorbic acid, co-pigments, enzymatic activity, and exposure to light, oxygen, metallic ions, and sugars, along with the resultant degradation products, can contribute to the degradation of anthocyanins and, as a result, impact the color of the food (Ekici et al., 2014).

6.1.4 Benefits of Anthocyanins

Anthocyanins have strong antioxidant properties and can offer an extensive array of health advantages via antioxidant or alternative mechanisms. However, to fully understand the role of anthocyanins in disease prevention, it is necessary to know the metabolism, extent, and rate of their absorption and tissue or cellular dispersion.

Flavonols and anthocyanins possess strong ultraviolet (UV) and visible light absorption capabilities, providing protection against the harmful effects of UV radiation when exposed to an excess of photons (Araguirang and Richter, 2022). Additionally, anthocyanins encompass other protective mechanisms that do not involve photophysical reactions (Gitelson et al., 2007). The detection of anthocyanins is crucial because it serves as an indicator of the aging and stress response of many plant species, offering valuable insights into how plants respond to and adapt to environmental stressors.

Anthocyanins are important for various functions within plants (Holton and Cornish, 1995). Apart from serving as color pigments, they play roles in pollination and seed distribution in plants. Additionally, when synthesized, they assist in seed dispersal.

Anthocyanins are present in reproductive structures and vegetative parts of plants. In vegetative tissues, their presence aids in safeguarding the photosynthetic machinery by absorbing surplus light. Additionally, they enhance tolerance to environmental conditions such as cold, drought, nutrient deficiencies, and high salinity while providing defense against pathogens (LaFountain and Yuan, 2021).

6.1.5 Foods Containing Anthocyanins

Anthocyanins, which offer antiviral and antimicrobial activities, can be found in various fruits like plums, grapes, peaches, and cherries as well as in vegetables such as black beans, red onions, and lettuce (Wrolstad et al., 2004). Red wine can also serve as a valuable source of anthocyanins. Anthocyanins found in raspberries and blackberries are significant for their health-related benefits, associated with anti-inflammatory, antioxidant, and chemopreventive properties. It has been shown that anthocyanins in black raspberries have biological activity against colon, esophageal, and oral cancers (Bowen-Forbes et al., 2010). According to some studies by Bowen-Forbes et al. (2010), black raspberries have a significantly higher anthocyanin content, while red raspberries have lower levels compared to others.

In addition to fruits and vegetables, a significant portion of anthocyanin intake can be obtained from fruit juices that are conducive to the digestion of health-promoting phytochemicals. Certain research studies have indicated that consumption of antioxidant fruit juices containing white grape, blackcurrant, elderberry, cherry, and blackberry juice can

TABLE 6.1

Concentrations of Total Anthocyanins (TA) (mg CGE/L)

	Total Anthocyanins (mg/L)
Black currant	1543.89 ± 5.5
Red Raspberry	217.39 ± 5.2
Blackberry	739.93 ± 37.5
Sour cherry	369.36 ± 2.4
Sweet cherry	256.6 ± 2.5
Strawberry	205.98 ± 2.2
Chokeberry	3042.2 ± 196.3
Elderberry	4188.63 ± 257

Source: Bowen-Forbes et al. (2010).

enhance plasma antioxidant capacity, indicating the beneficial effects of the antioxidant components in fruit juice (Table 6.1). It has also been observed that anthocyanins reduce lipid oxidation in the plasma compartment.

Because of the importance of color quality in both processed and fresh vegetables and fruits, agricultural experts and food scientists have conducted extensive research on these pigments. Anthocyanin-based colorants serve as not only natural color additives but are also manufactured for food applications using agricultural products cultivated for a particular intention, as well as from processing byproducts. In recent times, there has been a substantial rise in interest in anthocyanin pigments because of their potential to mitigate the risk of cancer, coronary heart disease, and stroke. Hence, apart from their functional purposes as color enhancers, anthocyanin extracts are thought to augment the nutritional value of food and beverages.

6.1.6 Natural Colorants Based on Anthocyanins

The regulations concerning the utilization of colorants based on anthocyanins exhibit notable variations from one country to another. Four out of the 26 colorants that are exempt from certification and authorized for food use in the United States are derived from anthocyanins, namely, grape color extract, grape skin extract, vegetable juice, and fruit juice. Besides grapes, more common sources of fruits and vegetables include red cabbage, radish, blackberry, currant, black raspberry, and blackcurrant.

In the European Union, all anthocyanin derivative colorants fall under the E 163 classification and are considered natural colorants. The regulations regarding anthocyanin use in specific food products vary between countries. Some countries, including Canada, Australia, Cyprus, Japan, Finland, Norway, Sweden, South Africa, New Zealand, Switzerland, and the United States, limit the use of anthocyanins in certain food products. On the other hand, countries like Chile, Malta, Iran, Israel, Saudi Arabia, Colombia, South Korea, Peru, and the United Arab Emirates have granted general approval for the use of anthocyanin colorants in food. It can be said that the United States has among the most stringent regulations regarding the utilization of anthocyanin-based colorants (Wrolstad, 2004).

6.1.7 Bioactivity of Anthocyanin Pigments

Over the past few years, there has been an increasing interest in anthocyanin pigments owing to their potential health advantages. These benefits include reduced coronary heart disease risk, anticarcinogenic activity, decreased stroke risk, improved visual acuity, anti-inflammatory effects, and enhanced cognitive behavior (Wrolstad, 2004).

Experiments have provided compelling evidence regarding the bioactivity of anthocyanin pigments. Hagiwara et al. (2002) showcased that anthocyanins extracted from purple sweet potatoes and red cabbage could inhibit tumor development in rats when incorporated into their diet. Joseph et al. (1999) showed that the supplementation of blueberry and strawberry extracts improved cognitive behavior and memory in elderly rats. Clinical investigations conducted in Italy revealed that diabetic patients, accounting for 79% of the participants, demonstrated an amelioration in diabetic retinopathy after consuming blackberry extract throughout the trial period. McGhie et al. (2003) determined intact anthocyanin glycosides after consumption of berry extracts in the plasma and urine of humans. The retrieved anthocyanins accounted for less than 1% of the dietary intake, and these pigments were eliminated from the body within a span of 4 hours.

In this chapter, we describe the definitions and sources of anthocyanin and the health effects of anthocyanin intake by consuming anthocyanin-rich sources.

6.2 Health Effects of Anthocyanin

6.2.1 Effects of Anthocyanins on Antidiabetic and Antiobesity Diseases

Insulin suppresses glucose uptake or output from the liver and keeps blood sugar levels at normal levels by stimulating metabolism. Insulin resistance is a disease in which insulin inadequately stimulates the transport of glucose in adipose tissue and skeletal muscle. Insufficient insulin level causes blood glucose to increase. Increased triglyceride levels also have a significant role in the deterioration of insulin resistance (Shipp and Abdel-Aal, 2010).

According to analyses conducted worldwide, obesity has increased approximately three times since 1975. Factors affecting obesity include diseases such as appetite disorders, genetics, endocrine disorders, and diabetes (Gomes et al., 2019). Diabetes is a metabolic disease related to high sugar concentration in the blood. It is generally caused by resistance to the peripheral effects of insulin and impairment of insulin secretion (Oliveira et al., 2020).

It is suggested that a diet low in fat, contained in sufficient amounts of vegetables and fruits, and rich in polyphenols affects insulin resistance and obesity. It may also reduce the risk of diabetes. In the treatment of these diseases, hypoglycemic drugs or sulfonylurea-based drugs that stimulate the release of insulin from beta cells secreted in the pancreas are used to bring blood sugar levels to normal levels (Shipp and Abdel-Aal, 2010).

Phytochemicals found in fruits, including anthocyanins, have various such pharmacological effects as anti-inflammatory and antioxidant effects. It is stated that anthocyanins are effective in reducing fat tissue and weight loss. It is also thought to be effective against diabetes because they have an antioxidant effect and decrease the production of reactive oxygen species (ROS) (Azzini et al., 2017; Oliveira et al., 2020). Compared to medications taken for weight control or to control blood sugar levels, a diet rich in vegetables and fruits is more economical and reliable (Shipp and Abdel-Aal, 2010).

Recent studies have been carried out investigating the effects of anthocyanins on antidiabetic and antiobesity. Herrera-Balandrano et al. (2021) examined the hypolipidemic and hypoglycemic effects of malvidin, malvidin derivatives (malvidin-3-galactoside and malvidin-3-glucoside), and blueberry anthocyanin extract. Their effects on both streptozotocin

(STZ)-induced diabetic mice and the human hepatocarcinoma cell line (HepG2) were examined. As a result of the research, it was stated that anthocyanin extract and malvidin derivatives reduced ROS and increased cell viability by 79–88%. They inhibited hyperlipidemia and hyperglycemia by strengthening enzymes that play a significant role in glycogenolysis and lipolysis in HepG2 cells. Additionally, diabetic mice were given anthocyanin extract in low concentration (100 mg/kg) and high concentration (400 mg/kg) for five weeks. It has been stated that anthocyanin has a significant effect on body weight, reduces the amount of total cholesterol and triglycerides, and is a functional agent for hypolipidemia and hypoglycemia.

Chen et al. (2018) researched the *in vitro* hypolipidemic and hypoglycemic effects of black soybean seed coat extract (BSSCE) in diabetic mice induced by STZ and a high-fat diet. It has been observed that BSSCE protects HepG2 cells damaged by oxidative stress. It also inhibits lipid accumulation in HepG2 cells. It was stated that body weight, insulin level, and blood sugar were significantly reduced in BSSCE-fed mice compared to diabetic mice. Additionally, as a result of this study, enzyme activities were observed to be improved. It has also been shown to decrease total triglyceride, total cholesterol, and non-esterified fatty acid content, increase high-density lipoprotein (HDL) cholesterol content, and contribute to hypolipidemia and hypoglycemia.

Wu et al. (2013) examined the anti-obesity effects of sweet cherry anthocyanins in mice fed a high-fat diet. Cherry anthocyanins have been shown to obstruct obesity in high-fat-fed mice. In the *in vivo* experiment, it was stated that body weight decreased by 5.2% in mice fed with 40 mg/kg cherry anthocyanin and by 11.2% in mice fed with 200 mg/kg cherry anthocyanin. Additionally, it has been observed to reduce blood glucose, low-density lipoprotein (LDL) cholesterol, total cholesterol, and triglycerides, and slow down obesity.

Yang et al. (2017) studied the cardiometabolic effects of purified anthocyanins from black currant and blueberry in adults (aged 40–75) with hyperglycemia. Purified anthocyanins were administered to 160 participants with early untreated diabetes or prediabetes. As a result of the research, it was stated that anthocyanins reduce the amount of sugar in the blood and LDL cholesterol, and also the consumption of anthocyanins are effective in improving insulin resistance and glycemic control.

Wu et al. (2013) examined the effects of honeysuckle anthocyanins (HA) on obesity in mice induced by a high-fat diet. Mice were fed either a high-fat diet or a low-fat diet during eight weeks. Different amounts of HA (50, 100, and 200 mg/kg) were given to mice fed a high-fat diet. As a result of the research, it was reported that 100 and 200 mg/kg HA prevented body weight gain, improved liver functions, reduced lipids in the blood and liver, and reduced serum insulin levels. Thus, it has been stated that HA may be effective against obesity by preventing lipid accumulation.

6.2.2 Effects of Anthocyanins on Eye Health

The innermost layer of the eye is the retina. It is one of the most metabolically active tissues in the body (Liu et al., 2011). High metabolic rate causes sensitivity to oxidative stress and high oxygen demand. There are some parameters that make the retina defenseless against oxidative stress. Examples of these parameters include light exposure on photoreceptor membranes, high amounts of PUFA, and phagocytosis leading to reactive intermediates in the retinal pigment epithelium (RPE). Damage caused by oxidative stress is seen in the macula and usually occurs with increasing age (Kalt et al., 2010). Free radical scavengers can conserve the retina from light-induced damage (Liu et al., 2011).

According to studies conducted in recent years, it has been stated that flavonoids may be effective in visual signal transmission and eye diseases such as macular degeneration caused by oxidative stress (Kalt et al., 2010). The first reported health effect of anthocyanins is their benefit on vision. The first effect of anthocyanins on eye health emerged during World War II, when British Royal Airmen used blueberries to improve night vision (Shipp and Abdel-Aal, 2010). It was stated that foods rich in anthocyanins protected against oxidative stress damage and improved eye health (Wattanathorn et al., 2023).

Miyake et al. (2012) investigated the effect of blueberry extract containing anthocyanins on the retina of mice. In the study, an endotoxin-induced uveitis (EIU) model was created in addition to retinal inflammation by injecting lipopolysaccharide. The extract, rich in anthocyanins, was measured by electroretinogram and appeared to prevent the deterioration of photoreceptor cell function. It has also been stated that rhodopsin associated with EIU is reduced and the shortening of the outer segments of photoreceptor cells is suppressed. Additionally, the extract obstructed STAT3 activation, which causes rhodopsin reduction, and the rise in interleukin-6 expression (which activates STAT3). As a result, blueberry extract has been shown to protect visual function against retinal inflammation.

Kang et al. (2019) studied the anthocyanin oligomers effects on dry eye disease in mice. In the study, 40 mice (80 eyes) were used as subjects. Dry eyes were created using 0.02% benzalkonium chloride (BAC). Anthocyanin oligomers were administered once a day at two different doses such as 10 and 50 mg/kg. Eye drops (0.1% cyclosporine A) were applied to the control group twice a day. The corneal erosion degree (National Eye Institute (NEI) grading system), phenol red test, and tear break-up time (TBUT) were observed under slit lamp microscopy. In the group administered 50 mg/kg anthocyanin oligomer, the wet length of the phenol red thread and TBUT indicated an important increase compared to the BAC group. The NEI score was lower compared to the BAC-treated group. Analysis of corneal tissue indicated that keratinization and thickening of the corneal epithelial layer were not more pronounced in the group applied with 50 mg/kg anthocyanin oligomer compared to the group applied with cyclosporine A eye drops. As a result, it was revealed that anthocyanin oligomers had a positive effect on the clinical index of dry eyes in mice.

Kosehira et al. (2020) examined the effects of blueberry extract intake during 12 weeks on the tonic spasm of the ciliary muscle induced by the visual display terminal (VDT). A total of 109 healthy adults between the ages of 20 and 60 were divided into two groups: placebo and blueberry-extract-consuming. While the capsule without blueberry extract was administered to the placebo groups, the capsule containing 240 mg of blueberry extract was administered to the other group. The subjects were tested before and after VDT at weeks 0, 4, 8, and 12. The high-frequency component (HCF) of the accommodative microfluctuation was measured and found that the values of HCF were significantly improved in the group consuming blueberry extract compared with the placebo group at weeks 8–12 ($p<0.05$). Oral consumption of the extract during 12 weeks relieves the tonic spasm of the ciliary muscle induced by near vision and VDT tasks.

Sekikawa et al. (2021) observed the effects of consuming a supplement including blueberry anthocyanin on eye health. Randomized, placebo-controlled, double-blind study was carried out on Japanese subjects (32 healthy adults) who used VDT and subsequently experienced eye fatigue. Subjects were separated into a placebo group or an active group containing blueberry anthocyanin (43.2 mg anthocyanin per capsule). One capsule was consumed (either anthocyanin-containing or placebo) daily for six weeks. Percentage of pupillary response, tear film breakup time, muscle stiffness, and Schirmer value were observed before and after using VDT. It was stated that the active group using anthocyanin indicated an important improvement in the percentage of pupillary response. Additionally, it prevented the reduction in accommodative function induced by oxidative stress with the use of VDT.

Meiqi Fan et al. (2023) examined the effects of grape peel extract anthocyanin oligomers on dry eye disease. The 108 patients with dry eyes were equally divided into treatment and placebo groups. A total of 800 mg/day of anthocyanin oligomers was given to the treatment group, while 800 mg/day of maltodextrin was given to the placebo group. The occurrence of adverse reactions, clinical indexes, and clinical activities were compared between the two groups. In the result of this study, anthocyanin oligomers have been observed to be effective in dry eye disease, improving ocular surface disease, intraocular pressure, and tear breakup time.

6.2.3 Effects of Anthocyanins on Cardiovascular Diseases

Cardiovascular diseases influence the heart and blood vessels. Factors such as excessive alcohol and cigarette consumption, metabolic syndrome, and diet have an important role in the onset and progression of cardiovascular diseases. One of the origins of morbidity and mortality in the world is cardiovascular disease (Krga and Milenkovic, 2019; Gonçalves et al., 2021). The genesis of cardiovascular diseases is related to the production of free radicals from sources such as the angiotensin II-mediated nicotinamide adenine dinucleotide phosphate (NADPH) oxidase system, inducible nitric oxide synthase, xanthine oxidase, and the mitochondrial electron transport chain. ROS and free radicals can damage endothelial cells in blood vessels and cause cardiomyocyte loss in the heart (Alam et al., 2021). Persistent inflammation in the walls of large and medium arteries causes the onset of atherosclerosis that is linked to peripheral artery disease, ischemic stroke, hypertension, and coronary artery disease (Mattioli et al., 2020; Samota et al., 2022).

According to clinical and epidemiological studies, it is thought that a diet rich in vegetables and fruits slows down the development of cardiovascular diseases (Krga and Milenkovic, 2019). The biggest reason for the biological effects of anthocyanins is their antioxidant properties. It has an antioxidant effect against ROS and can be effective in preventing cardiovascular diseases caused by mortality (Krga and Milenkovic, 2019; Alam et al., 2021). Cardiovascular diseases like hypertension and atherosclerosis are associated with reduced blood flow. Anthocyanins have a positive effect on cholesterol levels, regulating blood pressure and preventing the development of hypertension and atherosclerosis (Hair et al., 2021). In addition, it also shows improving arterial stiffness, inhibiting enzymes (COX-1 and COX-2), antiglycation, anti-inflammatory, and antithrombotic effects (Khoo et al., 2017; Mozos et al., 2021).

Miyazaki et al. (2008) investigated the *in vitro* protective effect of purple sweet potato anthocyanin (PSPA) against *in vitro* LDL oxidation. Additionally, the development of atherosclerotic lesions was examined in mice with Apolipoprotein E deficiency and fed with a diet enriched in fat and cholesterol, without or with 1% PSPA for four weeks. PSPA protected LDL against oxidation more strongly than l-ascorbic acid and other anthocyanins *in vitro*. PSPA significantly reduced the plasma level of soluble vascular cell adhesion molecule-1 (sVCAM-1), atherosclerotic plaque area, and liver level of thiobarbituric acid-reactive substances ($p<0.05$). It was determined that the effects on plasma lipid and cholesterol levels and body weight were insignificant ($p<0.05$). It has also been stated that it suppresses the development of atherosclerotic lesions.

The study examining the effects of a food supplement consisting of blueberry and black currant anthocyanins on cardiovascular risk factors, inflammation, and oxidative stress in prehypertensive men was conducted by Hassellund et al. (2013). Subjects were given 640 mg of anthocyanin per day. Compared to the placebo group, blood sugar and HDL cholesterol were higher in the anthocyanin group. It has also been stated that various plasma polyphenols increased after anthocyanin intake (1–3 hours).

Johnson et al. (2015) examined the effects of blueberry consumption during eight weeks on arterial stiffness and blood pressure in postmenopausal and premenopausal women with hypertension (stage 1). As a result of the studies, it was observed that consuming blueberries for eight weeks significantly reduced brachial-ankle pulse wave velocity, diastolic blood pressure, and systolic blood pressure. In addition, it was stated that nitric oxide amounts were higher in the group consuming blueberries compared with the initial value and the control group.

Tabart et al. (2018) observed the relationship between the induction of endothelium-dependent relaxation of blackcurrant juices in isolated coronary arteries and the polyphenol content, antioxidant capacity, and anthocyanin content in their study. Consumption of six different commercial blackcurrant juices has been shown to induce relaxation ranging from 21 to 100%. It was observed that this caused relaxation was not related to the amount of ascorbic acid, total phenolic content, antioxidant capacity, and total flavanol content. It has been stated that relaxation amplitude has a significant relationship with anthocyanin concentration and total anthocyanin content.

The anti-inflammatory activities of anthocyanin-rich extract (ArCC) obtained from red Chinese cabbage were determined by Joo et al. (2018). Anti-inflammatory effects were observed in hyperlipidemic apolipoprotein E-deficient mice and cultured endothelial cells. Administration of ArCC suppressed aortic inflammation in hyperlipidemic apolipoprotein E-deficient mice. Compared to control mice, the aortas of mice undergoing ArCC had lower blood inflammatory cytokine concentrations, reduced plaque formation, and lower leukocyte infiltration. It has been determined that the consumption of red Chinese cabbage has a significant role in decreasing the risk of vascular inflammatory disease.

6.2.4 Effects of Anthocyanins on Cancer

Cancer is an uncontrolled cell cycle. The mechanism of cancer prevention can be grouped under three headings: inhibition of signaling pathways, regulation of protein expression, and other signaling pathways (Lin et al., 2017). The initial stage of the cancer cycle is closely related to the level of the disease. Therefore, diets with antioxidant and anti-inflammatory properties are critical (Chen et al., 2022). Anthocyanin intake is highly effective in preventing cancer. There is no specific recommended intake of anthocyanins for optimal health benefits. Human health can be promoted by intake of anthocyanins from different foods (Diaconeasa et al., 2020). After the intake of anthocyanins into the body, it can prevent cancer through pathways such as electron donation and Nrf2 activation. It can also prevent cancer by inhibiting proinflammatory gut bacteria and cytokines (Figure 6.2) (Chen et al., 2022).

Anthocyanins show anti-carcinogenic effects due to their high antioxidant activity. The damage caused by oxidative stress is prevented by scavenging free radicals. Anthocyanins also bind directly to proteins by metal chelating (Wang and Stoner, 2008). Jing et al. (2015) examined the antioxidant properties of eggplant. Eggplant contains delphinidin and nasunin anthocyanins. In this study on colon cancer, they observed that eggplant supplementation reduced ROS and promoted glutathione reductase protein expression. Grimes et al. (2018) examined colon and breast cancer. The cyanidin and delphinidin anthocyanins of grapes enabled the hydroxylation of polyphenols at the B ring.

Inflammatory proteins (NF-κB and COX-2) and other cytokine cause abnormal cell growth in several types of cancer. Anthocyanins inhibit the expression levels of mRNA

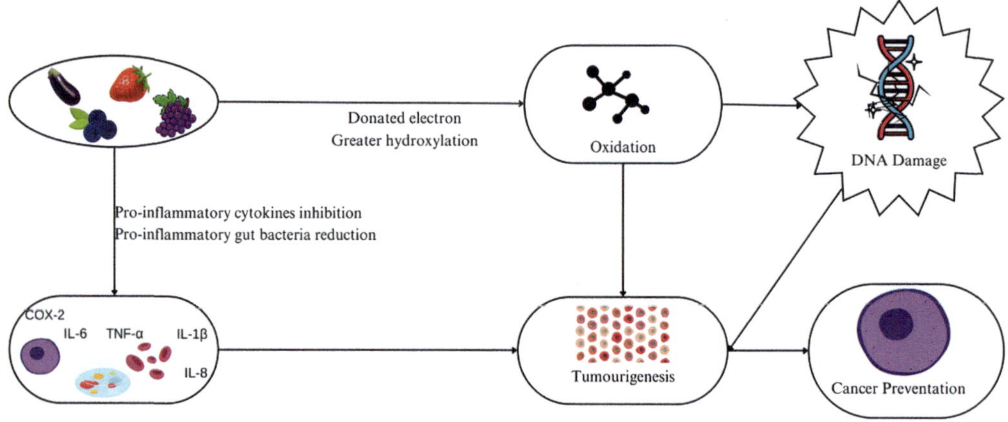

FIGURE 6.2
Main mechanism of the cancer prevention of anthocyanins (Chen et al., 2022).

and proteins. Many studies reveal the anti-inflammatory effect of anthocyanins against colorectal cancer. In addition, berries such as blackberries and raspberries support apoptosis of hepatocellular carcinoma, prostate cancer, and cervical cancer *in vitro* and *in vivo* (Ma et al., 2021). In a study conducted *in vivo*, blueberry and blackcurrant fruits were used. It was revealed that it reduced inflammation and inhibited NF-κB transactivation at the end of one month in the subject group. It was also found to improve glucose and lipid metabolism. Thus, it was concluded that tumor formation could be prevented (Aboonabi and Aboonabi, 2020). Chen et al. (2018) examined the properties of black raspberry anthocyanins on colorectal and colon cancer cells. In the study, anthocyanins were found to prevent the growth of cancer cells by showing anti-inflammatory effects against IL-1B, IL-6, and COX-2 inhibitors.

Other recent studies on cancer using anthocyanin sources are given in Table 6.2.

6.2.5 Effects of Anthocyanins on Neurodegenerative Diseases

Neurodegenerative diseases are defined as the rapid death of neurons and rapid brain damage. These diseases are caused by the loss of myelin, which causes dysfunction (Li et al., 2021). Neurodegenerative diseases are divided into acute and chronic. Acute diseases quickly cause permanent damage. Diseases such as Parkinson's, Alzheimer's, and Huntington's are chronic and reduce people's quality of life over time (dos Santos et al., 2019; Gardener et al., 2021). Although the origin of many diseases varies, factors contributing to the neurodegenerative process include oxidative stress, endoplasmic reticulum stress, excitotoxicity, nitrosative stress, protein aggregation, and neuroinflammation (Winter and Bickford, 2019).

The neuroprotective functions of anthocyanins are explained by their high oxygen radical absorbance capacity. Anthocyanins act as antioxidants by preventing the formation of ROS in cells (Alam et al., 2021). Anthocyanins exert neuroprotective effects by suppressing oxidative stress and neuroinflammation (Hair et al., 2021). Zhang et al. (2019) reported that anthocyanins demonstrated neuroprotective effects on the brain by suppression of N-terminal kinase activation, activation of neurotrophic factor signaling, and maintenance of calcium and zinc homeostasis. The anti-inflammatory effect of anthocyanins may be explained by their inhibition of some inflammatory biomarkers (interleukin-8 (IL-8)).

TABLE 6.2

Studies on Cancer Using Anthocyanin Sources

Source of Anthocyanins	Disease	Health Effects	Reference
Black rise anthocyanins	Breast cancer	Black rice anthocyanins have been shown to suppress metastasis by targeting the tumor stimulatory pathway (RAS/RAF/MAPK) on breast cancer cells by increasing antimetastatic activity.	Chen et al. (2015)
Blueberry anthocyanins	Breast cancer	Blueberry anthocyanins behaved as chemo inhibitors by inhibiting proliferation in breast cancer cells. In the study, anthocyanins promoted apoptosis by increasing caspase-3 activity.	Faria et al. (2010)
Black raspberries	Esophageal Tumors	*In vitro* study, black raspberry was found to prevent tumor cell proliferation by decreasing the expression of NF-κB and activator protein-1. It was revealed that apoptosis was induced in preneoplastic and papillomatous esophageal tissues.	Wang et al. (2009)
Amachi and Leona potatoes anthocyanins	Colonic cancer	Both potatoes showed cytotoxicity effect on tumor cells (Caco-2). Leona potato was reported to contain seven times more anthocyanins than Amachi potato. Therefore, Leona potato was found to have higher activity on cell viability protection.	Kubow et al. (2017)
Red Jasmine Rice (*Oryza sativa* L.) anthocyanins	Colonic cancer	Red jasmine rice with high antioxidant activity was found to induce DNA fragmentation through both extrinsic and intrinsic apoptosis in Caco-2 cells. The study demonstrated that anthocyanins in rice have high inhibition of free radicals.	Suantai et al. (2022)
Acai pulp (*Euterpe oleracea*)	Colonic cancer	Antitumor activity has been studied *in vitro* with a diet containing anthocyanins of the known acai fruit. In the study, anthocyanins appear to reduce the proliferation of tumor cells and the viability of tumors with high degrees of dysplasia.	Fragoso et al. (2018)
Cyanidin-3-*O*-glucoside	Colorectal cancer	Cyanidin-3-*O*-glucoside anthocyanin reduced cell motility against cells that acquired resistance at the epithelial-mesenchymal transition. It was concluded that it could improve oxaliplatin resistance by reversing the transition mechanism.	Kurter et al. (2023)
Hibiscus anthocyanins	Colorectal cancer	In the study, hibiscus anthocyanins promoted the activation of multiple signaling pathways by activating protein kinase. Anthocyanins were found to promote caspase-3 cleavage and apoptosis activation.	Tsai et al. (2023)
Pomegranate anthocyanins	Oral cancer	In the study, pomegranate anthocyanins have been shown to reduce metalloproteinase (MMP-9 and MMP-2) activity and expression. Additionally, pomegranate anthocyanins demonstrated antimetastatic treatment effects on oral cancer by reducing epithelial-mesenchymal transition signals.	Peng et al. (2020)

For example, pomegranate anthocyanins inhibit inflammatory biomarkers in the brain by activating NF-κB and inhibit the expression of proinflammatory cytokines and other biomolecules (Alam et al., 2021). Additively, cherry and blackberry anthocyanins have been shown to support the synthesis of prostacyclins in the vascular endothelium (Rehman et al., 2017).

Apoptosis is one of the main causes of neurological disorders. Lipopolysaccharides and glutamate can cause neuronal responses through mitochondria-dependent apoptotic pathways (Huo et al., 2021). In some cases, abnormal protein aggregation can be observed with accumulation of proteins. Intracellular calcium influx increases. Calcium imbalance causes mitochondrial oxidative stress. In these cases, the combination of glutamate and lipopolysaccharides induces apoptosis (Zhong et al., 2023). When anthocyanins are taken into the body, they rapidly reach the brain by way of P-glycoprotein transporters. It directly inhibits apoptosis by reducing oxidative stress and microglia (macrophages in the central nervous system) viability (Winter and Bickford, 2019). Ullah et al. (2014) studied mouse hippocampal cell lines and rat neurons. They reported that anthocyanins decreased calcium dysregulation, ROS accumulation, and apoptotic cells in the activation of adenosine monophosphate (AMP), which acts as a homeostasis regulator.

Li et al. (2020), studied 44 mice with Alzheimer's disease. Blueberry anthocyanin extracts were used in the study. 20 mg of extract was given by gavage every day for three months. At the end of three months, it has been observed that it reduced the expression of inflammatory factors and chemokine receptors in mice. It is seen that anthocyanin intake reduces hippocampal neuroinflammatory responses. It stimulates phagocytosis of β-amyloid protein plaques by regulating the signaling pathways of microglia. It has been demonstrated to reverse deficiency in cognitive function.

An important part of neuroinflammation is perioperative neurocognitive impairment. It is a common complication due to the lack of effective prevention and treatment measures. Zhang et al. (2020) studied the anthocyanins of black goji (*Lycium ruthenicum* Murr.) plant. In this study on mice, it was observed that neuroinflammation and microglia activation were significantly reduced. It was revealed that learning and memory functions improved in mice after treatment.

Hyperphosphorylation of microtubule-associated tau protein may result from peripheral insulin resistance. These mechanisms are closely associated with the development of Alzheimer's disease. Batista et al. (2017), examined anthocyanins (delphinidin-3-glucoside, cyanidin-3-glucoside, and petunidin-3-glucoside) in the peel of Jaboticaba fruit (*Myrciaria jaboticaba* Vell.). Mice with insulin sensitivity and Alzheimer's disease were treated with 24 mg/kg/day for 14 days. At the end of the study, mice showed lower phosphorylation in the hippocampus by reducing peripheral insulin resistance. Hippocampal inflammatory markers (such as pNF-κB, p-JNK, TNF-α, IL-1β) were reduced and neuronal viability was preserved. Oxidative stress and neuronal apoptosis decreased by increasing antioxidant activity in the brain. Connection between neurons (synaptic functions) increased.

Poulose et al. (2017) studied the neurochemical properties of anthocyanins (cyanidin-3-rutinoside, delphinidin-3-glycoside, cyanidin-3-glycoside, peonidin-3-glycoside, and malvidin-3-glycoside) of freeze-dried açaí powder. The diets of aged Fischer rats were supplemented with 20 g/kg anthocyanin every day for 45 days. At the end of the study, an increase in antioxidant enzymes (glutathione transferases (GST) and superoxide dismutase (SOD)) was observed. Significant decreases in prooxidant NADPH-oxidoreductase-2 and proinflammatory transcription factor (NF-κB) were observed in rats with treatment ($p<0.05$). In addition, regulation of neuroinflammation autophagy was achieved at the end of 45 days.

Oxidative stress, mitochondrial dysfunction, and loss of dopaminergic neurons lead to Parkinson's disease. A lot of studies are being conducted to prevent these processes. Qian et al. (2019) studied mice with Parkinson's disease. Blueberry anthocyanins were used in the study. After three weeks of treatment, glutathione peroxidase and SOD enzyme activities decreased. Anthocyanin intake showed a high antioxidant effect in scavenging free radicals. It was reported that oxidative damage in neurons decreased, and a significant neuroprotective effect was observed in Parkinson's disease processes ($p<0.05$).

6.2.6 Effects of Anthocyanins on Skin Health

The skin comprises three layers such as epidermis, dermis, and subcutaneous tissue. All layers undergo degenerative changes with aging (Shin et al., 2019). Skin aging is divided into two categories: intrinsic and extrinsic. Internal aging is associated with the epidermis. Extrinsic aging is usually related to sun exposure and decreased collagen (Kohl et al., 2011). There are many molecular mechanisms of skin aging and skin damage. Many treatments related to these mechanisms continue to be researched (Xia et al., 2015).

The health benefits expected from cosmetic products are increasing day by day. Researchers and producers are focused on both health benefits and sustainability (Câmara et al., 2022). Anthocyanins are the color sources of many fruits and vegetables. They draw attention with their pharmacological profiles, pharmacological mechanisms, and health-promoting properties (Correia et al., 2021; Zhao et al., 2021) Anthocyanins are an indispensable source for skin health. Rose et al. (2018) have been conducted the anthocyanins of black currant skins obtained as waste from industry. A hair dye formulation was created by extracting it with water. The stability of the hair color was high and industrial waste was evaluated.

The protective mechanism of anthocyanin pigments in human skin can be explained by reducing solar UV radiation penetration, the availability of ROS, biofilm formation, and the activation of skin barrier degrading enzymes (Figure 6.3) (Correia et al., 2021).

Anthocyanins continue to be researched for their benefits to human skin with their strong antioxidant capacity. In some applications, undesirable situations may arise due to changes in pH, temperature, etc. (Silva et al., 2020).

Poomanee et al. (2021) investigated the extraction of purple glutinous rice (*Oryza sativa* L.) anthocyanins and their effect on aging. High antioxidant capacity and high amounts of phytochemical components were observed in the extracts obtained. It was obtained that purple glutinous rice, which is a good source of cyanidin-3-*O*-glucoside component, showed a high anti-aging effect with skin fibroblast proliferation without cellular toxicity.

Purple sweet potato (*Dioscorea alata* L.) is known to have high radical scavenging ability, total phenolic content, and strong reducing ability. Chan et al. (2010) investigated the ability of PSPAs to protect against UV radiation. A cosmetic cream containing anthocyanins was studied. It was determined that the cream contained 0.61 mg of anthocyanins per 100 g of cream. As a result of the study, it was found that it showed high absorption ability against UV-A and UV-B radiation and was successful against 46% of UV radiation.

Abdellatif et al. (2021) have been conducted a study on skin protection of pomegranate (*Punica granatum*) fruit in cream form. A cream containing pomegranate anthocyanins was used by the fusion method. The study showed that the cream did not cause any toxicity on the skin. It was reported that the anthocyanin-containing cream improved skin barrier functions and showed improvements in many photodamage parameters.

Vaccinium uliginosum is a species of blueberry that grows in temperate regions of the northern hemisphere. Jo et al. (2020) extracted anthocyanins using this blueberry. It was given orally to hairless mice exposed to UV-B radiation. At the end of the study, the level

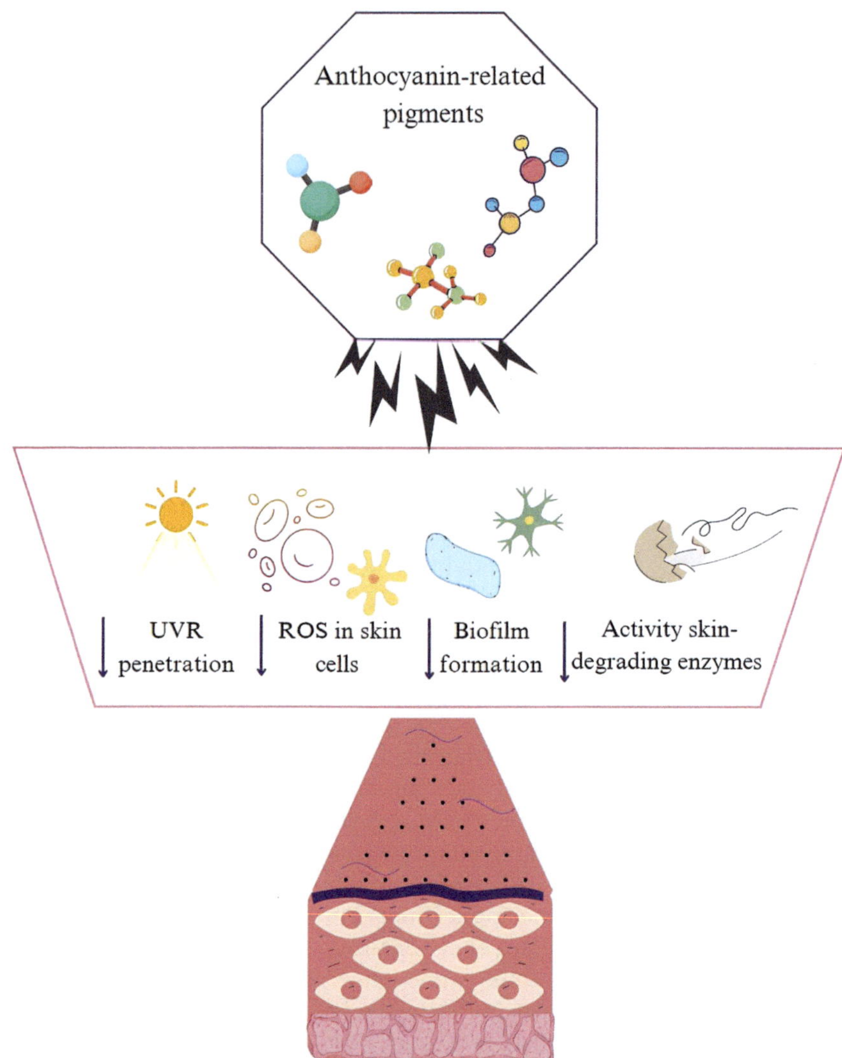

FIGURE 6.3
The effect of anthocyanin on factors that cause skin aging (Correia et al., 2021).

of metalloproteinase tissue inhibitor increased with high antioxidant activity. The level of UVB-affected phosphorylation was significantly decreased ($p<0.05$). It was stated that *V. uliginosum* may have a photoprotective effect against skin damages.

Papaioanou et al. (2018) examined the effects of anthocyanins of two varieties of *Rubus idaeus* L. fruit, Ruvi, and Cayuga on skin health. They reported that Ruvi fruit cultivated under low light conditions had high antioxidant activity and high anthocyanin content. A cream was developed with anthocyanin extract obtained from these fruits. The cream showed a high inhibitory effect against skin regulatory enzymes, tyrosinase, and elastase enzymes. In the study, it was mentioned that agricultural applications should be developed to maximize the cosmeceutical effect.

6.3 Conclusion

Anthocyanins are a member of the flavonoids that provide the blue, purple, and red pigment in vegetables and fruits. The chemical structure of anthocyanins shows that they have important effects on the suppression of diseases. These compounds, naturally found in many food sources, are one of the health-protecting components of our daily diet. In addition to their food applications, they have also been the subject of remarkable research for their therapeutic effects. It is stated that they have antioxidative effects, anti-inflammatory effects, prevention of oxidative stress, and protection against such diseases as cardiovascular diseases, neurodegenerative, obesity, and cancer. In addition, a diet containing anthocyanin-rich fruits and vegetables is more reliable and economical instead of medications used to control blood sugar and weight. Anthocyanins should be consumed more as natural coloring agents, dietary supplements, and nutraceuticals. Therefore, more research should be conducted on the antioxidant activities of anthocyanin-rich food sources, their effects on health, and the bioavailability of extracts. Additionally, more studies are needed to improve the stability of anthocyanins during digestion.

Acknowledgments

The authors are thankful to Eda Nurko, Selen Seyhan, and Gizem Tiryaki for their valuable contributions in the preparation of this chapter.

References

Abdellatif, A. A., Alawadh, S. H., Bouazzaoui, A., Alhowail, A. H., & Mohammed, H. A. (2021). Anthocyanins rich pomegranate cream as a topical formulation with anti-aging activity. *Journal of Dermatological Treatment*, 32(8), 983–990. https://doi.org/10.1080/09546634.2020.1721418

Aboonabi, A., & Aboonabi, A. (2020). Anthocyanins reduce inflammation and improve glucose and lipid metabolism associated with inhibiting nuclear factor-kappaB activation and increasing PPAR-γ gene expression in metabolic syndrome subjects. *Free Radical Biology and Medicine*, 150, 30–39. https://doi.org/10.1016/j.freeradbiomed.2020.02.004

Alam, M. A., Islam, P., Subhan, N., Rahman, M. M., Khan, F., Burrows, G. E., … & Sarker, S. D. (2021). Potential health benefits of anthocyanins in oxidative stress related disorders. *Phytochemistry Reviews*, 20(4), 705–749. https://doi.org/10.1007/s11101-021-09757-1

Araguirang, G. E., & Richter, A. S. (2022). Activation of anthocyanin biosynthesis in high light – What is the initial signal? *New Phytologist*, 236(6), 2037–2043. https://doi.org/10.1111/nph.18488

Azzini, E., Giacometti, J., & Russo, G. L. (2017). Antiobesity effects of anthocyanins in preclinical and clinical studies. *Oxidative Medicine and Cellular Longevity*. https://doi.org/10.1155/2017/2740364

Batista, Â. G., Soares, E. S., Mendonça, M. C., da Silva, J. K., Dionísio, A. P., Sartori, C. R., … & Marostica Junior, M. R. (2017). Jaboticaba berry peel intake prevents insulin-resistance-induced tau phosphorylation in mice. *Molecular Nutrition & Food Research*, 61(10), 1600952. https://doi.org/10.1002/mnfr.201600952

Bowen-Forbes, C. S., Zhang, Y., & Nair, M. G. (2010). Anthocyanin content, antioxidant, anti-inflammatory and anticancer properties of blackberry and raspberry fruits. *Journal of Food Composition and Analysis*, 23, 554–560. https://doi.org/10.1016/j.jfca.2009.08.012

Câmara, J. S., Locatelli, M., Pereira, J. A., Oliveira, H., Arlorio, M., Fernandes, I., … & Bordiga, M. (2022). Behind the scenes of anthocyanins—From the health benefits to potential applications in food, pharmaceutical and cosmetic fields. *Nutrients*, 14(23), 5133. https://doi.org/10.3390/nu14235133

Chan, C. F., Lien, C. Y., Lai, Y. C., Huang, C. L., & Liao, W. C. (2010). Influence of purple sweet potato extracts on the UV absorption properties of a cosmetic cream. *Journal of Cosmetic Science*, 61(5), 333–341.

Chen, J., Xu, B., Sun, J., Jiang, X., & Bai, W. (2022). Anthocyanin supplement as a dietary strategy in cancer prevention and management: A comprehensive review. *Critical Reviews in Food Science and Nutrition*, 62(26), 7242–7254. https://doi.org/10.1080/10408398.2021.1913092

Chen, L., Jiang, B., Zhong, C., Guo, J., Zhang, L., Mu, T., … & Bi, X. (2018). Chemoprevention of colorectal cancer by black raspberry anthocyanins involved the modulation of gut microbiota and SFRP2 demethylation. *Carcinogenesis*, 39(3), 471–481. https://doi.org/10.1093/carcin/bgy009

Chen, X. Y., Zhou, J., Luo, L. P., Han, B., Li, F., Chen, J. Y., … & Yu, X. P. (2015). Black rice anthocyanins suppress metastasis of breast cancer cells by targeting RAS/RAF/MAPK pathway. *BioMed Research International*. https://doi.org/10.1155/2015/414250

Chen, Z., Wang, C., Pan, Y., Gao, X., & Chen, H. (2018). Hypoglycemic and hypolipidemic effects of anthocyanins extract from black soybean seed coat in high fat diet and streptozotocin-induced diabetic mice. *Food & Function*, 9(1), 426–439. https://doi.org/10.1039/C7FO00983F

Correia, P., Araújo, P., Ribeiro, C., Oliveira, H., Pereira, A. R., Mateus, N., … & Fernandes, I. (2021). Anthocyanin-related pigments: Natural allies for skin health maintenance and protection. *Antioxidants*, 10(7), 1038. https://doi.org/10.3390/antiox10071038

de la Rosa, L. A., Moreno-Escamilla, J. O., Rodrigo-García, J., & Alvarez-Parrilla, E. (2019). Phenolic compounds. In *Postharvest Physiology and Biochemistry of Fruits and Vegetables* (pp. 253–271). Woodhead Publishing. https://doi.org/10.1016/B978-0-12-813278-4.00012-9

Diaconeasa, Z., Ştirbu, I., Xiao, J., Leopold, N., Ayvaz, Z., Danciu, C., … & Socaciu, C. (2020). Anthocyanins, vibrant color pigments, and their role in skin cancer prevention. *Biomedicines*, 8(9), 336. https://doi.org/10.3390/biomedicines8090336

dos Santos, N. M., Batista, P. B., Batista, A. G., & Júnior, M. R. M. (2019). Current evidence on cognitive improvement and neuroprotection promoted by anthocyanins. *Current Opinion in Food Science*, 26, 71–78. https://doi.org/10.1016/j.cofs.2019.03.008

Ekici, L., Simsek, Z., Ozturk, I., Sagdic, O., & Yetim, H. (2014). Effects of temperature, time, and pH on the stability of anthocyanin extracts: Prediction of total anthocyanin content using nonlinear models. *Food Anal Methods*, 7, 1328–1336. https://doi.org/10.1007/s12161-013-9753-y

Fan, M., Kim, S. A., Choi, Y. J., Tang, Y., Yang, H. P., & Kim, E. K. (2023). Anthocyanin oligomer (grape skin extract) administration improves dry eye disease: A randomised, double-blind, placebo-controlled study. *Clinical & Experimental Ophthalmology*, 51(2), 122–130. https://doi.org/10.1111/ceo.14207

Fragoso, M. F., Romualdo, G. R., Vanderveer, L. A., Franco-Barraza, J., Cukierman, E., Clapper, M. L., … & Barbisan, L. F. (2018). Lyophilized açaí pulp (*Euterpe oleracea* Mart) attenuates colitis-associated colon carcinogenesis while its main anthocyanin has the potential to affect the motility of colon cancer cells. *Food and Chemical Toxicology*, 121, 237–245. https://doi.org/10.1016/j.fct.2018.08.078

Gardener, S. L., Rainey-Smith, S. R., Weinborn, M., Bondonno, C. P., & Martins, R. N. (2021). Intake of products containing anthocyanins, flavanols, and flavanones, and cognitive function: A narrative review. *Frontiers in Aging Neuroscience*, 13, 640381. https://doi.org/10.3389/fnagi.2021.640381

Gitelson, A. A., Merzlyak, M. N., & Chivkunova, O. B. (2007). Optical properties and nondestructive estimation of anthocyanin content in plant leaves. *Photochem Photobiol*, 74, 38–45. https://doi.org/10.1562/0031-8655(2001)0740038opaneo2.0.co2

Gomes, J. V. P., Rigolon, T. C. B., da Silveira Souza, M. S., Alvarez-Leite, J. I., Della Lucia, C. M., Martino, H. S. D., & Rosa, C. D. O. B. (2019). Antiobesity effects of anthocyanins on mitochondrial biogenesis, inflammation, and oxidative stress: A systematic review. *Nutrition*, 66, 192–202. https://doi.org/10.1016/j.nut.2019.05.005

Gonçalves, A. C., Nunes, A. R., Falcão, A., Alves, G., & Silva, L. R. (2021). Dietary effects of anthocyanins in human health: A comprehensive review. *Pharmaceuticals*, 14(7), 690. https://doi.org/10.3390/ph14070690

Grimes, K. L., Stuart, C. M., McCarthy, J. J., Kaur, B., Cantu, E. J., & Forester, S. C. (2018). Enhancing the cancer cell growth inhibitory effects of table grape anthocyanins. *Journal of Food Science*, 83(9), 2369–2374. https://doi.org/10.1111/1750-3841.14294

Hagiwara, A., Yoshino, H., Ichihara, T., Kawabe, M., Tamano, S., Aoki, H., ... & Shirai, T. (2002). Prevention by natural food anthocyanins, purple sweet potato color and red cabbage color, of 2-amino-1-methyl-6-phenylimidazo [4, 5-B] pyridine (phip)-associated colorectal carcinogenesis in rats. *The Journal of Toxicological Sciences*, 27(1), 57–68. https://doi.org/10.2131/jts.27.57

Hair, R., Sakaki, J. R., & Chun, O. K. (2021). Anthocyanins, microbiome, and health benefits in aging. *Molecules*, 26(3), 537. https://doi.org/10.3390/molecules26030537

Hassellund, S. S., Flaa, A., Kjeldsen, S. E., Seljeflot, I., Karlsen, A., Erlund, I., & Rostrup, M. (2013). Effects of anthocyanins on cardiovascular risk factors and inflammation in pre-hypertensive men: A double-blind randomized placebo-controlled crossover study. *Journal of Human Hypertension*, 27(2), 100–106. http://dx.doi.org/10.1038/jhh.2012.4

Herrera-Balandrano, D. D., Chai, Z., Hutabarat, R. P., Beta, T., Feng, J., Ma, K., ... & Huang, W. (2021). Hypoglycemic and hypolipidemic effects of blueberry anthocyanins by AMPK activation: In vitro and in vivo studies. *Redox Biology*, 46, 102100. https://doi.org/10.1016/j.redox.2021.102100

Holton, T. A., & Cornish, E. C. (1995). Genetics and biochemistry of anthocyanin biosynthesis. *The Plant Cell*, 7(7), 1071. https://doi.org/10.2307/3870058

Huo, L., Du, X., Li, X., Liu, S., & Xu, Y. (2021). The emerging role of neural cell-derived exosomes in intercellular communication in health and neurodegenerative diseases. *Frontiers in Neuroscience*, 15, 738442. https://doi.org/10.1105%2Ftpc.7.7.1071

Jing, P., Qian, B., Zhao, S., Qi, X., Ye, L., Giusti, M. M., & Wang, X. (2015). Effect of glycosylation patterns of Chinese eggplant anthocyanins and other derivatives on antioxidant effectiveness in human colon cell lines. *Food Chemistry*, 172, 183–189. https://doi.org/10.1016/j.foodchem.2014.08.100

Jo, K., Bae, G. Y., Cho, K., Park, S. S., Suh, H. J., & Hong, K. B. (2020). An anthocyanin-enriched extract from *Vaccinium uliginosum* improves signs of skin aging in UVB-induced photodamage. *Antioxidants*, 9(9), 844. https://doi.org/10.3390/antiox9090844

Johnson, S. A., Figueroa, A., Navaei, N., Wong, A., Kalfon, R., Ormsbee, L. T., ... & Arjmandi, B. H. (2015). Daily blueberry consumption improves blood pressure and arterial stiffness in post-menopausal women with pre-and stage 1-hypertension: A randomized, double-blind, placebo-controlled clinical trial. *Journal of the Academy of Nutrition and Dietetics*, 115(3), 369–377. https://doi.org/10.1016/j.jand.2014.11.001

Joo, H. K., Choi, S., Lee, Y. R., Lee, E. O., Park, M. S., Park, K. B., ... & Jeon, B. H. (2018). Anthocyanin-rich extract from red Chinese cabbage alleviates vascular inflammation in endothelial cells and Apo E$^{-/-}$ mice. *International Journal of Molecular Sciences*, 19(3), 816. https://doi.org/10.3390/ijms19030816

Joseph, J. A., Shukitt-Hale, B., Denisova, N. A., Bielinski, D., Martin, A., McEwen, J. J., & Bickford, P. C. (1999). Reversals of age-related declines in neuronal signal transduction, cognitive, and motor behavioral deficits with blueberry, spinach, or strawberry dietary supplementation. *Journal of Neuroscience*, 19(18), 8114–8121. https://doi.org/10.1523/JNEUROSCI.19-18-08114.1999

Kalt, W., Hanneken, A., Milbury, P., & Tremblay, F. (2010). Recent research on polyphenolics in vision and eye health. *Journal of Agricultural and Food Chemistry*, 58(7), 4001–4007. https://doi.org/10.1021/jf903038r

Kang, K., Jeung, J. G., Afzal, R., Yang, H., & Hwang, H. B. (2019). Effects of grape skin extract, anthocyanin oligomer, on a murine dry eye model. *Journal of the Korean Society of Food Science and Nutrition*, 48(4), 395–402. https://doi.org/10.3746/jkfn.2019.48.4.395

Khoddami, A., Wilkes, M. A., & Roberts, T. H. (2013). Techniques for analysis of plant phenolic compounds. *Molecules*, 18(2), 2328–2375. https://doi.org/10.3390/molecules18022328

Khoo, H. E., Azlan, A., Tang, S. T., & Lim, S. M. (2017). Anthocyanidins and anthocyanins: Colored pigments as food, pharmaceutical ingredients, and the potential health benefits. *Food & Nutrition Research*, 61(1), 1361779. https://doi.org/10.1080/16546628.2017.1361779.

Kohl, E., Steinbauer, J., Landthaler, M., & Szeimies, R. M. (2011). Skin ageing. *Journal of the European Academy of Dermatology and Venereology*, 25(8), 873–884. https://doi.org/10.1111/j.1468-3083.2010.03963.x.

Kosehira, M., Machida, N., & Kitaichi, N. (2020). A 12-week-long intake of bilberry extract (*Vaccinium myrtillus* L.) improved objective findings of ciliary muscle contraction of the eye: A randomized, double-blind, placebo-controlled, parallel-group comparison trial. *Nutrients*, 12(3), 600. https://doi.org/10.3390/nu12030600

Krga, I., & Milenkovic, D. (2019). Anthocyanins: From sources and bioavailability to cardiovascular-health benefits and molecular mechanisms of action. *Journal of Agricultural and Food Chemistry*, 67(7), 1771–1783. https://doi.org/10.1021/acs.jafc.8b06737.

Kubow, S., Iskandar, M. M., Melgar-Bermudez, E., Sleno, L., Sabally, K., Azadi, B., ... & Zum Felde, T. (2017). Effects of simulated human gastrointestinal digestion of two purple-fleshed potato cultivars on anthocyanin composition and cytotoxicity in colonic cancer and non-tumorigenic cells. Nutrients, 9(9), 953.

Kurter, H., Basbinar, Y., Ellidokuz, H., & Calibasi-Kocal, G. (2023). The role of cyanidin-3-*O*-glucoside in modulating oxaliplatin resistance by reversing mesenchymal phenotype in colorectal cancer. *Nutrients*, 15(22), 4705. https://doi.org/10.3390/nu15224705.

LaFountain, A. M., & Yuan, Y. W. (2021). Repressors of anthocyanin biosynthesis. *New Phytologist*, 231(3), 933–949. https://doi.org/10.1111/nph.17397

Li, B., Wang, L., Bai, W., Chen, W., Chen, F., & Shu, C. (2022). *Anthocyanins: Chemistry, Processing and Bioactivity* (pp. 1–15). Springer Nature. https://doi.org/10.1007/978-981-16-7055-8

Li, J., Zhao, R., Jiang, Y., Xu, Y., Zhao, H., Lyu, X., & Wu, T. (2020). Bilberry anthocyanins improve neuroinflammation and cognitive dysfunction in APP/PSEN1 mice via the CD33/TREM2/TYROBP signaling pathway in microglia. *Food & Function*, 11(2), 1572–1584. https://doi.org/10.1039/C9FO02103E

Li, P., Feng, D., Yang, D., Li, X., Sun, J., Wang, G., ... & Bai, W. (2021). Protective effects of anthocyanins on neurodegenerative diseases. *Trends in Food Science & Technology*, 117, 205–217. https://doi.org/10.1016/j.tifs.2021.05.005

Liang, Z., Liang, H., Guo, Y., & Yang, D. (2021). Cyanidin 3-*O*-galactoside: A natural compound with multiple health benefits. *International Journal of Molecular Sciences*, 22(5), 2261. https://doi.org/10.3390/ijms22052261

Lin, B. W., Gong, C. C., Song, H. F., & Cui, Y. Y. (2017). Effects of anthocyanins on the prevention and treatment of cancer. *British Journal of Pharmacology*, 174(11), 1226–1243. https://doi.org/10.1111/bph.13627

Liu, Y., Song, X., Han, Y., Zhou, F., Zhang, D., Ji, B., ... & Jia, X. (2011). Identification of anthocyanin components of wild Chinese blueberries and amelioration of light-induced retinal damage in pigmented rabbit using whole berries. *Journal of Agricultural and Food Chemistry*, 59(1), 356–363. https://doi.org/10.1021/jf103852s

Ma, Z., Du, B., Li, J., Yang, Y., & Zhu, F. (2021). An insight into anti-inflammatory activities and inflammation related diseases of anthocyanins: A review of both in vivo and in vitro investigations. *International Journal of Molecular Sciences*, 22(20), 11076. https://doi.org/10.3390/ijms222011076

Mattioli, R., Francioso, A., Mosca, L., & Silva, P. (2020). Anthocyanins: A comprehensive review of their chemical properties and health effects on cardiovascular and neurodegenerative diseases. *Molecules*, 25(17), 3809. https://doi.org/10.3390/molecules25173809

McGhie, T. K., Ainge, G. D., Barnett, L. E., Cooney, J. M., & Jensen, D. J. (2003). Anthocyanin glycosides from berry fruit are absorbed and excreted unmetabolized by both humans and rats. *Journal of Agricultural and Food Chemistry*, 51(16), 4539–4548. https://doi.org/10.1021/jf026206w

Miyake, S., Takahashi, N., Sasaki, M., Kobayashi, S., Tsubota, K., & Ozawa, Y. (2012). Vision preservation during retinal inflammation by anthocyanin-rich bilberry extract: Cellular and molecular mechanism. *Laboratory Investigation*, 92(1), 102–109. https://doi.org/10.1038/labinvest.2011.132

Miyazaki, K., Makino, K., Iwadate, E., Deguchi, Y., & Ishikawa, F. (2008). Anthocyanins from purple sweet potato *Ipomoea batatas* cultivar Ayamurasaki suppress the development of atherosclerotic lesions and both enhancements of oxidative stress and soluble vascular cell adhesion molecule-1 in apolipoprotein E-deficient mice. *Journal of Agricultural and Food Chemistry*, 56(23), 11485–11492. https://doi.org/10.1021/jf801876n

Mozos, I., Flangea, C., Vlad, D. C., Gug, C., Mozos, C., Stoian, D., ... & Atanasov, A. G. (2021). Effects of anthocyanins on vascular health. *Biomolecules*, 11(6), 811. https://doi.org/10.3390/biom11060811

Oliveira, H., Fernandes, A., F. Brás, N., Mateus, N., de Freitas, V., & Fernandes, I. (2020). Anthocyanins as antidiabetic agents—In vitro and in silico approaches of preventive and therapeutic effects. *Molecules*, 25(17), 3813. https://doi.org/10.3390/molecules25173813

Papaioanou, M., Chronopoulou, E. G., Ciobotari, G., Efrose, R. C., Sfichi-Duke, L., Chatzikonstantinou, M., ... & Labrou, N. E. (2018). Cosmeceutical properties of two cultivars of red raspberry grown under different conditions. *Cosmetics*, 5(1), 20. https://doi.org/10.3390/cosmetics5010020

Peng, S. Y., Hsiao, C. C., Lan, T. H., Yen, C. Y., Farooqi, A. A., Cheng, C. M., ... & Chang, H. W. (2020). Pomegranate extract inhibits migration and invasion of oral cancer cells by downregulating matrix metalloproteinase-2/9 and epithelial-mesenchymal transition. *Environmental Toxicology*, 35(6), 673–682. https://doi.org/10.1002/tox.22903

Poomanee, W., Wattananapakasem, I., Panjan, W., & Kiattisin, K. (2021). Optimizing anthocyanins extraction and the effect of cold plasma treatment on the anti-aging potential of purple glutinous rice (*Oryza sativa* L.) extract. *Cereal Chemistry*, 98(3), 571–582. https://doi.org/10.1002/cche.10399

Poulose, S. M., Bielinski, D. F., Carey, A., Schauss, A. G., & Shukitt-Hale, B. (2017). Modulation of oxidative stress, inflammation, autophagy, and expression of Nrf2 in hippocampus and frontal cortex of rats fed with açaí-enriched diets. *Nutritional Neuroscience*, 20(5), 305–315. https://doi.org/10.1080/1028415X.2015.1125654

Prior, R. L., & Wu, X. (2006). Anthocyanins: Structural characteristics that result in unique metabolic patterns and biological activities. *Free Radical Research*, 40(10), 1014–1028. https://doi.org/10.1080/10715760600758522

Qian, F., Wang, M., Wang, J., & Lu, C. (2019). Anthocyanin-rich blueberry extract ameliorates the behavioral deficits of MPTP-induced mouse model of Parkinson's disease via anti-oxidative mechanisms. *Yangtze Medicine*, 3(1), 72–78. https://doi.org/10.4236/ym.2019.31008

Ray, S., Barman, A. K., Roy, P. K., & Singh, B. K. (2017). Chicken eggshell powder as dietary calcium source in chocolate cakes. *The Pharma Innovation Journal*, 6(9), 01–04.

Rehman, S. U., Shah, S. A., Ali, T., Chung, J. I., & Kim, M. O. (2017). Anthocyanins reversed D-galactose-induced oxidative stress and neuroinflammation mediated cognitive impairment in adult rats. *Molecular Neurobiology*, 54, 255–271. https://doi.org/10.1007/s12035-015-9604-5

Rose, P. M., Cantrill, V., Benohoud, M., Tidder, A., Rayner, C. M., & Blackburn, R. S. (2018). Application of anthocyanins from blackcurrant (*Ribes nigrum* L.) fruit waste as renewable hair dyes. *Journal of Agricultural and Food Chemistry*, 66(26), 6790–6798. https://doi.org/10.1021/acs.jafc.8b01044

Samota, M. K., Sharma, M., Kaur, K., Yadav, D. K., Pandey, A. K., Tak, Y., ... & Rani, H. (2022). Onion anthocyanins: Extraction, stability, bioavailability, dietary effect, and health implications. *Frontiers in Nutrition*, 9, 917617. https://doi.org/10.3389/fnut.2022.917617

Sekikawa, T., Kizawa, Y., Takeoka, A., Sakiyama, T., Li, Y., & Yamada, T. (2021). The effect of consuming an anthocyanin-containing supplement derived from Bilberry (*Vaccinium myrtillus*) on eye function: A randomized, double-blind, placebo-controlled parallel study. *Functional Foods in Health and Disease*, 11(3), 116–146. https://doi.org/10.31989/ffhd.v11i3.782

Shin, J. W., Kwon, S. H., Choi, J. Y., Na, J. I., Huh, C. H., Choi, H. R., & Park, K. C. (2019). Molecular mechanisms of dermal aging and antiaging approaches. *International Journal of Molecular Sciences*, 20(9), 2126. https://doi.org/10.3390/ijms20092126

Shipp, J., & Abdel-Aal, E. S. M. (2010). Food applications and physiological effects of anthocyanins as functional food ingredients. *The Open Food Science Journal*, 4(1). http://dx.doi.org/10.2174/1874256401004010007

Silva, G. T. M., Silva, K. M., Silva, C. P., Goncalves, J. M., & Quina, F. H. (2020). Hybrid pigments from anthocyanin analogues and synthetic clay minerals. *ACS Omega*, 5(41), 26592–26600. https://doi.org/10.1021/acsomega.0c03354

Suantai, B., Jantakee, K., Kaewkod, T., Sangboonruang, S., Chitov, T., & Tragoolpua, Y. (2022). Anthocyanins in red jasmine rice (*Oryza sativa* L.) Extracts and efficacy on inhibition of herpes simplex virus, free radicals and cancer cell. *Nutrients*, 14(9), 1905. https://doi.org/10.3390/nu14091905

Tabart, J., Auger, C., Kevers, C., Dommes, J., Pollet, B., Defraigne, J. O., … & Pincemail, J. (2018). The potency of commercial blackcurrant juices to induce relaxation in porcine coronary artery rings is not correlated to their antioxidant capacity but to their anthocyanin content. *Nutrition*, 51, 53–59. https://doi.org/10.1016/j.nut.2018.01.009

Tsai, M. C., Chen, C. C., Tseng, T. H., Chang, Y. C., Lin, Y. J., Tsai, I. N., … & Wang, C. J. (2023). Hibiscus anthocyanins extracts induce apoptosis by activating AMP-activated protein kinase in human colorectal cancer cells. *Nutrients*, 15(18), 3972. https://doi.org/10.3390/nu15183972

Ullah, I., Park, H. Y., & Kim, M. O. (2014). Anthocyanins protect against kainic acid-induced excitotoxicity and apoptosis via ROS-activated AMPK pathway in hippocampal neurons. *CNS Neuroscience & Therapeutics*, 20(4), 327–338. https://doi.org/10.1111/cns.12218

Wang, L. S., Hecht, S. S., Carmella, S. G., Yu, N., Larue, B., Henry, C., … & Stoner, G. D. (2009). Anthocyanins in black raspberries prevent esophageal tumors in rats. *Cancer Prevention Research*, 2(1), 84–93. https://doi.org/10.1158/1940-6207.CAPR-08-0155

Wang, L. S., & Stoner, G. D. (2008). Anthocyanins and their role in cancer prevention. *Cancer Letters*, 269(2), 281–290. https://doi.org/10.1016/j.canlet.2008.05.020

Wattanathorn, J., Tong-Un, T., Thukham-Mee, W., Paholpak, P., & Rangseekhajee, P. (2023). A randomized, double-blind, placebo-controlled study of an anthocyanin-rich functional ingredient on cognitive function and eye dryness in late adulthood volunteers: Roles of epigenetic and gut microbiome modulations. *Nutrients*, 15(16), 3499. https://doi.org/10.3390/nu15163499

Winter, A. N., & Bickford, P. C. (2019). Anthocyanins and their metabolites as therapeutic agents for neurodegenerative disease. *Antioxidants*, 8(9), 333. https://doi.org/10.3390/antiox8090333

Wrolstad, R. E. (2004). Symposium 12: Interaction of natural colors with other ingredient, anthocyanin pigments-bioactivity and coloring properties. *Journal of Food Science*, 69, 419–421.

Wu, T., Yu, Z., Tang, Q., Song, H., Gao, Z., Chen, W., & Zheng, X. (2013). Honeysuckle anthocyanin supplementation prevents diet-induced obesity in C57BL/6 mice. *Food & Function*, 4(11), 1654–1661. https://doi.org/10.1039/C3FO60251F

Xia, W., Quan, T., Hammerberg, C., Voorhees, J. J., & Fisher, G. J. (2015). A mouse model of skin aging: Fragmentation of dermal collagen fibrils and reduced fibroblast spreading due to expression of human matrix metalloproteinase-1. *Journal of Dermatological Science*, 78(1), 79–82. https://doi.org/10.1016/j.jdermsci.2015.01.009

Yang, L., Ling, W., Yang, Y., Chen, Y., Tian, Z., Du, Z., … & Yang, L. (2017). Role of purified anthocyanins in improving cardiometabolic risk factors in Chinese men and women with prediabetes or early untreated diabetes—A randomized controlled trial. *Nutrients*, 9(10), 1104. https://doi.org/10.3390/nu9101104

Zhang, C., Wu, W., Zhou, L., Cheng, H., Ye, X., & He, Y. (2020). Developing deep learning based regression approaches for determination of chemical compositions in dry black goji berries (Lycium ruthenicum Murr.) using near-infrared hyperspectral imaging. Food chemistry, 319, 126536.

Zhang, J., Wu, J., Liu, F., Tong, L., Chen, Z., Chen, J., … & Huang, C. (2019). Neuroprotective effects of anthocyanins and its major component cyanidin-3-*O*-glucoside (C3G) in the central nervous system: An outlined review. *European Journal of Pharmacology*, 858, 172500. https://doi.org/10.1016/j.ejphar.2019.172500

Zhao, J., Yu, J., Zhi, Q., Yuan, T., Lei, X., Zeng, K., & Ming, J. (2021). Anti-aging effects of the fermented anthocyanin extracts of purple sweet potato on *Caenorhabditis elegans*. *Food & Function*, 12(24), 12647–12658. https://doi.org/10.1039/D1FO02671B

Zhong, H., Xu, J., Yang, M., Hussain, M., Liu, X., Feng, F., & Guan, R. (2023). Protective effect of anthocyanins against neurodegenerative diseases through the microbial-intestinal-brain axis: A critical review. *Nutrients*, 15(3), 496. https://doi.org/10.3390/nu15030496

7

Anthocyanins as Functional Foods

Semih Ötleş and Gozde Turkoz Bakirci

7.1 Introduction

Anthocyanins are water-soluble plant pigments with a particle molecular structure belonging to the flavonoid class (Mattioli et al., 2020). Although the term anthocyanin was first used in 1835, it has been known that these substances are pH indicators since 1664 (Harborne and Williams, 2001). It is considered one of the most important pigment groups commonly found in plants (Kong et al., 2003). These compounds are water-soluble pigments derived from natural sources and are responsible for most of the red, blue and purple colors of fruits, vegetables, flowers and other plant tissues or products (Mazza et al., 2004).

In terminology (anthos = flower, kyanos = blue), it was originally used to describe the pigment in the blue cornflower (*Centaurea cyanus*) (Pascual-Teresa et al., 2010; Tsuda, 2012). They usually have one or more hydroxyl or methoxy groups at different positions of two benzene rings linked by different sugar moieties (Kay et al., 2017). To date, more than 600 natural anthocyanins have been discovered, varying in the number and location of hydroxyl and methoxy groups, the number and type of sugar moieties, and the extent and type of sugar acylation (Shen et al., 2022).

The fact that antioxidant substances taken with plants create a protective shield against the effects of oxygen and other harmful substances entering the body, which causes cells to deform, increases the interest in such natural products (Etherton et al., 2002). Anthocyanins act as natural antioxidants and protect against free radicals and harmful substances. It prevents molecules such as proteins, DNA and lipids from oxidizing. It reduces the risks of chronic diseases and relieves inflammation. Anthocyanins are also used in industry and cosmetics. It has been found to repair DNA damage, induce apoptosis in tumor cells, normalize lipid profiles, improve vascular endothelial function, reduce platelet reactivity and contribute to the improvement of neurotoxicity. In addition to adding color appeal to foods, they also increase the oxidative stability of the foods to which they are added due to their activities against radical substances (Espin et al., 2000). In addition to the broad biological activity developed through preclinical and clinical studies, the low-temperature properties of anthocyanins make them very safe to consume (Salehi et al., 2020). In recent years, interest in anthocyanins has been increasing due to their positive effects on health.

7.2 Functional Foods

Functional products are defined as products and long-term functional foodstuffs that are offered in one or more forms or have been found to be effective in maintaining or improving the appearance and performance of human beings thanks to their characteristics (Wildman, 2000).

People want to reach healthier and better quality of life depending on their knowledge and behavioral developments. They prefer to take preventive measures instead of treating their health problems in order to increase their life expectancy and quality. Nutritional style and preference are the most important of these preventive measures. While eating, functional foods that protect and improve well-being and reduce the risk of disease are preferred (Roberfroid, 2000).

Probiotics, prebiotics, synbiotics, dietary fibers, essential fatty acids, carotenoids, phenolic acids and flavonoids, phytoestrogens, phytosterols and phytostanols, vitamins and minerals are functional food components.

Probiotics are live microbial food additives that have beneficial effects on host health by regulating the microbial balance of the intestinal system. Today, the definition of probiotic covers all microbial preparations that support (strengthen) human and animal health and are added to food, feed or food additives. Lactic acid bacteria constitute the most important group of probiotic microorganisms. Among these, *Bifidobacterium* and *Lactobacillus* species are the most commonly used probiotic microorganisms. In addition, some bacterial species, yeast and mold species are also used in the preparation of probiotic products (Lee and Salminen, 1995; Billoo et al., 2006; Kim et al., 2006).

Prebiotics are short-chain carbohydrates that cannot be digested by digestive enzymes in humans and are called resistant short-chain carbohydrates. Prebiotics resist digestion in the colon and reach the colon undigested (Li et al., 2008). They are generally nondigestible oligosaccharides that can be dissolved in 80% ethanol. Prebiotics, which are inactive food ingredients, ferment in the colon. Its benefit to the host is due to its contribution to the bacterial flora (Al-Sherajia et al., 2013).

When content or foods contain both prebiotics and probiotics, they are called synbiotics (Gibson and Roberfroid, 1995). Dietary fiber is defined as carbohydrate polymers having ten or more monomeric units that are not hydrolyzed by endogenous enzymes in the small intestine of humans (Codex Alimentarius, 2010). Essential fatty acids are fatty acids that the body cannot produce and must be obtained through food (Harris et al., 2007).

Carotenoids, phenolic acids, flavonoids and phytosterols are phytochemicals that contain disease-preventing and protective compounds. Fruits and vegetables are rich sources of nutrients and phytochemicals. Some agents in the phytochemical group, such as polyphenols (flavonoids, anthocyanins), isoprenoids, soluble compounds, soluble and insoluble fibers, are used in humans. It helps fight cancer and some cardiovascular chronic diseases. Research shows that fruits and vegetables for a healthy life protect from the mentioned diseases and positively promote a healthy life (Oke and Paliyath, 2012). Anthocyanins are used in functional foods due to their antioxidant properties.

7.3 Anthocyanins

Anthocyanins are members of the flavonoid group of phytochemicals. Cyanidin, pelargonidin, petunidin are examples of anthocyanins (Skibola and Smith, 2000). The colorful anthocyanins are the members of the bioflavonoid phytochemicals and most recognized, or visible (Barros et al., 2016). Naturally occurring anthocyanins are in their glycosylated form and have sugar moieties such as glucose, arabinose, rhamnose, xylose or galactose conjugated to the anthocyanidin backbone (Huang and Zhou, 2019). Anthocyanins are largely responsible for the attractive pale yellow, orange, red, magenta, violet and blue color of a wide variety of plant tissues, especially flowers, fruit, storage organs, stems and grains (Chemler et al., 2009; Martín et al., 2017).

It is estimated that there are more than 600 anthocyanins in nature (Clifford, 2000). Factors such as the number of hydroxyl and hydrogen groups in the anthocyanin molecule, the location of these groups, the type, number and bonding style of sugars bonded to the molecule, the structure and number of aliphatic and aromatic acids bonded to these sugars reveal the different properties among anthocyanins (Eiro and Heinonen, 2002).

The concentration of the pigment is very important on determining the color expressed by anthocyanin. In lesser concentration, the color is blue. Meanwhile in higher concentration, the color is red. In medium concentration, the color is purple. The increase of the hydroxyl group tends to strengthen the color to be more bluish (Farahmandazad, 2015). Six anthocyanins found frequently in plants are pelargonidin, cyanidin, peonidin, delphinidin, petunidin and malvidin (Giusti and Wrolstad, 2003). Orange and pink colors are from pelargonidin individuals, red colors are from cyanidin individuals and purple and blue colors are from delphinidin individuals (Harborne and Williams, 2001).

Anthocyanins have higher antioxidant capacities (ACs) with free radical scavenging potential than other antioxidants. Anthocyanins have also been observed to reduce the risk of chronic disease by increasing antioxidant defenses and modulating antioxidant and inflammatory signaling pathways (Yang et al., 2011). Various studies have found that anthocyanin intake prevents the rise of blood sugar levels and increases insulin sensitivity in a Type 2 diabetes model (Prior et al., 2008).

Cardiovascular disorders, the largest cause of death in the world, are a class of diseases that affect the heart and blood vessels. Disease progression is closely related to platelet aggregation, hypertension, high LDL cholesterol concentration and vascular endothelial dysfunction. Dietary antioxidants, including anthocyanins, play a potential role in preventing or partially reversing these pathological conditions. Many studies have been done in this field (Hassellund et al., 2013; Wallace et al., 2016).

The preventive effects of anthocyanins on body fat accumulation were first reported in 2003 (Tsuda et al., 2003). To date, most studies evaluating the antiobesity role of anthocyanins have relied on anthocyanin-rich crude extracts from plant sources, and it has not been fully defined which molecular structures of anthocyanins are responsible for antiobesity (Tsuda, 2012).

Studies have shown that anthocyanins have effects on glucose modulation and insulin resistance associated with metabolic syndrome (Yamauchi et al., 2001; Russo et al., 2005; Shih et al., 2010).

The health benefits of anthocyanins on visual functions were one of the first health properties attributed to them (Ghosh and Konishi, 2007). Some in vitro and in vivo studies have been conducted focusing on food-derived anthocyanins regarding their effects on visual function (Smeriglio et al., 2016). Antioxidant effect is one of the physiological functions of anthocyanins that have been identified in studies. In studies, this antioxidant effect was observed in the retinal pigment epithelium (Milbury et al., 2007).

Inflammation is a complex biological response to tissue damage, associated with the onset, development and progression of cancer or tumors, provided there is a suitable microenvironment. Flavonoids have been found to have immunomodulatory properties in vitro. Anthocyanins in particular have been reported to have a number of anti-inflammatory effects (Miller et al., 2013).

It has been observed that anthocyanin-rich foods may have beneficial effects against cognitive decline and age-related neurodegeneration and may modulate neuronal functions (Nakaishi et al., 2000).

Anthocyanins have been found to support various cell cycle stages, affecting regulatory proteins and selectively inhibiting cancer cell growth (Smeriglio et al., 2016).

7.4 Anthocyanin-Rich Foods

Important sources of anthocyanins include black grapes, red cabbage, black carrots, currants, cherries, purple asparagus, strawberries, eggplant, purple cabbage, onions, red radishes and sweet potatoes, purple corn, elderberries and blueberries. The high anthocyanin content found in fruits shows that a daily intake can be achieved by regular consumption of fruits and fruit-derived drinks (Wu et al., 2006). To date, more than 90% of anthocyanins isolated in nature are based on six anthocyanidins: pelargodin, cyanidin, peonidin, delphinidin, petunidin and malvidin (Andersen and Jordheim, 2010).

Grapes have been consumed as dessert around the world for 2000 years and have a wide range of natural properties (Vouillamoz and Grando, 2006). As a taxonomic classification, it can be expressed as species (*Vinifera*) belonging to the Vitaceae family and the *Vitis* genus, in the *Rhamnales* order, under the Thalamiflorae group. A lot of research has been done on its fruits since 1950 (Yadav et al., 2009). Grapes contain phenolic compounds with very different structures, such as phenolic acids and polymeric proanthocyanidins. Grape seeds are rich in procyanidins, known as polymers of catechins. Grape skin and juice contain anthocyanin and phenolic acids, which are responsible for its color. These bioactive compounds vary depending on the grape variety, growing conditions and harvest time (Yu et al., 2005). Anthocyanin peonidin 3-glucoside, cyanidin 3-glucoside, malvidin 3-glucoside, petunidin 3-glucoside and delphinidin 3-glucoside compounds have been detected in four red (Red Globe, Flame Seedless, Crimson Seedless and Napoleon) and three white (Superior Seedless, Dominga and Moscatel Italica) grape varieties (Cantos et al., 2002). Five anthocyanin-3-O-glucosides named malvidin-3-O-glucoside, peonidin-3-O-glucoside and cyanidin-3-O-glucoside were isolated by extraction of grapes (Chiou et al., 2014).

Onion and onion peel contain high amounts of flavonols with antioxidant properties (Albishi et al., 2013). There are two main flavonoids found in onions: anthocyanins/derivatives, some of which give a red/purple color, and flavonols, which are quercetin/derivatives which are found in all onions and give a yellow/brown color. Benítez et al. (2011) and Zhang et al. (2014) examined the anthocyanin content in red onion and determined that it predominantly contained delphinidin 3,5-diglucoside, cyanidin 3,5-diglycoside and cyanidin 3-glycoside. It has been stated that the amounts of delphinidin 3-glycosyl glycoside, cyanidin 3-(6″-malonyl glucoside) and cyanidin 3-(6″-malonyl-3″glycosylglycoside) in the peel of the red onion are much higher than in the inner part and contain more cyanidin derivatives (Gennaro et al., 2002).

Generally, 54 phenolic compounds have been identified in black elderberry species. The predominant polyphenols in black elderberry are anthocyanins. Anthocyanins are found in black elderberry fruits in the form of cyanidin glycosides. The main anthocyanins in black elderberry fruits are cyanidin-3-sambubioside, cyanidin-3,5 diglycoside, cyanidin-3-sambubioyl-5-glycoside, cyanidin-3-O-glycoside and cyanidin-3-(E)-pcoumaroyl-sambubioside-5-glycoside (Mikulic-Petkovsek et al., 2015). Cyanidin-3-O-glycoside is a compound with anticancer activity (Marczylo et al., 2009). In addition to cyanidin glycosides, quercetin and caffeic acid derivatives have also been detected in black elderberry fruits. Although black elderberry extracts have an inhibitory effect on microbial activity, they also have antiradical activity with their anthocyanin content (Pliszka, 2017).

Pomegranate fruit is one of the foods with high AC. There are many components in pomegranate fruit that positively affect human health, such as phenolic substances, vitamins, minerals, organic acids and anthocyanins (Viuda-Martos et al., 2010; Shirode et al., 2014). With these features, pomegranate fruit is effective in protecting against many common diseases such as cancer, cardiovascular diseases and diabetes (Mphahlele et al., 2016). It is also useful in the treatment of bacterial/fungal/parasitic infections, periodontal diseases, food poisoning, ulcers, diarrhea, dysentery and hemorrhoids (Lee et al., 2010). Some of the anthocyanins detected in pomegranate juice are cyanidin-3-O-glucoside, cyanidin-3,5-di-O-glucoside, delphinidin-3-O-glucoside, delphinidin-3,5-di-O-glucoside, pelargonidin-3-O-glucoside and pelargonidin-3,5-di-O-glucoside (Lansky and Newman, 2007). It has been reported that consumption of pomegranate juice reduces LDL cholesterol in humans by preventing lipoprotein oxidation and prevents atherosclerosis by reducing platelet activation. It has been stated that these positive effects of pomegranate juice are due to its high AC and the polyphenolic compounds it contains (Aviram et al., 2000).

Anthocyanins are pigments naturally found in eggplant. Agronomic and genetic factors are concentrated within the fruit peel due to the intensity and type of light, temperature, processing and storage (Taher et al., 2020). Delphinidin glucosides (delphinidin anthocyanidin derivatives) are one of the major anthocyanins of eggplant peel and give a dark purple color. Some studies have shown that extracts from eggplant fruits have excellent therapeutic effects on warts, burns and many inflammatory diseases such as stomatitis, arthritis and gastritis. There are several studies proving that eggplant's high fiber content helps the body get rid of waste materials and harmful toxins, thus reducing the risk of colon and stomach cancer (Rodriguez-Jimenez et al., 2018).

Flavonoids in sorghum include quercetin, luteolin, apigenin, catechin, taxifolin, glycitein, naringenin, ononin and hispidulin. In addition, 3-deoxyanthocyanidin is an anthocyanin specific to sorghum and is found mostly in its colorful species (Irondi et al., 2019; Ofosu et al., 2020).

There are many studies showing the anti-inflammatory effects of *Brassica* vegetables due to their phytochemical properties. Anthocyanins, among the phytochemicals they contain, may have a wide range of biological activities, including antioxidant properties, anti-tumor activity, cardio- and hepato-protective effects, glucose control in diabetic individuals, reducing symptoms of gastrointestinal system diseases such as gastritis, ulcers, irritable bowel syndrome, immunomodulation and effects on aging, is stated (Kapusta-Duch et al., 2012). Many studies have reported that *Brassica* anthocyanins have high AC and thus reduce oxidative stress and oxidative damage (Wiczkowski et al., 2013).

It is stated that red cabbage has many health benefits, especially with the anthocyanin it contains, such as lowering cholesterol, protecting against cardiovascular system diseases, helping in weight loss, protecting against cancer and improving cognitive function (Zhang and Jing, 2020). Purple asparagus (*Asparagus officinalis* L.) contains more anthocyanins than green and white asparagus (Sonoda et al., 2005).

It is stated that the addition of anthocyanin improves the color of processed foods and provides an alternative to synthetic colorants. It has been determined that adding anthocyanin-rich extracts to bakery products, especially cookies, biscuits and macaroons, can protect the food from damage that may occur during cooking without affecting the acceptability of the food, and its AC may be greater than that of additives. In addition, products such as kefir and yogurt can be considered as foods suitable for anthocyanin supplementation, as anthocyanins show high stability during storage. However, the produced functional foods have positive effects on health thanks to their rich anthocyanin content (Echegaray et al., 2020).

Biscuits are a good choice for functional combination as biscuit is one of the popular, frequently consumed foods with a long shelf life. Since purple pigments are found in the outer layers of the pericarp, biscuit making experiments have been made with whole wheat flour. Many functional biscuits are formulated with antioxidant and/or prebiotic properties (Pasqualone et al., 2015). From the point of view of anthocyanin, the addition of the extract obtained from Italian black rice, rich in biscuits and cookies, is highly technological and the strengthening of its phenolic fortified content has been observed. In addition, the addition of anthocyanin obtained from different sources has made the color of foods such as cookies, wafers and pasta more desirable. Products with high AC in red or purple color such as kefir, yoghurt, cream, grape skins, black carrots, blueberries and red cabbage have been obtained (Echegaray et al., 2020). An artistic, functional baking product made from traditional Chinese bread, which is obtained from more sweet potatoes, has been achieved as it is suitable for launch (Zhu and Sun, 2019).

Due to their antioxidant properties, anthocyanins not only color the food product they are added to but also help protect it against spoilage. Anthocyanins are used as dyestuffs to color various foods due to their high dyeing power. Since anthocyanins are water-soluble, they can be easily added to aqueous food systems. It has been determined that anthocyanins not only provide attractive color properties to foods but also protect them against spoilage by increasing the oxidative stability of the foods they are added to, due to their high AC (Katsube et al., 2006; Palonen and Weber, 2019).

Although anthocyanins are among the best natural food dyes, their use and prevalence are not increasing sufficiently because they are difficult to obtain and do not have a stable structure. Anthocyanins can be used to impart red color in ready-to-drink beverages. For this purpose, the use of anthocyanins in very low concentrations is sufficient. They can also be used as colorants in fruit yoghurts and candies. Since anthocyanins are natural coloring substances, they are seen as an important alternative to synthetic dyes in the coloring of many foods (Khoo et al., 2017).

References

Albishi T, John JA, Al-Khalifa AS, Shahidi F. Antioxidative phenolic constituents of skins of onion varieties and their activities. Journal of Functional Foods. 2013; 5: 1191–1203.

Codex Alimentarius. Guidelines on nutrition labelling CAC/GL 2-1985 as last amended 2010. Joint FAO/WHO Food Standards Programme, Secretariat of the Codex Alimentarius Commission 2010; FAO, Rome.

Al-Sherajia SH, Ismail A, Manap MY, Mustafa S, Yusof RM, Hassan FA. Prebiotics as functional foods: A review. Journal of Functional Foods. 2013; 5: 1542–1553.

Andersen MO, Jordheim M. Anthocyanins. 2010; Chichester: John Wiley & Sons.

Aviram M, Rosenblat M, Volkova N, Kaplan M, Coleman R, Hayek T, Presser D, Fuhrman B. Pomegranate juice consumption reduces oxidative stress, atherogenic modifications to LDL, and platelet aggregation: Studies in humans and in atherosclerotic apolipoprotein E-deficient mice. The American Journal of Clinical Nutrition. 2000; 71: 1062–1076.

Barros J, Serrani-Yarce JC, Chen F, Baxter D, Venables BJ, Dixon RA. Role of bifunctional ammonia-lyase in grass cell wall biosynthesis. Nature Plants. 2016; 2(6): 16050.

Benítez V, Mollá E, Martín-Cabrejas MA, Aguilera Y, López-Andréu FJ, Cools K, Terry LA, Esteban RM. Characterization of industrial onion wastes (*Allium cepa* L.): Dietary fibre and bioactive compounds. Plant Foods for Human Nutrition. 2011; 66(1):48–57. doi: 10.1007/s11130-011-0212-x. PMID: 21318305.

Billoo AG, Memon MA, Khaskheli SA, Murtaza G, Iqbal K, Shekhani MS, Siddiqi AQ. Role of a probiotic (*Saccharomyces boulardii*) in management and prevention of diarrhoea. World Journal of Gastroenterology. 2006; 12: 4557–4560.

Cantos E, Espin JC, Tomás-Barberán FA. Varietal differences among the polyphenol profiles of seven table grape cultivars studied by LC–DAD–MS–MS. Journal of Agricultural and Food Chemistry. 2002; 50(20): 5691–5696.

Chemler JA, Leonard E, Koffas MAG Flavonoid Bio Transformations in Microorganisms. In Anthocyanins: Biosynthesis, Functions, and Applications. Edited by Gould K, Davies K, Winefield C, Springer. 2009; 191–238.

Chiou A, Panagopoulou EA, Gatzali F, De Marchi S, Karathanos VT. Anthocyanins content and antioxidant capacity of Corinthian currants (*Vitis vinifera* L., var. Apyrena). Food Chemistry. 2014; 146: 157–165.

Clifford MN. Review: Anthocyanins-nature, occurrence and dietary burden. Journal of the Science of Food and Agriculture. 2000; 80: 1063–1072.

Echegaray N, Munekata PES, Gullón P, Dzuvor CKO, Gullón B, Kubi F, Lorenzo JM. Recent advances in food products fortification with anthocyanins. Critical Reviews in Food Science and Nutrition. 2020;1–15. 10.1080/10408398.2020.1844141

Eiro MJ, Heinonen M. Anthocyanin color behavior and stability during storage: Effect of intermolecular copigmentation. Journal of Agricultural and Food Chemistry. 2002; 50: 7461–7466.

Espin JC, Soler-Rivas C, Wichers H, Garcia-Viguera C. Anthocyanin based natural colorants: A new source of foodstuff. Journal of Agricultural and Food Chemistry. 2000; 48(5): 1588–1592.

Etherton PMK, Hecker KD, Bonanome A, Coval SM, Binkoski AE, Hilpert KF, Griel AE, Etherton TD. Bioactive compounds in foods: Their role in the prevention of cardiovascular disease and cancer. The American Journal of Medicine. 2002; 113: 71–85.

Farahmandazad H. Recovery and Purification of Anthocyanins from Purple-Blue Potato. [Thesis]. 2015; Lappeenranta: Lappeenranta University of Technology.

Gennaro L, Leonardi C, Esposito F, Salucci M, Maiani G, Quaglia G, Fogliano V. Flavonoid and carbohydrate contents in tropea red onions: Effects of homelike peeling and storage. Journal of Agricultural and Food Chemistry. 2002; 50(7): 1904–1910.

Ghosh D, Konishi T. Anthocyanins and anthocyanin-rich extracts: Role in diabetes and eye function. Asia Pacific Journal of Clinical Nutrition. 2007; 16: 200–208.

Gibson GR, Roberfroid MB. Dietary modulation of the human colonic microbiota. Introducing the concept of prebiotics. Journal of Nutrition. 1995; 125: 1401–1412.

Harborne JB, Williams CA. Anthocyanins and other flavonoids. Natural Product Reports. 2001; 18: 310–333.

Harris WS, Miller M, Tighe AP, Davidson MH, Schaefer EJ. Omega-3 fatty acids and coronary heart disease risk: Clinical and mechanistic perspectives. Atherosclerosis. 2007; 197: 12–24.

Hassellund SS, Flaa A, Kjeldsen SE, et al. Effects of anthocyanins on cardiovascular risk factors and inflammation in pre-hypertensive men: A double-blind randomized placebo-controlled crossover study. Journal of Human Hypertension. 2013; 27(2): 100–106.

Huang Y, Zhou W. Microencapsulation of anthocyanins through two-step emulsification and release characteristics during in vitro digestion. Food Chemistry. 2019; 278: 357–363.

Irondi EA, Adegoke BM, Effion ES, Oyewo SO, Alamu EO, Boligon AA. Enzymes inhibitory property, antioxidant activity and phenolics profile of raw and roasted red sorghum grains in vitro. Food Science and Human Wellness. 2019; 8(2): 142–148.

Kapusta-Duch J, Kopec A, Piatkowska E, Borczak B, Leszczynska T. The beneficial effects of *Brassica* vegetables on human health. Roczniki Państwowego Zakladu Higieny. 2012; 63: 389–395.

Katsube T, Imawaka N, Kawano Y, Yamazaki Y. antioxidant flavonol glycosides in mulberry (*Morus alba* L.) leaves isolated based on LDL antioxidant activity. Food Chemistry. 2006; 97: 25–31.

Kay CD, Pereira-Caro G, Ludwig IA, Clifford MN, Crozier A. Anthocyanins and flavanones are more bioavailable than previously perceived: A review of recent evidence. Annual Review of Food Science and Technology. 2017; 8: 155–180.

Khoo HE, Azlan A, Tang ST, Lim SM. Anthocyanidins and anthocyanins: Colored pigments as food, pharmaceutical ingredients, and the potential health benefits. Food & Nutrition Research. 2017; 61: 1361779.

Kim HS, Park H, Cho IY, Paik HD, Park E. Dietary supplementation of probiotic Bacillus polyfermenticus, Bispan strain, modulates natural killer cell and T cell subset populations and immunoglobulin g levels in human subjects. Journal of Medicinal Food. 2006; 9: 321–327.

Kong JM, Chia LS, Goh NK, Chia TF, Brouillard R. Analysis and biological activities of anthocyanins. Phytochemistry. 2003; 64(5): 923–933.

Lansky EP, Newman RA. Punica granatum (pomegranate) and its potential for prevention and treatment of inflammation and cancer. Journal of Ethnopharmacology. 2007; 109: 177–206.

Lee CJ, Chen LG, Liang WL, Wanga CC. Anti-inflammatory effects of *Punica granatum* Linne in vitro and in vivo. Food Chemistry. 2010; 118: 315–322.

Lee YK, Salminen S. The coming age of preobiotics. *Trends in Food Science and Technology*. 1995; 6(7): 241–245.

Li D, Kim JM, Jin Z, Zhou J. Prebiotic effectiveness of inulin extracted from edible burdock. Anaerobe. 2008; 14: 29–34.

Marczylo TH, Cooke D, Brown K, Steward WP, Gescher AJ. Pharmacokinetics and metabolism of the putative cancer chemopreventive agent cyanidin-3-glucoside in mice. Cancer Chemotherapy and Pharmacology. 2009; 64(6): 1261–1268.

Martín J, Navas MJ, Jiménez-Moreno AM, Asuero AG. Anthocyanin pigments: Importance, sample preparation and extraction. Phenolic Compounds-Natural Sources, Importance and Applications. 2017; 117–152.

Mattioli R, Francioso A, Mosca L, Silva P. Anthocyanins: A comprehensive review of their chemical properties and health effects on cardiovascular and neurodegenerative diseases. Molecules. 2020; 25(17): 3809.

Mazza G, Cacace J, Kay C. Methods of analysis for anthocyanins in plants and biological fluids. Journal of AOAC International. 2004; 87(1): 129–145.

Mikulic-Petkovsek M, Ivancic A, Todorovic B, Veberic R, Stampar F. Fruit phenolic composition of different elderberry species and hybrids. Journal of Food Science. 2015; 80: 2180–2190.

Milbury PE, Graf B, Curran-Celentano JM, et al. Bilberry (*Vaccinium myrtillus*) anthocyanins modulate heme oxygenase1 and glutathione S-transferase-pi expression in ARPE-19 cells. Investigative Ophthalmology & Visual Science. 2007; 48(5): 2343–2349.

Miller R, Putnam S, Edwards M. et al. Potential Health Benefits of Blackcurrants: Bioactives in Fruit, Health Benefits and Functional Foods, 1st ed. New Jersey: John Wiley & Sons; 2013: 215–250.

Mphahlele RR, Fawole OA, Mokwena LM, Opara UL. Effect of extraction method on chemical, volatile composition and antioxidant properties of pomegranate juice. South African Journal of Botany. 2016; 103: 135–144.

Nakaishi H, Matsumoto H, Tominaga S, et al. Effects of blackcurrant anthocyanosides intake on dark adaptation and VDT work-induced transient refractive alternation in healthy humans. Alternative Medicine Review. 2000; 5: 553–562.

Ofosu FK, Elahi F, Daliri EBM, Tyagi A, Chen XQ, Chelliah R, … Oh DH. UHPLC-ESI-QTOF-MS/MS characterization, antioxidant and antidiabetic properties of sorghum grains. Food Chemistry. 2020; 337: 127788.

Oke M, Paliyath G. Fertilizer Application and Nutraceutical Content in Health-Functional Foods. In Fertilizing Crops to Improve Human Health: a Scientific Review Vol:2, Functional Foods. Edited by T.W. Bruulsema, P. Heffer, R. M. Welch, I., Çakmak, K., Moran, IPNI, Norcross, GA; IFA, Paris, France, July, 2012; 175p

Palonen P, Weber C. Fruit color stability, anthocyanin content, and shelf life were not correlated with ethylene production rate in five primocane raspberry genotypes. Scientia Horticulturae. 2019; 247: 9–16.

Pascual-Teresa D, Moreno DA, García-Viguera C. Flavanols and anthocyanins in cardiovascular health: A review of current evidence. International Journal of Molecular Sciences. 2010; 11(4): 1679–1703.

Pasqualone A, Bianco A, Paradiso V, Summo C, Gambacorta G, Caponio F, Blanco A. Production and characterization of functional biscuits obtained from purple wheat. Food Chemistry. 2015; 180, 64–70. doi: 10.1016/j.foodchem.2015.02.025

Pliszka B. Polyphenolic content, antiradical activity, stability and microbiological quality of elderberry (*Sambucus nigra* L.) extracts. Acta Scientiarum Polonorum Technologia Alimentaria. 2017; 16(4): 393–401.

Prior RL, Wu X, Gu L, et al. Whole berries versus berry anthocyanins: Interactions with dietary fat levels in the C57BL/6J mouse model of obesity. Journal of Agricultural and Food Chemistry. 2008; 56: 647–653.

Roberfroid MB. A European consensus of scientific concepts of functional foods. Nutrition. 2000; 16: 689–691.

Rodriguez-Jimenez JR, Amaya-Guerra CA, Baez-Gonzalez JG, Aguilera-Gonzalez C, Urias-Orona V, Nino-Medina G. Physicochemical, functional, and nutraceutical properties of eggplant flours obtained by different drying methods. Molcules. 2018; 23(12): 3210.

Russo A, La Fauci L, Acquaviva R, et al. Ochratoxin A-induced DNA damage in human fibroblast: Protective effect of cyanidin 3-O-beta-d-glucoside. Journal of Nutritional Biochemistry. 2005; 16(1): 31–37.

S.L. Lactic acid bacteria in health and disease. In: Lactic Acid Bacteria: Microbiology and Functional Aspects. Edited by Salminen S., von Wright, A. 2nd ed. New York: Marcel Dekker Inc. 1998; pp. 211–254.

Salehi B, et al., The therapeutic potential of anthocyanins: Current approaches based on their molecular mechanism of action. Frontiers in Pharmacology. 2020; 11: 1300.

Shen Y, Zhang N, Tian J, Xin G, Liu L, Sun X, Li B. Advanced approaches for improving bioavailability and controlled release of anthocyanins. Journal of Controlled Release. 2022; 341: 285–299.

Shih PH, Chan YC, Liao JW, et al. Antioxidant and cognitive promotion effects of anthocyanin-rich mulberry (*Morus atropurpurea* L.) on senescence-accelerated mice and prevention of Alzheimer's disease. Journal of Nutritional Biochemistry. 2010; 21(7): 598–605

Shirode AB, Kovvuru P, Chittur SV, Henning SM, Heber D, Reliene R. Antiproliferative effects of pomegranate extract in MCF-7 breast cancer cells are associated with reduced DNA repair gene expression and induction of double strand breaks. Molecular Carcinogenesis. 2014 Jun;53(6):458–70. doi: 10.1002/mc.21995. Epub 2013 Jan 28. PMID: 23359482.

Skibola C, Smith M. Potential health impacts of excessive flavonoid intake. Free Radical Biology and Medicine. 2000; 29(3-4): 375–383.

Smeriglio A, Barreca D, Bellocco E. Chemistry, pharmacology and health benefits of anthocyanins. Phytotherapy Research. 2016; 30(8): 1265–1286.

Sonoda T, Motoki S, Maekawa K, Suzuki T, Oosawa K, Maeda T, Kakuta H. Differences in antioxidative polyphenols contents of asparagus related to cultivars and seasonal change under various cultural conditions of the mother-fern culture. In XI International Asparagus Symposium. 2005; 776 (pp. 227–234).

Taher D, Solberg S, Prohns J, Chou YY, Rakha M, Wu T. World vegetable center eggplant collection: Origin, composition, seed dissemination and utilization in breeding. Frontiers in Plant Science. 2020; 8: 1484.

Tsuda T, Horio F, Uchida K, et al. Dietary cyanidin 3-o-beta-dglucoside-rich purple corn color prevents obesity and ameliorates hyperglycemia in mice. Journal of Nutrition. 2003; 133(7): 2125–2130.

Tsuda T. Dietary anthocyanin-rich plants: Biochemical basis and recent progress in health benefits studies. Molecular Nutrition & Food Research. 2012; 56: 159–170.

Viuda-Martos M, Fernandez-Lopez J, PerezAlvarez JA. Pomegranate and its many functional components as related to human health: A review. Comprehensive Reviews in Food Science and Food Safety. 2010; 9: 635–654.

Vouillamoz JF, Grando MS. Genealogy of wine grape cultivars: 'Pinot' is related to 'Syrah. Heredity (Edinb). 2006; 97(2): 102.

Wallace TC, Slavin M, Frankenfeld CL. Systematic review of anthocyanins and markers of cardiovascular disease. Nutrients. 2016; 8(1): 32–45.

Wiczkowski W, Szawara-Nowak D, Topolska J. Red cabbage anthocyanins: Profile, isolation, identification, and antioxidant activity. Food Research International. 2013; 51, 303–309. DOI: https://doi.org/10.1016/j.foodres.2012.12.015

Wildman RE. C. Nutraceuticals: A Brief Review of Historical and Teleological Aspects. In: Handbook of Nutraceuticals and Functional Foods. Edited by Wildman REC. Boca Raton, FL: CRC Press; 2000.

Wu X, Beecher GR, Holden JM, et al. Concentrations of anthocyanins in common foods in the United States and estimation of normal consumption. Journal of Agricultural and Food Chemistry. 2006; 54(11): 4069–4075.

Yadav M, Jain S, Bhardwaj A, Nagpal R, Puniya M, Tomar R, Singh V, Parkash O, Prasad TR, Marotta F, Yadav H. Biological and medicinal properties of grapes and their bioactive constituents: An update. Journal of Medicinal Food. 2009; 12(3): 473–484.

Yamauchi T, Kamon J, Waki H, et al. The fat-derived hormone adiponectin reverses insulin resistance associated with both lipoatrophy and obesity. Nature Medicine. 2001; 7: 941–946.

Yang M, Koo SI, Song WO, et al. Food matrix affecting anthocyanin bioavailability. Current Medicinal Chemistry. 2011; 18(2): 291–300.

Yu J, Ahmedna M, Goktepe İ. Effects of processing methods and extraction solvents on concentration and antioxidant activity of peanut skin phenolics. Food Chemistry. 2005; 90: 199–206.

Zhang N, Jing P. Anthocyanins in Brassicaceae: Composition, stability, bioavailability, and potential health benefits. Critical Reviews in Food Science and Nutrition. 2022; 62(8): 2205–2220. doi: 10.1080/10408398.2020.1852170. Epub 2020 Dec 1. PMID: 33256437.

Zhang X, Chen F, Wang M. Antioxidant and antiglycation activity of selected dietary polyphenols in a cookie model. Journal of Agricultural and Food Chemistry. 2014; 62(7): 1643–1648.

Zhu F, Sun J. Physicochemical and sensory properties of steamed bread fortified with purple sweet potato flour. Food Bioscience. 2019; 30, 100411. doi: 10.1016/j.fbio.2019.04.012

Section II

Extraction Methods and Analysis of Anthocyanins

8

Extraction and Purification of Anthocyanins: A Review

Jiaqi Tan, Yanmei Han, Bo Han, Xiangmei Qi,
Xu Cai, Shaoqin Ge, and Hongkun Xue

8.1 Introduction

Anthocyanins are a kind of flavonoid natural water-soluble pigments widely existing in vacuoles of plant flowers, fruits, stems, and leaves [1]. Anthocyanins exhibit potential health benefits owing to their particular chemical structure and special fragrance. Extensive studies indicate that anthocyanins exhibited a variety of biological activities such as free radical scavenging [2], antibacterial [3], inhibition of cancer cell increment [4], and protection of eyesight [5]. Moreover, low toxicity and high safety of anthocyanins were also confirmed using *in vitro* and *in vivo* studies [6]. Consequently, anthocyanins are widely used in food, medicine, cosmetics, and other fields.

Presently, the extraction of anthocyanins mostly comes from fruits, vegetables, colored grains, by-products of fruit, and vegetable processing [7]. Fruit residues, such as grape, mulberry, and raspberry, are rich in anthocyanins. China is a large fruit-producing country with an annual output of 24 million tons and nearly 10 million tons of fruit residues [8]. There are abundant anthocyanins resources in fruit and residues. The high-value utilization of fruit residue waste is a hot spot in food processing and development. Therefore, utilizing fruit residues to extract anthocyanins can not only improve the comprehensive utilization rate of fruit residues, solve the problem of industrial waste treatment, and reduce the pressure of environmental protection but also obtain high value-added products to improve their economic benefits.

According to existing literature, anthocyanins can be used as functional active ingredients and therapeutic agents for disease prevention and treatment [9, 10]. Efficient extraction and purification of anthocyanins are the premise of its application, and Figure 8.1 summarizes different extraction and purification methods of anthocyanins. Therefore, the research progress of extraction and purification and future prospects of anthocyanins must be systematically reviewed to better understand anthocyanins from natural resources. Herein, the aim of this review was to present the research progress of anthocyanins extraction and purification and discuss the advantages and disadvantages of each method. These findings provide an important scientific basis for the further development and utilization of anthocyanins.

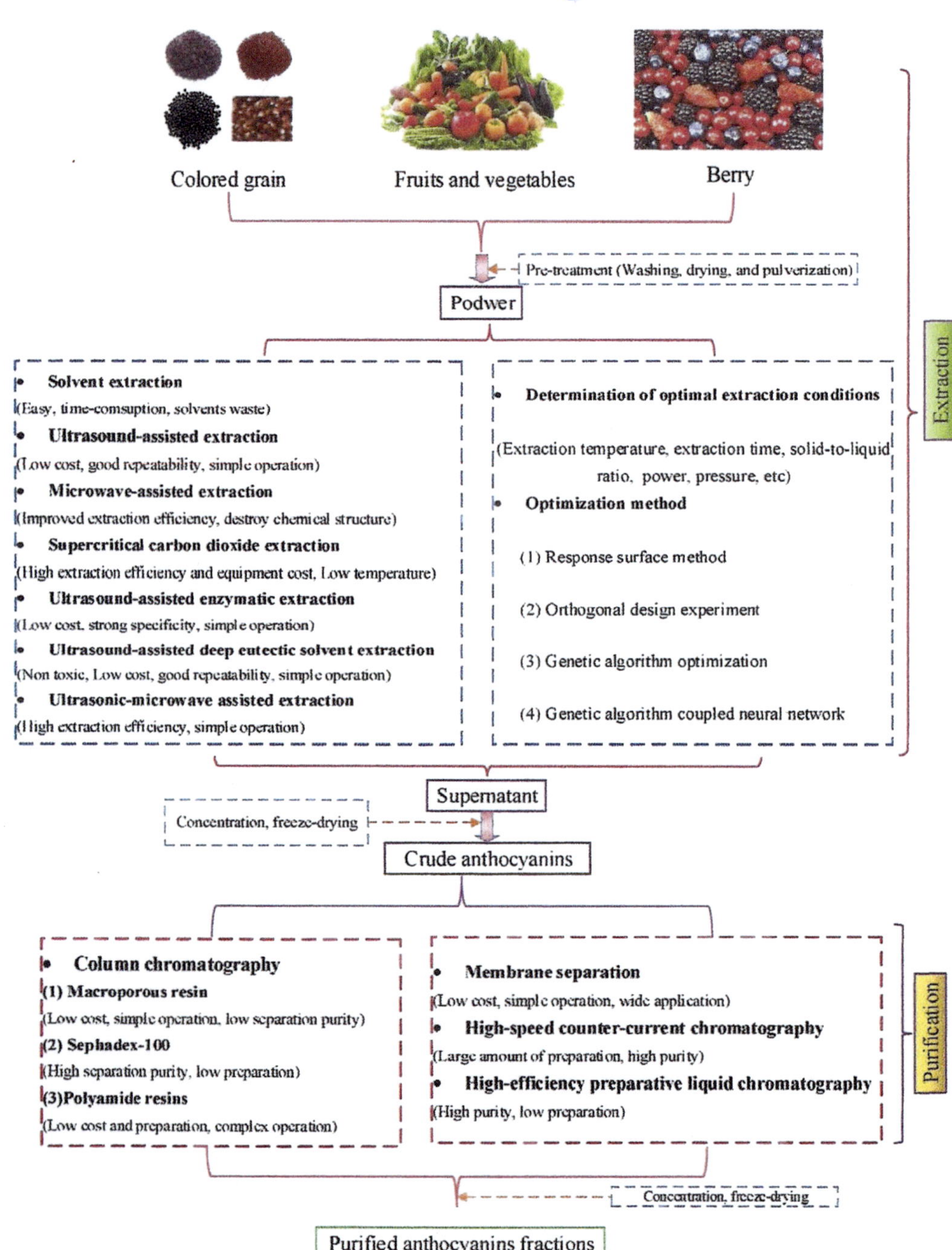

FIGURE 8.1

The available methods extraction and purification methods of anthocyanins.

8.2 Structure and Stability of Anthocyanins

Anthocyanins, as the most common and widely distributed flavonoids, are water-soluble pigments in flowers, seeds, fruits, and leaves of many plants. Figure 8.2 shows the basic skeleton characteristics of anthocyanins C6-C3-C6, and 2-phenylbenzopyran is its basic structure [11]. Generally, the properties of anthocyanins are extremely unstable and rarely exist alone under natural conditions, and they are often formed by glycosidic bonds with sugars such as glucose, galactose, rutose, and rhamnose [12]. There are six common anthocyanins: cyanidin, petunidin, petonidin, pelargonidin, delphinidin, and malvidin (Figure 8.2). The different types of anthocyanins are mainly due to the differences in the position of glycosidic bond binding, the type and number of sugars, the number of hydroxyl groups, the type and number of aromatic acids, and so on [13]. Anthocyanins contain unsaturated double bonds and easily oxidized groups, which lead to their extremely unstable properties. The external environment can easily affect their stability. The main external factors affecting the degradation of anthocyanins are pH, temperature, and O_2 concentration, and the secondary factors are enzymes, metal ions, ascorbic acid, and its degradation products [14, 15]. Moreover, the structure of anthocyanins, such as the number and position of hydroxyl groups, the degree of glycosylation, methylation, and acylation, affects the stability of anthocyanins [16]. Therefore, extraction factors (solid-to-liquid ratio, ultrasonic power, microwave power, extraction temperature, and extraction time) should be reasonably controlled to optimize the extraction rate of anthocyanins by using response surface analysis, orthogonal test, genetic algorithm, and genetic algorithm coupled neural network.

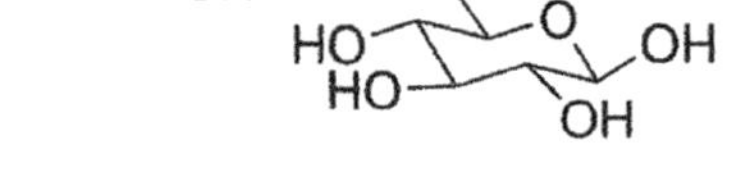

R_1=H	R_2=H	Pelargonidin	(Pg)
R_1=OH	R_2=H	Cyanidin	(Cy)
R_1=OH	R_2=OH	Delphinidin	(Dp)
R_1=OCH$_3$	R_2=H	Peonidin	(Pn)
R_1=OCH$_3$	R_2=OH	Petunidin	(Pt)
R_1=OCH$_3$	R_2=OCH$_3$	Malvidin	(Mv)

FIGURE 8.2
Basic structure of anthocyanins.

8.3 Extraction Methods of Anthocyanins

With the development of extraction technology, the nutritional and medical values of anthocyanins have increasingly attracted attention. All kinds of traditional and potential extraction methods have already been established and proposed. Table 8.1 summarizes the different raw materials, extraction methods, extraction process, and the corresponding anthocyanins yield.

8.3.1 Solvent Extraction Method (SEM)

The technical principle of the solvent extraction method (SEM) is close to the principle of similar solubility [17]. It is very important to reasonably select organic solvents. The most commonly used solvents for anthocyanins extraction are methanol, ethanol, acidified water, or acidified ethanol [18]. Based on a thorough review of previous literature, it was found that anthocyanins could be extracted by SEM under the following conditions: Extraction time of 5 min–4.2 h, extraction temperature of 34.7–52.03°C, solid-to-liquid ratio of 1:15–1:30 g/mL [19–23]. To sum up, SEM has certain advantages, such as convenient operation, simple equipment, and easy to implement, whereas it has significant disadvantages, such as long time-consuming, low efficiency, large solvent consumption, and high temperature. Hence, the applications of SEM are limited.

To solve the limitations of traditional solvent extraction, some advanced and effective extraction methods were developed to improve the yield of anthocyanins. Especially, ultrasound-assisted extraction (UAE), microwave-assisted extraction (MAE), supercritical carbon dioxide extraction (SCDE), ultrasound-assisted enzymatic extraction (UAEE), ultrasound-assisted deep eutectic solvent extraction (UADESE), and ultrasonic-MAE (UMAE) methods have been widely used in the extraction of anthocyanins.

8.3.2 Ultrasound-Assisted Extraction (UAE)

UAE utilizes the cavitation effect and strong shear forces produced by ultrasound with frequencies between 20 kHz and 50 MHz to enhance the extraction ability and shorten the extraction time of anthocyanins [24]. Table 8.1 summarizes the process conditions of UAE anthocyanins from red rice bran, *Auricularia auricula*, purple corn bran, fresh purple eggplant parts, and blueberry [25–28]. By analyzing the literature of UAE anthocyanins at home and abroad, ultrasound has the effect of strengthening the extraction of natural anthocyanins to a certain extent when compared with traditional SEM. However, the cavitation and mechanical effects produced by ultrasound may destroy the structure of anthocyanins in the process of UAE. Therefore, the ultrasonic conditions (ultrasound power, extraction temperature, solid-to-liquid ratio, and extraction time) must be strictly controlled to give better play to the advantages of ultrasonic extraction.

8.3.3 Microwave-Assisted Extraction (MAE)

MAE has been used as a potential extraction method due to the higher recovery and efficiency in comparison to conventional SEM [30]. The main mechanisms of microwave extraction are inherent ion conduction and dipole relaxation in dielectric materials [31]. Microwave radiation rapidly elevates the temperature of the solvent, and the increasing temperature can obviously reduce the viscosity of the extract and promote the solubility

TABLE 8.1

Extraction Methods of Anthocyanins

Source	Extraction Solvent	Extraction Method	Extraction Process	Yield	References
Purple passion fruit peel	Potable water	SEM	pH 2.0, extraction temperature 52.03°C, solid-to-liquid ratio 29.79 g/mL, and extraction time 180 min	577.59 mg/100 g	[19]
Dried blackcurrant skins	Methanol/water	SEM	Extraction time 2 h, extraction temperature 50°C	1712.3 ± 56.1 mg/100 g	[20]
Blueberry bagasse	Acidulant (1% w/v) in water	SEM	Extraction temperature 50°C, extraction time 5 min	17.17 ± 0.70 mg/100 g	[21]
Black rice	Water	SEM	Extraction temperature 34.7°C, extraction time 80 min, and solid-to-liquid ratio 1:30 g/mL	16.58 mg/100 g	[22]
Aronia melanocarpa	Ethanol	SEM	Extraction temperature 46°C, extraction time 4.2 h, and solid-to-liquid ratio 1:23.8 g/mL	6.12 mg/g	[23]
Red rice bran	78.37% ethanol	UAE	Solid-liquid ratio 1:17.46 g/mL, ethanol concentration 78.37%, ultrasonication time 55.23 min, and pH 2.31	5.80 mg/g	[25]
Auricularia auricula	80% ethanol	UAE	Ultrasonication time 5 min, extraction temperature 90°C, and ethanol concentration 80%	382.4 μg/g	[26]
Purple corn bran	95% ethanol, 0.1 mol/L citric acid, and distilled water	UAE	Extraction time 35 min, ultrasonic time 90 s, solid-liquid ratio 1:8 g/mL, and ultrasonic power 400 W.	3.625 ± 0.105 g/kg	[27]
Fresh purple eggplant parts	Acidified water	UAE	Extraction temperature 75°C, pH 2.0, and extraction time 60 min	29.011 mg GAE/g DM	[28]
Blueberry	Ethanol	UAE	Ultrasonic power 512.7 W, extraction time 29.8 min, and liquid to solid ratio 9.5:1 mL/g	2.903 mg/g	[29]
Purple sweet potato	30% ethanol	MAE	Solid-to-liquid ratio 1:3 g/mL, ethanol concentration 30%, microwave irradiation power 320 W, extraction time 500 s	31.16 mg/100g	[33]
Red cabbage	50% ethanol	MAE	Extraction time 5 min, solid-solvent ratio 1:20 g/mL, and microwave power 200 W	241.20 mg/g	[34]

(Continued)

TABLE 8.1 *(Continued)*

Extraction Methods of Anthocyanins

Source	Extraction Solvent	Extraction Method	Extraction Process	Yield	References
Blackberry	52% ethanol	MAE	Microwave power of 469 W, solvent concentration of 52%, liquid-solid ratio of 25 g/mL, and microwave time of 4 min.	2.18 ± 0.06 mg/g	[35]
Cranberry	52% ethanol	MAE	Extraction temperature of 50°C, extraction time of 8 s, ethanol concentration of 52% and solid-to-liquid ratio of 1:28 g/mL	3.06 ± 0.05 mg/g	[36]
Lonicera edulis	85% ethanol	MAE	Microwave power 280 W, microwave time 90 s, material liquid ratio 1:25 g/mL, and ethanol volume fraction 85%	292.16 ± 1.25 mg/100 g	[37]
Sour cherry (*Prunus cerasus* L.) peels	80% ethanol	MAE	Microwave power 500 W, irradiation time 90 s, and ethanol solvent concentration 80%	12.47 mg/g	[38]
Boletus edulis	CO_2	SCDE	Extraction temperature of 65°C, pressure 45 MPa, static time 15 min, dynamic time 20 min	52.7%	[39]
Blueberry pomace	CO_2	SCDE	Extraction temperature 40°C, pressure 34.7 MPa, CO_2 velocity of flow 4.5 L/min, and extraction time 1.86 h	1.48 mg/g	[40]
Blueberry	CO_2	SCDE	Extraction temperature 40°C, pressure 28 MPa, solid-to-liquid ratio 1:7 g/mL, and extraction time 60 min	1.58 mg/g	[41]
Blueberry	CO_2	SCDE	Temperature, pressure, and solvent flow rate kept constant at 40°C, 20 MPa, and 10 mL/min	85 mg/100g	[42]
Grape skins	60% ethanol	UAEE	Extraction temperature of 50°C, ultrasonic power of 400 W, pectinase dosage of 0.16%, and extraction time of 28 min	3.01 ± 0.04 mg/g	[43]
Raspberry wine residues	60% ethanol	UAEE	Extraction temperature of 44°C, ultrasound power of 290 W, pectinase dosage of 0.16%, and extraction time of 30 min	0.853 ± 0.009 mg/g	[44]
Mulberry wine residues	Distilled water acidified to pH 3.5	UAEE	Extraction temperature of 52°C, ultrasound power of 315 W, enzyme of 0.22%, incubation time of 94 min incubation	5.98 mg/g	[45]

(Continued)

TABLE 8.1 *(Continued)*

Extraction Methods of Anthocyanins

Source	Extraction Solvent	Extraction Method	Extraction Process	Yield	References
Blueberry wine residues	Deep eutectic solvent	UADESE	Water content of 29%, ultrasonic power of 380 W, extraction temperature of 55°C, and extraction time of 40 min.	9.32 ± 0.08 mg/g	[46]
Raspberry	Deep eutectic solvent	UADESE	Water content of 29%, ultrasonic power of 210 W, extraction temperature of 51°C, and extraction time of 32 min	1.378 ± 0.009 mg/g	[47]
Black carrots	Deep eutectic solvent	UADESE	Ultrasound power of 78.4 W, extraction time of 19.8 min, solvent/solid ratio of 24.5:1 mL/g, and cycle of 0.195	16.84 mg/g	[48]
Lonicera caerulea residue	30% ethanol	UMAE	Microwave power 150 W, extraction time 2 min, solid/solvent ratio of 3:90 g/mL, and ethanol solvent concentration 30%	9.37 mg/g	[49]
Mulberry	30% ethanol	UMAE	Microwave power 150 W, ultrasound power of 360 W, extraction time 2 min, solid/solvent ratio of 1:70 g/mL, and ethanol solvent concentration 30%	13.57 ± 1.30 mg/g	[50]
Purple yam	60% ethanol	UMAE	Ultrasonic power of 50 W, ultrasonic frequency of 40 kHz, material-liquid ratio of 1:48 g/mL, extraction time of 283 s and extraction temperature of 46°C	4.73 mg/g	[51]
Grape	67.4% ethanol	UMAE	Ethanol concentration 67.4%, extract time 192 s, microwave power 478 W	3.920%	[52]

of the target components. In addition, microwave radiation destroys the microstructure of plant cells, which markedly decreases the mass transfer resistance of the target components and promotes the anthocyanins diffusion from the inside to the outside [32]. Thus, the extraction efficiency of the target components is improved by microwave radiation. Currently, MAE is widely used in the extraction of bioactive compounds from natural plant resources, for example, purple sweet potato, red cabbage, blackberry, cranberry, *Lonicera edulis*, and sour cherry [33–38]. However, UAE and MAE methods may lead to the structural destruction of anthocyanins, which are mainly attributed to local excessive vibration and local high temperature of the extract, respectively. Moreover, the extraction parameters (microwave power, extraction time, and solid-to-liquid ratio) should be controlled at a reasonable level to obtain high anthocyanin yield.

8.3.4 Supercritical Carbon Dioxide Extraction (SCDE)

In recent years, a growing number of study has used supercritical carbon dioxide extraction (SCDE) to extract anthocyanins from natural sources, such as *Boletus edulis*, blueberry pomace, and blueberry (Table 8.1) [39–42]. The nature of supercritical carbon dioxide is between gas and liquid. It has different physico-chemical properties from traditional solvents, strong solubility, large mass transfer coefficient, nontoxic, high cost performance, and easy access [40]. Hence, SCDE, as a novel extraction technology, has many advantages, such as high efficiency, green, safety, and no pollution. Moreover, SCDE method has low treatment temperature and is especially suitable for the extraction of heat sensitive substances, such as anthocyanins [41]. However, this technology is still far from large-scale and industrialized application in the food industry due to the high cost of equipment and large technical investment. In addition, extraction parameters (pressure and CO_2 velocity) are difficult to control.

8.3.5 Combined Extraction Method

With the development of extraction technology, some combined extraction technologies, such as UAEE and UADESE methods, have been proposed for the extraction of active components from natural sources. The addition of enzymatic method, including cellulase and pectinase, has also increase the anthocyanins yield as has ultrasound processing for anthocyanins extraction from natural plant resources. Hence, UAEE method was widely used to extract anthocyanins from grape skins, raspberry wine residues, and mulberry wine residues [43–45]. At present, it is common to extract anthocyanins with organic solvents, whereas the toxicity of organic solvents and their harm to the environment bring hidden dangers to their use. In addition, the waste treatment in the production process also needs high-cost investment, which brings a burden to enterprises. Therefore, a green and efficient anthocyanins extraction method must be explored to improve the yield of anthocyanins. Recently, more and more researchers try to use deep eutectic solvents (DESs) instead of organic solvents. DESs, as a new generation of liquid salts, are usually based on mixtures of relatively cheap and easily available components. Since their emergence, DESs as potential green solvents have attracted extensive attention in various industrial fields, including the extraction of bioactive compounds from various natural plant sources. The combination of UAE and DESs extraction methods has the merits of both methods, thereby providing effective and environmentally friendly alternatives to the traditional anthocyanins extraction methods. The UADESE method is used to extract anthocyanins from blueberry wine residues, raspberry, black carrots, and black rice bran to achieve the maximum yield of anthocyanins [46–48]. Moreover, UMAE is also one of the important methods of anthocyanins extraction. This method can give full play to the advantages of microwave and ultrasound. Therefore, UMAE is widely used in the extraction of anthocyanins in *Lonicera caerulea* residue, mulberry, purple yam, and grape [49–52]. However, although the combined extraction method can improve the yield of anthocyanins to a certain extent, the extraction process cannot be accurately controlled. Many problems should be solved to realize large-scale extraction of anthocyanins by using the combined extraction method in industry. How to develop a rapid, environmentally friendly, and efficient extraction method of anthocyanins from plant resources needs to be further studied.

8.4 Purification Methods of Anthocyanins

In the extraction process, anthocyanins and a large number of impurities (soluble sugar, protein, and organic acid) are extracted at the same time. Excessive impurities will have a significant impact on the physiological activities, stability, and final product quality of anthocyanins. Therefore, the separation and purification of the crude extract is an essential link to obtain anthocyanins with high stability, strong physiological activity, and high quality. At present, the purification methods of anthocyanins mainly include column chromatography, membrane separation, high-speed counter-current chromatography (HSCCC), and high-efficiency preparative liquid chromatography.

8.4.1 Column Chromatography Method

Column chromatography is the most common purification method for the separation and purification of anthocyanins. The principle is that the distribution coefficients of anthocyanins in solid and mobile phases are different, which can better separate anthocyanins and impurities [53]. The packed column usually contains macroporous resins, sephadex-100, and polyamide resins. Macroporous resin, as an adsorbent of high polymer, has a porous skeleton and does not contain ion exchange groups. In addition, macroporous resin has the advantages of fast adsorption rate, large adsorption capacity, low production cost, and recycling [54]. Therefore, this method has become a very rapid purification method, which is widely used in the separation and purification of plant active components. Chen et al. [55] compared the adsorption and desorption properties of five macroporous resins on anthocyanins from mulberry. It was found that XDA-7HP macroporous resin had stronger adsorption and desorption capacities than the other four macroporous resins. The purity of anthocyanins obtained by elution with 40% ethanol solution was 8.5 times higher than that of the crude extract [55]. Wang et al. [56] used macroporous resin to separate and purify the crude extract of anthocyanins from blueberry. The results show that the purity of anthocyanins increased from 7.5% to 32.0% after purification by macroporous resin [56]. Xue et al. [57] first dynamically adsorbed the crude extract of anthocyanins in blueberry by microporous resin, then dynamically desorbed via pH = 3 and 60% ethanol aqueous solution, and collected the desorbed solution. The anthocyanins content was only 4.58%, then separated the anthocyanins in the analytical solution by Sephadex LH-20, and finally collected the anthocyanins content in the powder to reach 90.96% [57]. Column chromatography is the most common method for the separation and purification of anthocyanins. However, this technology can not realize the large-scale purification of anthocyanins in industry due to the small amount of preparation.

8.4.2 Membrane Separation Method

Membrane separation technology is a method that uses artificial and natural synthetic membranes to separate and purify substances. The separation principle is based on different molecular weights to better separate impurities and target substances [58]. At present, the membranes used for separation and crude extraction mainly include microfiltration membrane (MF), ultra-filtration membrane (UF), and nanofiltration membrane (NF). The whole separation process of membrane separation technology belongs to physical process, which does not involve chemical reaction, and it has the advantages of mild action, no phase change during separation, acid and alkali resistance, low energy consumption, and so on. Therefore, this technology is widely used in the fields of biology, medicine, food, and water treatment. Woo et al. (2008) used membrane separation technology to recover anthocyanins from cranberry pulp wastes. After ultrafiltration, infiltration, and concentration, the concentrated powder contained 0.11% anthocyanins [59]. He et al. extracted anthocyanins from black rice [60]. Firstly, the membrane with a relative molecular weight of 3 kDa was used to intercept and remove proteins and polysaccharides. The anthocyanins were adsorbed with DM 301 resin, eluted with an ethanol aqueous solution with a volume fraction of 85%, and finally intercepted with a 200 Da membrane to obtain purified anthocyanins with a purity of 95.93%. We can consider improving the pretreatment technology of raw materials when using membrane separation. In purification, a variety of methods were used, such as membrane separation and resin adsorption, gel separation and membrane separation. A suitable and efficient method for the separation and preparation of anthocyanins was selected according to the feasibility of the experiment.

8.4.3 High-Speed Counter-Current Chromatography (HSCCC) Method

HSCCC, as a continuous liquid-liquid separation method, has been extensively used to prepare bioactive compounds from natural plant resources [61, 62]. HSCCC can avoid the irreversible adsorption of samples on solid phase support; thus, it can improve the sample loading capacity and facilitate the rapid and large-scale preparation of bioactive compounds compared with traditional column chromatography [63]. Moreover, the impurities of the target fraction will be obviously reduced and the success rate of the separation of active compounds would be highly improved after separation by HSCCC. Our research group successfully isolated and purified two high-purity anthocyanins monomers (cyanidin-3-glucoside and delphinidin-3-glucoside) from blueberry and cranberry by HSCCC, and their purity was more than 94% [64, 65]. In addition, we isolated and purified seven major kinds of anthocyanins from raspberry wine residues by HSCCC [44]. Moreover, Xiao et al. [53] isolated anthocyanins from *Perilla frutescens* by HSCCC. It was found that the purities of malonylshisonin and shisonin were 96.7% and 97.5%, respectively, indicating that HSCCC is a fast and efficient technique to prepare pure anthocyanins monomer from natural resources [53]. Liang et al. isolated cyanidin-3-glucoside from blue honeysuckle fruits by HSCCC. The results show that each injection of 100 mg crude extract yielded 22.8 mg of cyanidin-3-glucoside at 98.1% purity [66]. The selection of appropriate extraction solvent system is still an urgent problem to be solved in the technology of HSCCC. Therefore, the technology still needs further research.

8.4.4 High Performance Preparative Liquid Chromatography (HPPLC) Method

The principle of high performance preparative liquid chromatography (HPPLC) is based on the different physicochemical properties of each component in the crude extract of natural products, and the components are distributed in two immiscible phases to varying degrees and move relatively in the two phases at different speeds to be eluted from the column, which can achieve better separation effect [67]. When using HPPLC to separate and purify the crude extract, the main factors considered are the purity, yield, operating components, and production cycle of the target substance. HPPLC has the advantages of good separation and purification effect, high detection accuracy, wide application, and automatic continuous separation [68]. Therefore, it is widely used in the separation of high value-added products, such as organic acids, bioactive bases, and flavonoids. However, this technology has a large amount of equipment investment and a small amount of preparation. Therefore, it is still unable to realize large-scale industrial application.

In conclusion, when enriching and purifying anthocyanins, the pretreatment of samples can help to improve the extraction rate and purity of anthocyanins, save time, reduce the use of organic solvents, and protect the environment. According to the production requirements, the selection of appropriate and feasible equipment, and the combined use of purification technology are conducive to the rapid separation and purification of anthocyanins with high purity.

8.5 Concluding Remarks and Prospects

The main factors affecting the extraction of anthocyanins from berries and fruit residue are the characteristics of sample matrix (such as water activity of sample and rigidity of plant cell wall) and extraction process parameters (such as pH, solvent, temperature, and time). At present, the extraction method has developed from traditional solvent method to emerging technologies such as DES extraction, ultrasonic-assisted extraction, MAE, and supercritical fluid extraction. Compared with traditional solvent extraction, the novel technologies have obvious advantages in extraction rate, energy consumption, extraction time, and environmental protection. However, the new technologies also have some shortcomings, such as high equipment requirements and process optimization standards. From the perspective of industrialization, microwave and ultrasonic-assisted extraction methods have been used in industry, whereas the industrialization of supercritical fluid extraction and high voltage pulsed electric field-assisted extraction methods are difficult due to the problem of equipment cost. Currently, the research on the extraction of anthocyanins is mainly to optimize the specific extraction process, or select the optimal extraction method by comparing different extraction processes and taking the extraction rate of total anthocyanins as the index. Laboratory extraction process should be closely related to industrial production, and laboratory extraction equipment should be used to provide more data support for simulating industrial production. Anthocyanins purification needs to consider factors such as efficiency, production cost, and easy operation. The resin materials with high cost performance are used or combined with membrane separation and gel method. Chromatographic technology must be used to prepare high-purity anthocyanins.

Based on the development status at home and abroad, the research direction of anthocyanins from plant resources in China in the future can be considered from the following aspects: (1) ameliorate the extraction equipment with high cost and potential safety hazards, and improve the safety performance of the extraction equipment while reducing the production cost; (2) according to the characteristics of extraction technology, different extraction technologies are investigated to diversify the extraction process and provide technical and theoretical support for screening the best process; (3) improve the laboratory extraction equipment to better match the industrial production; (4) increase the optimization of sample pretreatment methods (such as fully grinding the sample, soaking and softening in advance, adding cellulase and pectinase to hydrolyze cellulose in raw materials, steam explosion, and pressurization), which will be conducive to the release of anthocyanins in raw materials; (5) different extraction and purification processes are formulated for anthocyanins products with different purities.

Authors Contribution

Hongkun Xue: Writing-original draft, Writing-review & editing, Data curation, Visualization. Yanmei Han, Bo Han, Shaoqin Ge, and Xu Cai: Formal analysis and Software. Jiaqi Tan: Conceptualization, Resources, Writing-review & editing, Supervision.

Declaration of Competing Interest

All authors have declared no conflict of interest.

Acknowledgment

The authors gratefully thank the financial support provided by Hebei University High-level Talent Scientific Research Start-up Project (521100221072) for this research project.

References

1 [a]. J.M. Alvarez-Suarez, C. Cuadrado, I. Ballesteros, F. Giampieri, C.S. Buelga, Novel approaches in anthocyanin research-plant fortification and bioavailability issues, Trends Food Sci. Technol. 117 (2021) 92–105. [b] X. Zhao, Z. Yuan, Anthocyanins from pomegranate (*Punica granatum* L.) and their role in antioxidant capacities) and their role in antioxidant capacities, Chem. Biodivers. 18 (2021), e2100399.

2. X. Jiao, B. Li, Q. Zhang, N. Gao, X. Zhang, X. Meng, Effect of *in vitro*-simulated gastrointestinal digestion on the stability and antioxidant activity of blueberry polyphenols and their cellular antioxidant activity towards HepG2 cells, Int. J. Food Sci. Technol. 53 (2018) 61–71.

3. P.B. Pertuzatti, M.T. Barcia, L.P.G. Rebello, S. Gómez-Alonso, R.M.T. Duarte, M.C.T. Duarte, I. Hermosin-Gutierrez, Antimicrobial activity and differentiation of anthocyanin profiles of rabbiteye and highbush blueberries using HPLC-DAD-ESI-MS^n and multivariate analysis, J. Funct. Foods 26 (2016) 506–516.

4. E. Eroglu Ozkan, M.F. Seyhan, O. Kurt Sirin, T. Yilmaz-Ozden, E. Ersoy, S.D. Hatipoglu Cakmar, A.C. Goren, H. Yilmaz Aydogan, O. Ozturk, Antiproliferative effects of *Turkish pomegranate* (*Punica granatum* L.) extracts on MCF-7 human breast cancer cell lines with focus on antioxidant potential and bioactive compounds analyzed by LC-MS/MS, J. Food Biochem. 45 (2021) e13904.

5. P. Zhang, Y. Li, T. Wang, Z. Cai, H. Cao, H. Zhang, Y. Cao, B. Chen, D. Yang, Statistics on the bioactive anthocyanin/proanthocyanin products in China online sales, Food. Sci. Nutr. 9 (2021) 5428–5434.

6. M. Zielinska, A. Michalska, Microwave-assisted drying of blueberry (*Vaccinium corymbosum* L.) fruits: Drying kinetics, polyphenols, anthocyanins, antioxidant capacity, colour and texture, Food Chem. 212 (2016) 671–680.

7. P. Ockermann, L. Headley, R. Lizio, J. Hansmann, A review of the properties of anthocyanins and their influence on factors affecting cardiometabolic and cognitive health, Nutrients 13 (2021) 2831.

8. P. Zhou, Z. Wu, C. Huang, Y. Zhang, J. Zheng, Research progress of extraction and purification of anthocyanins, Fine Chem. 37 (2020) 1513–1523.

9. A. Kozlowska, T. Dzierzanowski, Targeting inflammation by anthocyanins as the novel therapeutic potential for chronic diseases: An update, Molecules 26 (2021) 4380.

10. H. Wen, H. Tian, C. Liu, X. Zhang, Y. Peng, X. Yang, F. Chen, J. Li, Metformin and cyanidin 3-O-galactoside from *Aronia melanocarpa* synergistically alleviate cognitive impairment in SAMP8 mice, Food Funct. 12 (2021) 10994–11008.

11. X. Su, J. Xu, D. Rhodes, Y. Shen, W. Song, B. Katz, W. Wang, Identification and quantification of anthocyanins in transgenic purple tomato, Food Chem. 202 (2016) 184–188.

12. Y.W. Zhao, C.K. Wang, X.Y. Huang, D.G. Hu, Anthocyanin stability and degradation in plants, Plant Signal. Behav. 16 (2021) 1987767.

13. X.F. Yue, S.S. Jing, X.F. Ni, K.K. Zhang, Y.L. Fang, Z.W. Zhang, Y.L. Ju, Anthocyanin and phenolic acids contents influence the color stability and antioxidant capacity of wine treated with mannoprotein, Front. Nutr. 8 (2021) 691784.

14. B. Micó-Vicent, M. Ramos, V. Viqueira, F. Luzi, F. Dominici, A. Terenzi, E. Maron, M. Hamzaoui, S. Kohnen, L. Torre, A. Jiménez, D. Puglia, M.C. Garrigós, Anthocyanin hybrid nanopigments from pomegranate waste: Colour, thermomechanical stability and environmental impact of polyester-based bionanocomposites, Polymers 13 (2021) 1966.

15. S. Oancea, A review of the current knowledge of thermal stability of anthocyanins and approaches to their stabilization to heat, Antioxidants 10 (2021) 1337.

16. M.E. West, L.J. Mauer, Color and chemical stability of a variety of anthocyanins and ascorbic acid in solution and powder forms, J. Agric. Food Chem. 61 (2013) 4169–4179.

17. J. Johnson, T. Collins, K. Walsh, M. Naiker, Solvent extractions and spectrophotometric protocols for measuring the total anthocyanin, phenols and antioxidant content in plums, phenols and antioxidant content in plums, Chem. Pap. 9 (2020) 145.

18. R. Muangrat, P.T. Williams, P. Saengcharoenrat, Subcritical solvent extraction of total anthocyanins from dried purple waxy corn: Influence of process conditions, J. Food Process. Preserv. 41 (2017) 46–50.

19. J. Herrera-Ramirez, N. Meneses-Marentes, M.P. Tarazona Diaz, Optimizing the extraction of anthocyanins from purple passion fruit peel using response surface methodology, J. Food Meas. Char. 14 (2020) 72–79.

20. E.M. Azman, D. Charalampopoulos, A. Chatzifragkou, Acetic acid buffer as extraction medium for free and bound phenolics from dried blackcurrant (*Ribes nigrum* L.) skins, J. Food Sci. 85 (2020) 3745–3755.

21. A. Lff, B. Nmm, B. Rfr, A. Rp, C. Er, A. Te, D. Vcb, Citric acid water-based solution for blueberry bagasse anthocyanins recovery: Optimization and comparisons with microwave-assisted extraction (MAE), LWT – Food Sci. Tech. 133 (2020) 11257.

22. A.C. Pedro, D. Granato, N.D. Rosso, Extraction of anthocyanins and polyphenols from black rice (*Oryza sativa* L.) by modeling and assessing their reversibility and stability, Food Chem. 191 (2016) 12–20.

23. L. Tan, X. Xue, Y. Li, G. Shi, Extraction and purification techniques of anthocyanins from *Aronia melanocarpa*, North. Horti. 3 (2020) 111–119.

24. T. Belwal, H. Huang, L. Li, Z. Duan, X. Zhang, H. Aalim, Z. Luo, Optimization model for ultrasonic-assisted and scale-up extraction of anthocyanins from *Pyrus communis 'Starkrimson'* fruit peel, Food Chem. 297 (2019) 124993.

25. Y. Wang, L. Zhao, R. Zhang, X. Yang, Y. Sun, L. Shi, P. Xue, Optimization of ultrasound-assisted extraction by response surface methodology, antioxidant capacity, and tyrosinase inhibitory activity of anthocyanins from red rice bran, Food Sci. Nutr. 8 (2020) 921–932.

26. C.J. Lopez, C. Caleja, M.A. Prieto, M.F. Barreiro, L. Barros, F.R. Ferreira, Optimization and comparison of heat and ultrasound assisted extraction techniques to obtain anthocyanin compounds from *Arbutus unedo* L. fruits, Food Chem. 264 (2018) 81–91.

27. L. Chen, M. Yang, H. Mou, Q. Kong, Ultrasound-assisted extraction and characterization of anthocyanins from purple corn bran, J. Food Process. Preserv. 42 (2017) e13377.

28. S. Ferarsa, W. Zhang, N. Moulai-Mostefa, L. Ding, M.Y. Jaffrin, N. Grimi, Recovery of anthocyanins and other phenolic compounds from purple eggplant peels and pulps using ultrasonic-assisted extraction, Food Bioprod. Process. 7 (2018) 1836.

29. Y. Tian, Y. Yang, P. Gao, J.H. Wang, Y.Q. Xu, Z.Y. Yu, Optimization of ultrasonic-assisted extraction of flavonols and anthocyanins from blueberry using RSM, Adv. Mater. Res. 4 (2012) 2423–2430.

30. Y. Sun, X. Liao, Z. Wang, X. Hu, F. Chen, Optimization of microwave-assisted extraction of anthocyanins in red raspberries and identification of anthocyanin of extracts using high-performance liquid chromatography-mass spectrometry, Eur. Food Res. Technol. 225 (2007) 511–523.

31. Z. Yang, W. Zhai, Optimization of microwave-assisted extraction of anthocyanins from purple corn (*Zea mays* L.) cob and identification with HPLC-MS, Innov. Food Sci. Emerg. 11 (2010) 470–476.

32. H. Xue, H. Xu, X. Wang, L. Shen, H. Liu, C. Liu, Q. Li, Effects of microwave power on extraction kinetic of anthocyanin from blueberry powder considering absorption of microwave energy, J. Food Qual. 1(2018) 9680184.

33. W. Liu, C. Yang, C. Zhou, Z. Wen, X. Dong, An improved microwave-assisted extraction of anthocyanins from purple sweet potato in favor of subsequent comprehensive utilization of pomace, Food Bioprod. Process. 5 (2019) 1–9.

34. N. Yiit, E.T. Yolaaner, A. Hamzalolu, V. Gkmen, Optimization of microwave-assisted extraction of anthocyanins in red cabbage by response surface methodology, J. Food Process. Preserv. 6 (2021) 111–117.

35. Y. Wen, H. Chen, X. Zhou, Q. Deng, Y. Zhao, C. Zhao, X. Gong, Optimization of the microwave-assisted extraction and antioxidant activities of anthocyanins from blackberry using a response surface methodology, RSC Adv. 5 (2015) 19686–19695.

36. H. Xue, J. Tan, L. Fan, Q. Li, X. Cai, Optimization microwave-assisted extraction of anthocyanins from cranberry using response surface methodology coupled with genetic algorithm and kinetics model analysis, J. Food Process. Eng. 44 (2021) e13688.

37. L.I. Feng, X.L. Zhang, X.C. Liu, X.T. Zhang, H.T. Zhao, Optimization of microwave-assisted extraction of anthocyanins from *Lonicera edulis* and its antioxidant activity by response surface methodology, Sci.Tech. Food Ind. 5 (2019) 116–170.

38. E.K. Ahin, M. Bilgin, S. Ahin, Recovery of anthocyanins from sour cherry (*Prunus cerasus* L.) peels via microwave assisted extraction: Monitoring the storage stability, Prep. Biochem. Biotechnol. 5 (2020) 1–11.

39. G.L. Jiao, A.K. Pour, Extraction of anthocyanins from Haskap berry pulp using supercritical carbon dioxide: Influence of co-solvent composition and pretreatment, LWT–Food Sci. Tech. 6 (2018) 947.

40. G.W. Qin, H. Han, X.W. Ding, Y.L. Wang, Optimization of extracting technology of anthocyanins from blueberry pomace by supercritical carbon dioxide, Appl. Chem. Ind. 48 (2019) 109–112.

41. M.X. Tian, Y.D. Li, W.Z. Hu, Y.Y. Wang, A.L. Jiang, C.H. Liu, Optimization of supercritical CO_2 extraction of blueberry anthocyanins using response surface methodology, Sci. Tech. Food Ind. 37 (2016) 208–212.

42. J. Paes, R. Dotta, G.F. Barbero, J. Martínez, Extraction of phenolic compounds and anthocyanins from blueberry (*Vaccinium myrtillus* L.) residues using supercritical CO_2 and pressurized liquids, J. Supercrit. Fluids 95 (2014) 8–16.

43. J. Tan, Q. Li, H. Xue, J. Tang, Ultrasound-assisted enzymatic extraction of anthocyanins from grape skins: Optimization, identification, and antitumor activity, J. Food Sci. 85 (2020) 3731–3744.

44. H. Xue, J. Tan, Q. Li, J. Tang, X. Cai, Ultrasound-assisted enzymatic extraction of anthocyanins from raspberry wine residues: Process optimization, isolation, purification, and bioactivity determination, Food Anal. Methods 14 (2021) 1–18.

45. L. Zhang, G. Fan, M.A. Khan, Z. Yan, T. Beta, Ultrasonic-assisted enzymatic extraction and identification of anthocyanin components from mulberry wine residues, Food Chem. 323 (2020) 126714.

46. H. Xue, J. Tan, Q. Li, J. Tang, X. Cai, Ultrasound-assisted deep eutectic solvent extraction of anthocyanins from blueberry wine residues: Optimization, identification, and HepG2 antitumor activity, Molecules 25 (2020) 5456.

47. H. Xue, J. Tan, Q. Li, J. Tang, X. Cai, Optimization ultrasound-assisted deep eutectic solvent extraction of anthocyanins from raspberry using response surface methodology coupled with genetic algorithm, Foods 9 (2020) 1409.

48. D.A. Türker, M. Doan, Ultrasound-assisted natural deep eutectic solvent extraction of anthocyanin from black carrots: Optimization, cytotoxicity, *in vitro* bioavailability and stability, Food Bioprod. Process. 132 (2022) 99–113.

49. X. Liu, L. Meng, X. Xin, Optimization of ultrasonic-microwave assisted extraction of anthocyanin from *Lonicera caerulea* residue, Food Res. Dev. 42 (2021) 50–54.

50. C. Dong, X. Yan, X. Li, S. Liu, C. Liu, Optimization of ultrasonic-microwave assisted extraction of anthocyanins from mulberry and theirs antioxidant activities, China Cond. 45 (2020) 172–178.

51. Z.S. Pei, H.D. Zhang, L.M. Yuan, X.C. Xin, L.Z. Sheng, Ultrasonic/microwave-assisted extraction (UMAE) of anthocyanins from Purple Yam, Food Sci. (N. Y.) 33 (2012) 78–83.

52. L. Chao, W. Wang, Z. Yi, R. Yang, D. Fan, Optimization of ultrasonic-microwave synergistic extraction of anthocyanins by response surface methodology, Food Ferment. Ind. 35 (2009) 167–171.

53. H.U. Xiao, A.D. Sun, D.Q. Zhang, Separation of anthocyanins from *Perilla frutescens* by high speed counter-current chromatography, J. Chin. Med. Mater. 33 (2010) 1586.

54. X.X. Wang, C. Hansen, K. Allen, Identification of anthocyanins isolated from black bean canning waste water by macroporous resin using optimized conditions, Food Nutr. Sci. 4 (2013) 174–181.

55. Y. Chen, W. Zhang, T. Zhao, F. Li, M. Zhang, J. Li, L. Yang, Adsorption properties of macroporous adsorbent resins for separation of anthocyanins from mulberry, Food Chem. 194 (2016) 712–722.

56. E. Wang, Y. Yin, C. Xu, J. Liu, Isolation of high-purity anthocyanin mixtures and monomers from blueberries using combined chromatographic techniques, J. Chromatogr. A 1327 (2014) 39–48.

57. H.K. Xue, L.Y. Shen, X.R. Wang, C. Liu, C. Liu, H. Liu, X. Zheng, Isolation and purification of anthocyanin from blueberry using macroporous resin combined Sephadex LH-20 Techniques, Food Sci. Technol. Res. 25 (2019) 29–38.

58. J. Martin, J. Enrique, M. Diaz, A.G. Asuero, Recovery of anthocyanins using membrane technologies: A review, Crit. Rev. Anal. Chem. 48 (2017) 143–175.

59. A.H. Woo, J.H.V. Elbe, C.H. Amundson, Anthocyanin recovery from cranberry pulp wastes by membrane technology, J. Food Sci. 45 (1980) 875–879.

60. S. He, Q. Lou, J. Shi, H. Sun, M. Zhang, L. Qian, Water extraction of anthocyanins from black rice and purification using membrane separation and resin adsorption, J. Food Process. Preserv. 41 (2017) e13091.

61. F.X. Yang, P. Xu, J.G. Yang, J. Liang, M.H. Zong, W.Y. Lou, Efficient separation and purification of anthocyanins from Saskatoon berry by using low transition temperature mixtures, RSC Adv. 6 (2016) 104582–104590.

62. H. Zou, Y. Ma, Z. Xu, X. Liao, A. Chen, S. Yang, Isolation of strawberry anthocyanins using high-speed counter-current chromatography and the copigmentation with catechin or epicatechin by high pressure processing, Food Chem. 247 (2018) 81–88.

63. T. Chen, X. Yang, N. Wang, H. Li, J. Zhao, Y. Li, Separation of six compounds including two n-butyrophenone isomers and two stibene isomers from *Rheum tanguticum* Maxim by recycling high speed counter-current chromatography and preparative high-performance liquid chromatography, IEEE J. Solid State Circ. 41 (2018) 3660–3668.

64. H.K. Xue, X.H. Zhu, J.Q. Tan, L.L. Fan, Q. Li, J.T. Tang, X. Cai, Counter-current fractionation-assisted bioassay-guided separation of active compound from blueberry and the interaction between the active compound and α-glucosidase, Foods. 10 (2021) 509.

65. H.K. Xue, J.Q. Tan, Q. Li, J.T. Tang, X. Cai, Counter-current fractionation-assisted and bioassay-guided separation of active compounds from cranberry and their interaction with α-glucosidase, LWT – Food Sci. Technol. (Lebensmittel-Wissenschaft-Technol.) 145 (2021) 111374.

66. C.A. Liang, A. Xx, L.A. Rong, B. Qy, A. Xw, L.A. Ye, Isolation of cyanidin-3-glucoside from blue honeysuckle fruits by high-speed counter-current chromatography, Food Chem. 152 (2014) 386–390.

67. Y. Liu, M. Nobutoshi, L. Wang, S. Zhang, Preparative high-performance liquid chromatography for the purification of natural acylated anthocyanins from red radish (*Raphanus sativus* L.), J. Chromatogr. Sci. 5 (2008) 743–746.

68. A. Skalska, A. Matysik, M. Gerkowicz, M. Wójciak-Kosior, Preparative reversed-phase high-performance thin-layer chromatography for analysis of anthocyanins, JPC-J. Planar. Chromat. 19 (2006) 463–466.

69. Woo, A. H., Von Elbe, J. H., & Amundson, C. H. (1980). Anthocyanin recovery from cranberry pulp wastes by membrane technology. Journal of Food Science, 45(4), 875–879.

9

Extraction Methods of Anthocyanins

Kamran Javed Naquvi, Mohd. Javed Naim, and Javed Ahamad

9.1 Introduction

Anthocyanins are flavonoid glycosides with anthocyanidin as aglycone and these anthocyanins are abundantly distributed in plants (Byamukama *et al.*, 2006; Jordheim *et al.*, 2006). Over 500 various kinds of anthocyanins have been identified to date, and such anthocyanins are found in 72 genera and 27 distinct families of plants (de Pascual-Teresa and Sanchez-Ballesta, 2008). Anthocyanins are colored pigments mainly distributed in fruits, flowers, and vegetables. Anthocyanins have been the subject of extensive research due to their health benefits such as anti-inflammatory, antioxidant, antidiabetic, antimicrobial, analgesic, hepatoprotective, and anticancer (Kong *et al.*, 2003; Wallace and Giusti, 2019). The most predominant examples of anthocyanin molecules include malvidin, pelargonidin, peonidin, petunidin, cyanidin, and delphinidin (Castañeda-Ovando *et al.*, 2009). Through anti-oxidation, anthocyanins can prevent genomic damage from oxidative stress and can manage cancer in its early stages of tumor development and metabolic disorders like diabetes, obesity, and cardiovascular diseases (Wang and Stoner, 2008; Naseri *et al.*, 2018; Jiang *et al.*, 2019).

Due to the extensive use of anthocyanins as a colorant in foodstuffs, cosmetics, and pharmaceutical products as well as their potential health benefits, the interest in anthocyanins increased in recent years. Despite their potential application in pharmacological, pharmaceutical, food, and cosmetic industries, anthocyanin use is limited because of their relative instability, and low extraction yield (Castañeda-Ovando *et al.*, 2009). Anthocyanins occur in plants either in a free state (anthocyanidin) or in combination with sugar (flavonoid glycoside). Extraction and purification of anthocyanins are difficult tasks due to their possible break at higher temperatures and low quantity in natural sources (Fan *et al.*, 2008; Tan *et al.*, 2022). The present book chapter discusses the different conventional and modern methods for the extraction of different anthocyanin molecules.

9.2 Sources and Chemistry of Anthocyanins

Anthocyanins are coloring pigments predominantly found in fruits, berries, vegetables, legumes, and cereals. The important sources of anthocyanins include blueberries, raspberries, bilberries, elderberries, strawberries, chokeberries, pomegranates, apples, grapes, pistachios, red onions, and red cabbage (Jaiswal *et al.*, 2019; Oladzadabbasabadi *et al.*, 2022).

DOI: 10.1201/9781003453260-11

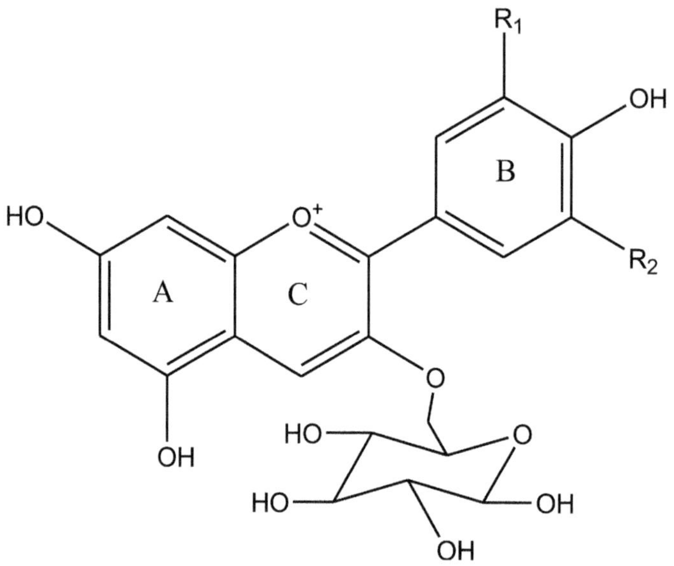

FIGURE 9.1
The basic chemical skeleton of anthocyanin.

Anthocyanins belong to the category of flavonoids, and the basic skeleton of anthocyanin consists of $C_6C_3C_6$. The anthocyanins are derived from anthocyanidins which are aglycone parts of anthocyanins. Anthocyanins rarely occur in the free state in plants, generally, they are combined with sugar parts, the most common sugars present in anthocyanins include glucose, galactose, and rhamnose. Many anthocyanins occur in plants, the different types of anthocyanins mainly due to the differences in the position of glycosidic bond binding, the number and types of sugars, the number of hydroxyl groups, and the number of aromatic groups. The most important anthocyanins that occur in plants include malvidin, pelargonidin, peonidin, petunidin, cyanidin, and delphinidin (Figure 9.1) (Silva et al., 2016; Su et al., 2016).

Anthocyanins	R_1	R_2
Malvidin	OCH_3	OCH_3
Pelargonidin	H	H
Peonidin	OCH_3	H
Petunidin	OCH_3	OH
Cyanidin	OH	H
Delphinidin	OH	OH

9.3 Approaches of Anthocyanin Extraction

Anthocyanins occur in the glycosidic form and most of them are unsaturated and in low content in plants. These conditions of anthocyanins in plants make difficulty in the extraction and purification process. Their thermal instability and possible degradation are due to high temperature, pH, oxygen concentration, and some other factors like enzymes, metal ions, and their degradation products. Moreover, the complex chemical structure of anthocyanins, such as the number and position of hydroxyl groups, degree of glycosylation, methylation, and acylation, affects the stability of anthocyanins (Oancea, 2021; Zhao *et al.*, 2021). To solve above-stated problems associated with anthocyanins, the extraction of anthocyanins is done at low temperatures using appropriate solvents (e.g., ethanol, ethyl acetate, water) using conventional methods (e.g., maceration, percolation, and Soxhlet extraction) and modern methods (microwave-assisted extraction (MAE), ultrasound-assisted extraction, supercritical fluid extraction (SFE), enzyme-assisted extraction (EAE), etc.). The extraction time, solid-to-solvent ratio, extraction temperature, microwave power, and ultrasonic power all affect the extraction of anthocyanins (West and Mauer, 2013; Tan *et al.*, 2022).

9.4 Methods of Anthocyanin Extraction

Anthocyanins are plant secondary metabolites from the flavonoid family, serving as water-soluble natural pigments (Constantin and Istrati, 2022). These compounds provide red, purple, and blue hues in many fruits and vegetables, including berries and eggplants. A significant drawback of anthocyanins is their low stability. Their stability is affected by several factors, including pH, light, temperature, co-pigmentation, sulfites, ascorbic acid, oxygen, and enzymes (Enaru *et al.*, 2021). Extracting and separating anthocyanins from plants is crucial due to their instability. Optimizing the process is important as anthocyanins are prone to degradation by factors such as pH, temperature, oxygen, water activity, co-pigments, and enzymes (Chung *et al.*, 2016; Weber *et al.*, 2017). The most used method for extracting anthocyanins is solvent extraction, also known as conventional solid-liquid extraction, due to its effectiveness and simplicity. Conventional solid-liquid extraction is a process used to extract compounds from solid materials using a liquid solvent. The solid is mixed with the solvent, dissolving the desired compounds into the liquid. The mixture is then filtered or centrifuged to separate the liquid extract from the solid residue, enabling compound isolation (Wallace and Giusti, 2013). Some most common conventional methods for the extraction of anthocyanins are discussed below.

9.4.1 Conventional Methods

The extraction method for anthocyanins is determined by the need to maintain the stability and prolong the shelf life of these compounds, which directly impacts their beneficial properties (Sharma *et al.*, 2016; Enaru *et al.*, 2021). Percolation, maceration, and Soxhlet extraction are three commonly used methods for extracting anthocyanins from plant materials. Percolation involves passing a solvent through a packed bed of the material, allowing the solvent to extract the anthocyanins as it percolates through. Maceration, on the other hand, relies on soaking the plant material in a solvent to allow the anthocyanins to diffuse into the liquid over time. Soxhlet extraction is a more intensive technique that involves continuously cycling a solvent between a heated flask containing the plant material and a condenser, effectively extracting the anthocyanins through repeated cycles of vaporization and condensation. Each of these methods has its advantages and limitations in terms of efficiency, yield, and the quality of the extracted anthocyanins, making it important to choose the most suitable method based on the specific requirements of the extraction process. These techniques utilize various solvents and/or heat. Following the principle that 'like dissolves like,' the solvents commonly used to extract anthocyanins include methanol, ethanol, water, acetone, or their mixtures (Tena and Asuero, 2022).

9.4.1.1 Percolation

Percolation involves passing a solvent through a tightly packed column of plant material, allowing the solvent to flow slowly and uniformly through the material. As the solvent percolates downward, it interacts with the plant tissues, dissolving the anthocyanins and other soluble compounds. This method ensures that the solvent has maximum contact with the material, facilitating efficient extraction of the desired anthocyanins. The process can be optimized by controlling the flow rate, solvent composition, and extraction time, ensuring a high yield and quality of the extracted anthocyanins (Ravanfar *et al.*, 2018; Wang *et al.*, 2020). The authors extracted anthocyanins from red cabbage (*Brassica oleracea* L. var. *capitata f. rubra*) using both ultrasound-assisted extraction and conventional percolation methods. For the conventional method, they percolated red cabbage leaves in water to obtain the crude extract, using water as the solvent. The researchers optimized and compared the efficiency of percolation and ultrasound-assisted extraction methods for extracting anthocyanins, utilizing ultrasonic irradiation in the latter method. For ultrasound-assisted extraction, they determined that the optimal extraction efficiency was achieved at an ultrasonic output power of 100 W, with a pulse mode of 30 seconds on and 30 seconds off, conducted at a temperature of 15°C, over 90 minutes. This approach allowed for a more effective extraction process by enhancing the interaction between the solvent and the plant material through ultrasonic waves, resulting in a higher yield of anthocyanins compared to conventional percolation (Ravanfar *et al.*, 2018). The extraction of anthocyanins from the bright red and dark red calyces of *Hibiscus sabdariffa* was carried out using the percolation method. This process was meticulously optimized to ensure the extracted anthocyanins are suitable for food applications. The optimization process focused on maximizing yield and preserving the stability and bioactivity of the anthocyanins, ensuring their efficacy and safety when incorporated into food products (Diessana *et al.*, 2015). The phenolic fraction was extracted from *Lonicera caerulea* L. (blue honeysuckle; Caprifoliaceae) berries using 0.1% H_3PO_4 for percolation and Sepabeads SP207 for solid-phase extraction (SPE). Cyanidin-3-glucoside was identified as the predominant anthocyanin in the extract. During the comparison of conventional and nonconventional phytochemical extraction methods from the heartwood of *Pterocarpus marsupium* Roxb. (also known as Malabar kino or Indian kino, family Fabaceae), it was found that the percolation method produced the highest yield compared to other conventional extraction methods (Devgun *et al.*, 2012).

9.4.1.2 Maceration

Maceration involves submerging coarse or powdered plant materials in an airtight container filled with a solvent (called menstruum). The mixture is kept at room temperature for at least three days and stirred regularly until the soluble matter dissolves (Manousi *et al.*, 2019; Samira *et al.*, 2024). The resulting mixture is then filtered to remove impurities, extracting flavors, colors, and active ingredients (Jovanović *et al.*, 2017; Samira *et al.*, 2024). Infusion and decoction use the same principle as maceration for extracting compounds from plant materials (Manousi *et al.*, 2019). Infusion involves shorter soaking periods with hot water, ideal for delicate anthocyanins. Decoction uses boiling, suitable for tougher materials but risks degrading anthocyanins. All methods aim to extract and preserve these pigments for their vibrant colors and antioxidant properties. Maceration advantages include simplicity and effectiveness for delicate compounds; disadvantages are time consumption and potential degradation (Zhang *et al.*, 2018). For anthocyanin extraction, maceration preserves these sensitive pigments, yielding vibrant colors and beneficial antioxidants, but may require long durations for optimal results. Although new technologies like ultrasound and MAEs enhance yield and extract stability, they require sophisticated instrumentation, increasing the cost of anthocyanin production (Tena and Asuero, 2022; Taghavi *et al.*, 2023).

Maceration is essential for extracting anthocyanins, which give highbush blueberries (*Vaccinium corymbosum* L.) their color and potential health benefits. It's interesting how similar processes occur in both highbush blueberries and red wine, affecting compound extraction from the skin. The duration of fermentation and maceration in red wine also impacts compound extraction, like highbush blueberries during maceration. Identifying specific anthocyanidins like peonidin and malvidin, along with compounds like malvidin-3-galactoside, offers valuable insights into highbush blueberry composition. The isolation and quantification of delphinidin, cyanidin, petunidin, peonidin, and malvidin demonstrate the variety of anthocyanins in these berries, each potentially contributing distinct health benefits or flavors (Varo *et al.*, 2022).

In a study comparing conventional and UAE extraction methods, anthocyanin content was measured in European red and green gooseberry (*Ribes uva-crispa* L.) extracts. A significant difference in anthocyanin content, which is responsible for the red and blue colors in the fruits, was observed. Conventional extraction with ethanol (E-CE-20) yielded the highest anthocyanin content (3.63 mg/100 g) in red gooseberries, while the lowest content (1.14 mg/100 g) was found in green gooseberry extract (E-UAE-20). Maximum anthocyanin extraction yield was observed using the E-CE method in green and red gooseberries. Incorporating the Graciano variety during pre-fermentative maceration seems to boost the extraction and retention of pigments in Tempranillo wines. This improvement likely stems from the better co-pigmentation abilities of flavanols in Graciano grape skins compared to those in Tempranillo. This correlation underscores the importance of maceration techniques in enhancing color and pigment characteristics in wine production (García-Marino *et al.*, 2010). Anthocyanins have been extracted from dried chokeberry (*Aronia melanocarpa*) fruits using 50% ethanol as the solvent (Ćujić *et al.*, 2016).

9.4.1.3 Soxhlation

Soxhlet extraction, an amalgamation of percolation and maceration, is pivotal in chemistry and biochemistry for extracting compounds from solids. Invented by Franz von Soxhlet in 1879, it employs a specialized apparatus with an extraction chamber, vapor duct, and siphon tube. The process begins with solid material in a thimble, soaked in a solvent. Heating vaporizes the solvent, which condenses and drips back, dissolving compounds from the material. The siphon tube maintains solvent levels, ensuring continuous extraction. This method's efficiency and selectivity in extracting target compounds make it indispensable across scientific and industrial domains, delivering reliable results (Malik and Mandal, 2022). Anthocyanins were extracted from *H. sabdariffa* petals using solvent, Soxhlet, and microwave methods. Solvent extraction yielded the highest anthocyanin content. The effect of these anthocyanins as natural antioxidants on extra virgin olive oil was examined, suggesting that anthocyanins could be a suitable natural additive for commercial olive oil production.

9.4.2 Modern Methods of Extraction

In recent years, interest in food quality has surged as consumers become more aware of the link between nutrition and health. Anthocyanins, known for their antioxidant properties, also serve as natural colorants and high-value compounds due to their role in enhancing sensory qualities, nutrition, and health benefits (Tena and Asuero, 2022). Various modern extraction techniques are employed to extract anthocyanins to harness these benefits efficiently. Anthocyanin extraction employs various advanced techniques for enhanced efficiency and sustainability. MAE employs microwave energy to heat solvents rapidly, increasing extraction rates and reducing solvent use. These methods collectively improve anthocyanin recovery, aligning with green chemistry principles (Marić *et al.*, 2018; Shen *et al.*, 2023). Ultrasound-assisted extraction (UAE) utilizes acoustic cavitation to improve solvent penetration and reduce extraction time. SFE utilizes supercritical CO_2, providing a green, solvent-free method with high efficiency. Carbon dioxide (CO_2) is the most frequently used supercritical fluid, particularly effective and widely employed in the food industry for extracting analytes from vegetables. SFE offers a viable alternative to conventional solvent extraction due to its nontoxic, cost-effective, relatively inert, recyclable, and nonflammable properties (Benchikh *et al.*, 2023). SPE provides high purity by using selective adsorbents. EAE enhances yield and preserves bioactivity by breaking down plant cell walls with specific enzymes. EAE is an energy-efficient and environment-friendly technique for the selective extraction of targeted products, without compromising the functional properties (Kumar and Tingirikari, 2023).

9.4.2.1 Microwave-Assisted Extraction

MAE is a significant technique in analytical chemistry, offering numerous advantages over traditional extraction methods. These benefits include improved extraction efficiency, shorter extraction times, reduced solvent consumption, and enhanced analyte recovery (Eskilsson and Björklund, 2000). By directly applying heat to the sample, microwaves enable the rapid and efficient extraction of target compounds by increasing their solubility and diffusion, thereby requiring less solvent. Consequently, MAE is a more environmentally friendly and cost-effective option, promoting greener and more sustainable analytical chemistry workflows (Ferrara *et al.*, 2023). The extraction of anthocyanins from saffron (*Crocus sativus*) flower tepals using MAE has been investigated. Optimal processing conditions were determined for cyanidin-3-glucoside, specifically focusing on extraction temperatures ranging from 35 to 75°C and extraction times between 5 and 15 minutes, with a maximum irradiation power of 360 W. Microstructural analysis revealed that MAE effectively disrupted cell walls under microwave irradiation, thereby demonstrating its rapid and efficient capability for extracting saffron anthocyanins (Jafari *et al.*, 2019).

The optimization of anthocyanins, specifically cyanidin-3-glucoside and cyanidin-3-rutinoside found in mulberry, was achieved using MAE. Subsequent identification of these anthocyanins in the extract was performed utilizing HPLC-ESI-MS (Zou *et al.*, 2012). A study developed a method for extracting anthocyanins from mulberry using MAE with ethanol/ammonium sulfate as a biphasic extractant. This approach, termed microwave-assisted aqueous two-phase system extraction, was applied to the extraction of functional natural products, specifically anthocyanins, from mulberry fruits (Wang *et al.*, 2024). A study on the stability of 3-deoxyanthocyanin pigment from sorghum compared to anthocyanins from grains, maize, and cowpea under MAE revealed that the enhanced extraction of 3-deoxyanthocyanin was attributed to its structural stability. Additionally, the improved diffusion from the cell matrix due to microwave-induced disruption of sorghum cell walls contributed to the efficiency of MAE (Herrman *et al.*, 2020).

9.4.2.2 Ultrasound-Assisted Extraction

Ultrasound, an evolving technology in the food industry, plays a pivotal role in enhancing processes such as the extraction of compounds like anthocyanins (Nurkhasanah *et al.*, 2023). Applying ultrasound to solid or fluid products generates acoustic waves and cavitation, triggering modifications in the sample matrix that lead to improved outcomes. This technique is particularly very effective in extracting valuable compounds such as natural pigments (Linares and Rojas, 2022), thereby boosting overall yield and sustainability. Furthermore, ultrasound alters the physical and chemical properties of plant materials, facilitating the release of extractable compounds by disrupting cell walls and enhancing mass transport, ultimately contributing to more efficient and sustainable food production (Chemat *et al.*, 2017; Linares and Rojas, 2022; Nurkhasanah *et al.*, 2023). An analytical system was created to optimize ultrasound-assisted extraction (UAE) conditions for anthocyanin extraction, employing Box-Behnken design (BBD). The analysis identified six anthocyanins in the samples, particularly in various pigmented corn samples, highlighting potential varieties that offer enhanced health benefits (Nurkhasanah *et al.*, 2023).

The major anthocyanin content analysis in *Erica australis* flowers was optimized using a UAE method. This optimization involved the application of BBD along with response surface methodology, focusing on solvents, pH, extraction temperature, and sample-to-solvent ratio as key variables (Carrera *et al.*, 2021). In a study, both heat- and ultrasound-assisted extraction methods were used to recover anthocyanin content. The research revealed higher levels of delphinidin-3-O-sambubioside and cyanidin-3-O-sambubioside anthocyanins in *H. sabdariffa* calyces compared to other extraction methods reported in the literature (Pinela *et al.*, 2019). UAE was utilized to extract anthocyanins from Malus 'Royalty' fruit, with optimized parameters such as temperature, solid-liquid ratio, solvents, extraction time, and ultrasonic power. The main anthocyanin identified was cyanidin-3-galactoside. This study indicates that the concentrated anthocyanin fraction from Malus 'Royalty' fruits may possess promising antitumor and supportive therapeutic properties for gastric cancer treatment (Liu *et al.*, 2022). A study optimized the extraction conditions for ultrasound-assisted enzymatic extraction (UAEE) of anthocyanins from grape skins using response surface methodology combined with a genetic algorithm. The highest yield of anthocyanins (3.01 ± 0.04 mg/g) from grape skins via UAEE was achieved with the optimized extraction parameters. Delphinidin-3,5-O-diglucoside and cyanidin-3-O-rutinoside were among the purified compounds. The results indicated that anthocyanin extracts obtained under the optimum extraction conditions (AEOEC) had a superior antitumor effect on breast cancer MCF-7 cells compared to two purified anthocyanins (Tan *et al.*, 2020).

9.4.2.3 Supercritical Fluid Extraction

SFE has gained attention as an innovative method for extracting bioactive compounds from plant materials due to its efficiency and environmental friendliness (Wen *et al.*, 2019). SFE has been explored extensively for the extraction of anthocyanins, which are natural pigments with various health benefits. Anthocyanins responsible for the colors, red, purple, and blue, are in fruits and vegetables. Berries, currants, grapes, and some tropical fruits have high anthocyanin content (Khoo *et al.*, 2017). This extraction technique utilizes supercritical fluids, such as carbon dioxide, at specific temperatures and pressures to extract anthocyanins effectively while minimizing solvent use and preserving the integrity of the extracted compounds (Machado *et al.*, 2013; Tyśkiewicz *et al.*, 2018; Nunes *et al.*, 2022). Phenolic compounds and anthocyanins were extracted from blueberry (*Vaccinium myrtillus* L.) residues using supercritical CO_2 and pressurized liquids. This study employed both supercritical CO_2 extraction and pressurized liquid extraction to recover bioactives from the blueberry residues. Interestingly, the concentrations of bioactive compounds in these extracts were found to be higher than those typically reported for fresh blueberries (Paes *et al.*, 2014). *Arrabidaea chica* (Humb. and Bonpl.) Verlot leaves, abundant in anthocyanins, have long been utilized as a medicinal plant in the Amazon. A sequential extraction in a fixed bed was conducted at 40°C and 300 bar using supercritical carbon dioxide ($scCO_2$) to obtain various extracts from this plant. The primary anthocyanin component, carajurin, was notably present in the highest concentration in the ethanolic extracts of *A. chica* (Paula *et al.*, 2013).

$scCO_2$ and water as a co-solvent were utilized to extract anthocyanin-rich compounds from Haskap berry (*L. caerulea* L.) pulp paste. The extraction parameters, including pressure, temperature, and water quantity, were further refined through the BBD. The optimal conditions resulted in the highest total anthocyanin yield of 52.7% achieved at 45 MPa pressure, 65°C temperature, using 5.4 g of water per 3.2 g of berry pulp paste, with 15 minutes of static time followed by 20 minutes of dynamic time (Jiao and Pour, 2018). Iridoids and anthocyanins were extracted from haskap (*L. caerulea* L.) berries using $scCO_2$ with ethanol as a modifier. Compared to conventional ethanolic extraction, the dynamic streamlined method resulted in a 97% yield of iridoids and approximately 7% more anthocyanins (Martinez *et al.*, 2021). The extraction yield and anthocyanin content of *H. sabdariffa* L. (Roselle) were investigated using the $scCO_2$ method. The results indicated that employing a low flow rate (4 ml/min), smaller particle size (200–355 µm), and a higher percentage ratio of modifier (10%) led to an increased anthocyanin content in the extract (Idham *et al.*, 2020). The study explored the supercritical carbon dioxide extraction ($scCO_2$) of total anthocyanin and phenolic compounds from *Syzygium cumini* fruit pulp, examining the impact of three independent variables (pressure, temperature, and co-solvent flow rate) using a Box-Behnken response surface design with three factors at three levels. The optimized conditions were assessed to evaluate the effects of these three independent variables (pressure, temperature, and co-solvent flow rate) on the extraction process (Maran *et al.*, 2014).

9.4.2.4 Solid-Phase Extraction

SPE is a widely used and cost-effective technique for purifying samples, crucial for analyzing both major and minor food components, including squalene, tocopherols, and anthocyanins (Ferreiro-González *et al.*, 2014; Grigoriadou *et al.*, 2007). These natural pigments, responsible for the red, purple, and blue hues in many fruits and vegetables, are often found in complex mixtures (He and Giusti, 2011; Ferreiro-González *et al.*, 2014). SPE involves conditioning the solid phase, loading the sample to adsorb anthocyanins, washing away impurities, and eluting the purified anthocyanins. Typically, reverse-phase materials like C_{18} or polymeric resins are used for their high affinity due to hydrophobic interactions, making SPE an efficient method for isolating and studying these valuable compounds (Badawy *et al.*, 2018; Badawy *et al.*, 2022). SPE was employed to obtain anthocyanin-rich extracts from five berry species: chokeberry (*A. melanocarpa*), elderberry (*Sambucus nigra*), black currant (*Ribes nigrum*), blackberry (*Rubus fruticosus*), and blueberry (*V. corymbosum* L.). This process effectively separated nearly all sugars and acids from the crude berry extracts. SPE achieved an anthocyanin recovery rate of 90–95.6% from these extracts (Denev *et al.*, 2010).

A simple SPE method was developed to fractionate and isolate di-acylated anthocyanins from red cabbage extracts, which contain cyanidin derivatives capable of exhibiting various colors from red to blue. These di-acylated anthocyanins are known for their greater stability and more desirable blue colorimetric properties. The SPE method achieved a high recovery rate of over 93% for the di-acylated pigments (Ahmadiani *et al.*, 2019). A new SPE method was developed to determine anthocyanins in grapes. Various SPE cartridges were evaluated, including C-18 and vinylbenzene-based cartridges. This method is particularly valuable for assessing phenolic maturity, as it facilitates the analysis of anthocyanins present in grapes (Ferreiro-González *et al.*, 2014). Capitani *et al.* employed aqueous and organic extracts of blueberries for untargeted nuclear magnetic resonance (NMR) metabolite profiling. They also conducted targeted NMR characterization of anthocyanins using SPE. This led to the identification of five anthocyanins: malvidin-3-glucoside, malvidin-3-galactoside, delphinidin-3-glucoside, delphinidin-3-galactoside, and petunidin-3-glucoside (Capitani *et al.*, 2014).

Traditional extraction methods from plant materials are non-selective, resulting in pigment solutions with many by-products like sugars, organic acids, and proteins, which can accelerate anthocyanin degradation and complicate processing steps like spray drying. To obtain a highly concentrated and purified anthocyanin pigment from black chokeberry (*A. melanocarpa* var. *nero*), juices and skin extracts were purified using SPE. SPE can purify anthocyanins in solution by removing impurities such as sugars, phenols, organic acids, proteins, salts, and flavonoids that co-extract during solid-liquid extraction. To optimize anthocyanin extraction from elderberry fruits (*S. nigra*), adsorbent C_{18} silica gel was used. After three extractions with 70% ethanol water and 1% citric acid for 30 minutes each, anthocyanins were obtained at approximately 0.98% of the fresh plant weight (Salamon *et al.*, 2015). A novel SPE-HPLC method for analyzing the anthocyanin and organic acid composition of Finnish cranberry (*Vaccinium oxycoccos* L.) was studied. The most abundant individual anthocyanins were cyanidin-3-galactoside (19.2%), cyanidin-3-arabinoside (23.1%), peonidin-3-galactoside (21.5%), and peonidin-3-arabinoside (14.1%), as analyzed by RP-HPLC (Huopalahti *et al.*, 2000).

9.4.2.5 *Enzyme-Assisted Extraction*

In EAE, the plant material undergoes pretreatment with enzymes like protease, pectinase, pectinesterase, cellulase, hemicellulase, cellobiase, α-amylase, and fructosyltransferase. These enzymes hydrolyze the cell walls, releasing phytochemicals bound to lipid and carbohydrate chains within the cells (Shinwari, 2021). EAE is an energy-efficient and environmentally friendly method for selectively extracting targeted products without compromising their functional properties (Kumar and Tingirikari, 2023). Anthocyanins obtained through EAE not only exhibit higher content but also demonstrate increased stability compared to those extracted by traditional methods. This enhanced stability encompasses a broader range of anthocyanin compounds, making the process more effective for preserving the vibrant colors and beneficial properties of these phytochemicals.

An aqueous solution of Pectinex (containing cellulase, hemicellulase, and pectinase) at concentrations of 1–10% and a temperature of 40°C was used to extract anthocyanins from saffron tepals over reaction times of 20–180 minutes, and the results were compared with those obtained using an ethanol solvent under similar conditions. The anthocyanin content in the Pectinex solution reached approximately 6.7 mg/g of tepal powder, which is over 40% higher than that achieved with the ethanol method, at enzyme concentrations of 5% and extraction times of 60 minutes. Anthocyanins extracted with aqueous enzymes showed degradation rates three times slower and had 50% more vibrant chroma color compared to those recovered using ethanol after 3 hours of extraction. HPLC analysis indicated that anthocyanins extracted with mixed enzymes contained about 80% more cyanidin-3-glucosides and 20% less pelargonidin 3,5-glucosides compared to the ethanol method (Lotfi *et al.*, 2015).

EAE is more efficient because enzymes effectively break down plant cell walls, leading to the release of intracellular proteins and other valuable compounds. This breakdown enhances the extraction process by allowing easier access to and release of phytochemicals and nutrients that are typically bound within the plant's cellular structure (José Aliaño González *et al.*, 2021). Consequently, this method increases yield and preserves the integrity and activity of the extracted compounds. In a study, EAE was found to be superior to conventional extraction methods for extracting anthocyanins from eggplant (*Solanum melongena* L.) peel. EAE yielded the highest total yields and anthocyanin content compared to conventional solvent extraction. The extraction process was optimized using response surface methodology, with enzyme concentration, extraction temperature, and maceration time being the key variables optimized through a central-composite design. The results confirmed that EAE is more effective than traditional extraction methods (Amulya and Ul Islam, 2023).

Modern extraction techniques, like EAE, are more efficient, eco-friendly, and cost-effective. They enhance the stability of extracted compounds, improve food quality and nutritional value, and maximize health benefits. This is especially true for anthocyanins, where these methods yield higher content and better preservation of beneficial properties. The examples of modern extraction methods used for anthocyanins are discussed in detail in Table 9.1.

9.5 Conclusion

In conclusion, this chapter has provided a comprehensive overview of the chemistry of anthocyanins and the various approaches for their extraction. Anthocyanins, natural pigments found in many fruits and vegetables, possess significant health benefits due to their antioxidant properties. Traditional extraction methods, such as maceration, percolation, and Soxhlet extraction, have been widely used but often lack efficiency and can be environmentally unfriendly. Modern extraction methods have emerged to address these shortcomings, offering more efficient, cost-effective, and environmentally friendly alternatives. Techniques such as MAE and UAE enhance the disruption of plant tissue, leading to higher yields of anthocyanins. SFE and SPE are also notable for their efficiency and ability to maintain the stability of the extracted compounds. EAE further improves yield and purity by breaking down cell walls more effectively than conventional methods.

These advanced techniques not only improve the efficiency and effectiveness of anthocyanin extraction but also ensure the stability of these sensitive compounds. The use of modern extraction methods preserves the functional properties of anthocyanins, thereby maximizing their health benefits. As the demand for natural pigments and nutraceuticals grows, the continued development and optimization of these extraction methods will be crucial for meeting industry needs while promoting sustainability.

TABLE 9.1

Extraction of Anthocyanins by Modern Extraction Techniques

Name of Plant	Modern Extraction Method	Optimized Parameters	Results	References
Rhodomyrtus tomentosa (Ait.) Hassk	MAE	Solvent concentration (30–70%), application ratio (10:1–1:1 ml/g), microwave power (100–500 W), and time (1–9 minutes).	The anthocyanin content reached its highest value of 137.54 mg/l when extracted using a 50% ethanol solvent.	Pham et al. (2022)
Clitoria ternatea petals	UAE	Optimum extraction parameters are 30 minutes extraction time, 50°C extraction temperature, and 10:1 solvent-to-solid ratio.	The glycerol-water system was identified as a green alternative for anthocyanin extraction, also serving effectively as a storage medium.	Shu et al. (2022)
Aronia melanocarpa	UMAE-NADES (ultrasonic-microwave-assisted natural deep eutectic solvent)	Eight different natural eutectic solvents were screened initially, and choline chloride-glycerol was selected as the extraction solvent.	Six anthocyanins were isolated and UMAE-NADES demonstrated significant efficiency in extracting bioactive substances, showing potential utility in both the food and pharmaceutical industries.	Lin et al. (2022)
Sour cherry (*Prunus cerasus* L.), peels	MAE	The optimized parameters for the extraction process were 500 W of microwave power, 90 seconds of irradiation time, and an 80% ethanol solvent concentration.	The maximum predicted yield of total anthocyanins was found to be 12.47 mg of cyanidin-3-glucoside per gram of fresh mass (mg-C3G/g-FM).	Kurtulbaş Şahin et al. (2021)
Raspberry (*Rubus Coreanus* Miq.)	MAE	Different solvents, including methanol, ethanol, and acetone, were used.	The canonical analysis estimated that the highest total anthocyanin content of raspberry was 17.93 mg cyanidin-3-O-glucoside equivalents per gram dry weight.	Teng et al. (2013)
Pyrus communis 'Starkrimson' fruit peel	UAE	Optimized UAE conditions: ultrasonic power, 162 W; temperature, 71°C; trifluoroacetic acid, 3%; ethanol, 57%; ultrasonication time, 11 minutes; and sample-to-solvent ratio = 1:30 g/ml).	Cyanidin-3-galactoside (Cy3-gal) yield was significantly higher (0.343 ± 0.005 mg/g) than from CE (0.266 ± 0.004 mg/g).	Belwal et al. (2019)
Raspberry (*Rubus idaeus* L. var. Heritage)	UAE	Optimized conditions: The ratio of solvents to materials was 4:1 (ml/g), extraction time was 200 seconds, and ultrasonic power was 400 W.	UAE was found to be more efficient and rapid for extracting anthocyanins from red raspberries, due to the strong disruption of fruit tissue structure caused by ultrasonic acoustic cavitation.	Chen et al. (2007)
Petals of *Chaenomeles sinensis*	SFE and HSCCC	SFE coupled with high-speed counter-current chromatography (HSCCC) was successfully used for the extraction and on-line isolation of the anthocyanidins.	Six compounds including delphinidin-3-O-glucoside, cyanidin-3-O-glucoside, peonidin-3-O-glucoside, delphinidin, peonidin, and malvidin (Mv) were successfully separated within 300 minutes.	Li et al. (2013)

References

Abdel-Aal ESM, Akhtar H, Rabalski I, Bryan M. Accelerated, microwave-assisted, and conventional solvent extraction methods affect anthocyanin composition from colored grains. Journal of Food Science. 2014; 79(2): C138–46.

Ahmadiani N, Sigurdson GT, Robbins RJ, Collins TM, Giusti MM. Solid phase fractionation techniques for segregation of red cabbage anthocyanins with different colorimetric and stability properties. Food Research International. 2019; 120: 688–96.

Amulya PR, Ul Islam R. Optimization of enzyme-assisted extraction of anthocyanins from eggplant (*Solanum melongena* L.) peel. Food Chemistry: X. 2023; 18: 100643.

Badawy ME, El-Nouby MA, Kimani PK, Lim LW, Rabea EI. A review of the modern principles and applications of solid-phase extraction techniques in chromatographic analysis. Analytical Sciences. 2022; 38(12): 1457–87.

Badawy ME, El-Nouby MA, Marei AE. Development of a solid-phase extraction (SPE) cartridge based on chitosan-metal oxide nanoparticles (CH-MO NPs) for extraction of pesticides from water and determination by HPLC. International Journal of Analytical Chemistry. 2018; 2018(1): 3640691.

Belwal T, Huang H, Li L, Duan Z, Zhang X, Aalim H, Luo Z. Optimization model for ultrasonic-assisted and scale-up extraction of anthocyanins from *Pyrus communis* 'Starkrimson' fruit peel. Food Chemistry. 2019; 297: 124993.

Benchikh Y, Bachir-bey M, Chaalal M, Ydjedd S, Kati DE. Extraction of phenolic compounds, Editor(s): Dr. Inamuddin, Tariq Altalhi, Green sustainable process for chemical and environmental engineering and science. Elsevier, 2023; 329–354. https://doi.org/10.1016/B978-0-323-95156-2.00008-8

Byamukama R, Jordheim M, Kiremire B, Namukobe J, Andersen Ø M. Anthocyanins from flowers of *Hippeastrum* cultivars. Scientia Horticulturae. 2006; 109(3): 262–6.

Capitani D, Sobolev AP, Delfini M, Vista S, Antiochia R, Proietti N, Bubici S, Ferrante G, Carradori S, Salvador FR, Mannina L. NMR methodologies in the analysis of blueberries. Electrophoresis. 2014; 35(11): 1615–26.

Carrera C, Aliaño-González MJ, Rodríguez-López J, Ferreiro-González M, Ojeda-Copete F, Barbero GF, Palma M. Optimization of an ultrasound-assisted extraction method for the analysis of major anthocyanin content in *Erica australis* flowers. Molecules. 2021; 26(10): 2884.

Castañeda-Ovando A, de Lourdes Pacheco-Hernández M, Páez-Hernández ME, Rodríguez JA, Galán-Vidal CA. Chemical studies of anthocyanins: A review. Food Chemistry. 2009; 113(4): 859–71.

Chemat F, Rombaut N, Meullemiestre A, Turk M, Perino S, Fabiano-Tixier AS, Abert-Vian M. Review of green food processing techniques, preservation, transformation, and extraction. Innovative Food Science & Emerging Technologies. 2017; 41: 357–77.

Chen F, Sun Y, Zhao G, Liao X, Hu X, Wu J, Wang Z. Optimization of ultrasound-assisted extraction of anthocyanins in red raspberries and identification of anthocyanins in extract using high-performance liquid chromatography-mass spectrometry. Ultrasonics Sonochemistry. 2007; 14(6): 767–78.

Chung C, Rojanasasithara T, Mutilangi W, McClements DJ. Stabilization of natural colors and nutraceuticals: Inhibition of anthocyanin degradation in model beverages using polyphenols. Food Chemistry. 2016; 212: 596–603.

Constantin OE, Istrati DI. Extraction, quantification and caxterization techniques for anthocyanin compounds in various food matrices – A review. Horticulturae. 2022; 8(11): 1084.

de Pascual-Teresa S, Sanchez-Ballesta MT. Anthocyanins: From plant to health. Phytochemistry Reviews. 2008; 7: 281–99.

Denev P, Ciz M, Ambrozova G, Lojek A, Yanakieva I, Kratchanova M. Solid-phase extraction of berries' anthocyanins and evaluation of their antioxidative properties. Food Chemistry. 2010; 123(4): 1055–61.

Devgun M, Nanda AR, Ansari SH. Comparison of conventional and non-conventional methods of extraction of heartwood of *Pterocarpus marsupium* Roxb. Acta Poloniae Pharmaceutica – Drug Research. 2012; 69(3): 475–85.

Diessana A, Parkouda C, Cissé M, Diawara B, Dicko M. Optimization of aqueous extraction of anthocyanins from *Hibiscus sabdariffa* L. calyces for food application. Food Science and Quality Management. 2015; 45: 23–32.

Enaru B, Drețcanu G, Pop TD, Stănilă A, Diaconeasa Z. Anthocyanins: Factors affecting their stability and degradation. Antioxidants (Basel). 2021; 10(12): 1967.

Eskilsson CS, Björklund E. Analytical-scale microwave-assisted extraction. Journal of Chromatography A. 2000; 902(1): 227–50.

Fan G, Han Y, Gu Z, Chen D. Optimizing conditions for anthocyanins extraction from purple sweet potato using response surface methodology (RSM). LWT – Food Science and Technology. 2008; 41(1): 155–60.

Ferrara D, Beccaria M, Cordero CE, Purcaro G. Microwave-assisted extraction in closed vessel in food analysis. Journal of Separation Science. 2023; 46(20): e2300390.

Ferreiro-González M, Carrera C, Ruiz-Rodríguez A, Barbero GF, Ayuso J, Palma M, Barroso CG. A new solid phase extraction for the determination of anthocyanins in grapes. Molecules. 2014; 19(12): 21398–410.

García-Marino M, Hernández-Hierro JM, Rivas-Gonzalo JC, Escribano-Bailón MT. Colour and pigment composition of red wines obtained from co-maceration of Tempranillo and Graciano varieties. Analytica Chimica Acta. 2010; 660(1–2): 134–42.

Grigoriadou D, Androulaki A, Psomiadou E, Tsimidou MZ. Solid phase extraction in the analysis of squalene and tocopherols in olive oil. Food Chemistry. 2007; 105(2): 675–80.

He J, Giusti MM. High-purity isolation of anthocyanins mixtures from fruits and vegetables – A novel solid-phase extraction method using mixed mode cation-exchange chromatography. Journal of Chromatography A. 2011; 1218(44): 7914–22.

Herrman DA, Brantsen JF, Ravisankar S, Lee KM, Awika JM. Stability of 3-deoxyanthocyanin pigment structure relative to anthocyanins from grains under microwave-assisted extraction. Food Chemistry. 2020; 333: 127494.

Huopalahti R, Järvenpää EP, Katina K. A novel solid-phase extraction-HPLC method for the analysis of anthocyanin and organic acid composition of Finnish cranberry. Journal of Liquid Chromatography & Related Technologies. 2000; 23(17): 2695–2701.

Idham Z, Zaini AS, Putra NR, Rusli NM, Mahat NS, Yian LN, Yunus MA. Effect of flow rate, particle size and modifier ratio on the supercritical fluid extraction of anthocyanins from *Hibiscus sabdariffa* (L), IOP conference series: Materials science and engineering. IOP Publishing, 2020, 932(1): 012031.

Jafari SM, Mahdavee Khazaei K, Assadpour E. Production of a natural color through microwave-assisted extraction of saffron tepal's anthocyanins. Food Science & Nutrition. 2019; 7(4): 1438–45.

Jaiswal YS, Guan Y, Moon KH, Williams LL. Anthocyanins: Natural sources and traditional therapeutic uses, Flavonoids – A coloring model for cheering up life. IntechOpen, 2019.

Jiang X, Li X, Zhu C, Sun J, Tian L, Chen W, Bai W. The target cells of anthocyanins in metabolic syndrome. Critical Reviews in Food Science and Nutrition. 2019; 59(6): 921–46.

Jiao G, Pour AK. Extraction of anthocyanins from haskap berry pulp using supercritical carbon dioxide: Influence of co-solvent composition and pretreatment. LWT. 2018; 98: 237–44.

Jordheim M, Fossen T, Andersen ØM. Characterization of hemiacetal forms of anthocyanidin 3-O-β-glycopyranosides. Journal of Agricultural and Food Chemistry. 2006; 54(25): 9340–6.

José Aliaño González M, Carrera C, Barbero GF, Palma M. A comparison study between ultrasound-assisted and enzyme-assisted extraction of anthocyanins from blackcurrant (*Ribes nigrum* L. Food Chemistry: X. 2021; 13: 100192.

Jovanović AA, Đorđević VB, Zdunić GM, Pljevljakušić DS, Šavikin KP, Gođevac DM, Bugarski BM. Optimization of the extraction process of polyphenols from *Thymus serpyllum* L. herb using maceration, heat-and ultrasound-assisted techniques. Separation and Purification Technology. 2017; 179: 369–80.

Khoo HE, Azlan A, Tang ST, Lim SM. Anthocyanidins and anthocyanins: Colored pigments as food, pharmaceutical ingredients, and the potential health benefits. Food & Nutrition Research. 2017; 61(1): 1361779.

Kong JM, Chia LS, Goh NK, Chia TF, Brouillard R. Analysis and biological activities of anthocyanins. Phytochemistry. 2003; 64(5): 923–33.

Kumar V, Tingirikari JM. Degree of methylation: A key to the functional properties of low methylated pectin derived from pectin-rich agrowaste. Food Bioscience. 2023; 103386.

Linares G, Rojas ML. Ultrasound-assisted extraction of natural pigments from food processing by-products: A review. Frontiers in Nutrition. 2022; 9: 891462.

Liu Y, Zhao Y, Zhuo Y, Li Y, Meng J, Wang Y, Li H. Ultrasound-assisted extraction of anthocyanins from Malus 'royalty' fruits: Optimization, separation, and antitumor activity. Molecules. 2022; 27(13): 4299.

Lotfi L, Kalbasi-Ashtari A, Hamedi M, Ghorbani F. Effects of enzymatic extraction on anthocyanins yield of saffron tepals (*Crocus sativus*) along with its color properties and structural stability. Journal of Food and Drug Analysis. 2015; 23(2): 210–8.

Machado BAS, Pereira CG, Nunes SB, Padilha FF, Umsza-Guez MA. Supercritical fluid extraction using CO_2: Main applications and future perspectives. Separation Science and Technology. 2013; 48: 2741–60.

Malik J, Mandal SC. Extraction of herbal biomolecules, Editor(s): Subhash C. Mandal, Amit Kumar Nayak, Amal Kumar Dhara, Herbal biomolecules in healthcare applications. Academic Press, 2022; 21–46. ISBN 9780323858526. https://doi.org/10.1016/B978-0-323-85852-6.00015-9

Manousi N, Sarakatsianos I, Samanidou V. 10 – Extraction techniques of phenolic compounds and other bioactive compounds from medicinal and aromatic plants, Editor(s): Grumezescu, A.M., Holban, A.M., Engineering tools in the beverage industry. Woodhead Publishing, 2019; 283–314. https://doi.org/10.1016/B978-0-12-815258-4.00010-X

Maran JP, Priya B, Manikandan S. Modeling and optimization of supercritical fluid extraction of anthocyanin and phenolic compounds from *Syzygium cumini* fruit pulp. Journal of Food Science and Technology. 2014; 51: 1938–46.

Marić M, Grassino AN, Zhu Z, Barba FJ, Brnčić M, Brnčić SR. An overview of the traditional and inno-vative approaches for pectin extraction from plant food wastes and by-products: Ultrasound-, microwaves-, and enzyme-assisted extraction. Trends in Food Science & Technology. 2018; 76: 28–37.

Martinez AS, Kornpointner C, Haselmair-Gosch C, Mikulic-Petkovsek M, Schröder K, Halbwirth H. Dynamic streamlined extraction of iridoids, anthocyanins and lipids from haskap berries. LWT. 2021; 138: 110633.

Naseri R, Farzaei F, Haratipour P, Nabavi SF, Habtemariam S, Farzaei MH, Khodarahmi R, Tewari D, Momtaz S. Anthocyanins in the management of metabolic syndrome: A pharmacological and biopharmaceutical review. Frontiers in Pharmacology. 2018; 9: 1310.

Nunes AN, Borges A, Matias AA, Bronze MR, Oliveira J. Alternative extraction and downstream purification processes for anthocyanins. Molecules. 2022; 27(2): 368.

Nurkhasanah A, Fardad T, Carrera C, Setyaningsih W, Palma M. Ultrasound-assisted anthocyanins extraction from pigmented corn: Optimization using response surface methodology. Methods and Protocols. 2023; 6(4): 69.

Oancea S. A review of the current knowledge of thermal stability of anthocyanins and approaches to their stabilization to heat. Antioxidants. 2021; 10(9): 1337.

Oladzadabbasabadi N, Nafchi AM, Ghasemlou M, Ariffin F, Singh Z, Al-Hassan AA. Natural anthocyanins: Sources, extraction, characterization, and suitability for smart packaging. Food Packaging and Shelf Life. 2022; 33: 100872.

Paes J, Dotta R, Barbero GF, Martínez J. Extraction of phenolic compounds and anthocyanins from blueberry (*Vaccinium myrtillus* L.) residues using supercritical CO_2 and pressurized liquids. The Journal of Supercritical Fluids. 2014; 95: 8–16.

Paula JT, Paviani LC, Foglio MA, Sousa IM, Cabral FA. Extraction of anthocyanins from *Arrabidaea chica* in fixed bed using CO_2 and CO_2/ethanol/water mixtures as solvents. The Journal of Supercritical Fluids. 2013; 81: 33–41.

Pham TN, Le XT, Pham VT, Le HT. Effects of process parameters in microwave-assisted extraction on the anthocyanin-enriched extract from *Rhodomyrtus tomentosa* (Ait.) Hassk and its storage conditions on the kinetic degradation of anthocyanins in the extract. Heliyon. 2022; 8(6): e09518.

Pinela J, Prieto MA, Pereira E, Jabeur I, Barreiro MF, Barros L, Ferreira ICFR. Optimization of heat- and ultrasound-assisted extraction of anthocyanins from *Hibiscus sabdariffa* calyces for natural food colorants. Food Chemistry. 2019; 275: 309–21.

Ravanfar R, Moein M, Niakousari M, Tamaddon A. Extraction and fractionation of anthocyanins from red cabbage: Ultrasonic-assisted extraction and conventional percolation method. Journal of Food Measurement and Characterization. 2018; 12: 2271–7.

Salamon I, Mariychuk R, Grulova D. Optimal extraction of pure anthocyanins from fruits of *Sambucus nigra*. Acta Horticulturae. 2015; 1061: 73–8.

Samira O, Laila B, Moussa NA, Mohamed I, Devkota K, Abdelhakim B, Maggi F, Caprioli G, El Hassan S, Said G. Recent advances in the extraction of bioactive compounds from plant matrices and their use as potential antioxidants for vegetable oils enrichment. Journal of Food Composition and Analysis. 2024; 105995.

Sharma RJ, Gupta RC, Singh S, Bansal AK, Singh IP. Stability of anthocyanins-and anthocyanidins-enriched extracts, and formulations of fruit pulp of *Eugenia jambolana* ('jamun'). Food Chemistry. 2016; 190: 808–17.

Shen L, Pang S, Zhong M, Sun Y, Qayum A, Liu Y, Rashid A, Xu B, Liang Q, Ma H, Ren X. A comprehensive review of ultrasonic assisted extraction (UAE) for bioactive components: Principles, advantages, equipment, and combined technologies. Ultrasonics Sonochemistry. 2023; 106646.

Shinwari KJ. Chapter 4 – Emerging technologies for the recovery of bioactive compounds from saffron species, Editor(s): Galanakis, Charis M., Saffron. Academic Press, 2021; 143–182.

Shu MH, Annamalai KK, Idris FN, Kamaruddin AH, Nadzir MM. Dataset of ultrasound-assisted extraction of anthocyanin from the petals of *Clitoria ternatea* using Taguchi method and effect of storage conditions on the anthocyanin stability. Data in Brief. 2022; 40: 107803.

Silva VO, Freitas AA, Maçanita AL, Quina FH. Chemistry and photochemistry of natural plant pigments: The anthocyanins. Journal of Physical Organic Chemistry. 2016; 29(11): 594–9.

Su X, Xu J, Rhodes D, Shen Y, Song W, Katz B, Tomich J, Wang W. Identification and quantification of anthocyanins in transgenic purple tomato. Food Chemistry. 2016; 202: 184–8.

Taghavi T, Patel H, Rafie R. Extraction solvents affect anthocyanin yield, color, and profile of strawberries. Plants. 2023; 12(9): 1833.

Tan J, Han Y, Han B, Qi X, Cai X, Ge S, Xue H. Extraction and purification of anthocyanins: A review. Journal of Agriculture and Food Research. 2022; 8: 100306.

Tan J, Li Q, Xue H, Tang J. Ultrasound-assisted enzymatic extraction of anthocyanins from grape skins: Optimization, identification, and antitumor activity. Journal of Food Science. 2020; 85(11): 3731–44.

Tena N, Asuero AG. Up-to-date analysis of the extraction methods for anthocyanins: Principles of the techniques, optimization, technical progress, and industrial application. Antioxidants. 2022; 11(2): 286.

Tyśkiewicz K, Konkol M, Rój E. The application of supercritical fluid extraction in phenolic compounds isolation from natural plant materials. Molecules. 2018; 23(10): 2625.

Varo MA, Serratosa MP, Martín-Gómez J, Moyano L, Mérida J. Influence of fermentation time on the phenolic compounds, vitamin c, color and antioxidant activity in the winemaking process of Blueberry (*Vaccinium corymbosum*) wine obtained by maceration. Molecules. 2022; 27(22): 7744.

Wallace TC, Giusti MM, editors. Anthocyanins in health and disease. CRC Press; 2013.

Wallace TC, Giusti MM. Anthocyanins-nature's bold, beautiful, and health-promoting colors. Foods. 2019; 8(11): 550.

Wang DQ, Cui HP, Zong KL, Hu HC, Yang JT. Extraction of functional natural products employing microwave-assisted aqueous two-phase system: Application to anthocyanins extraction from mulberry fruits. Preparative Biochemistry & Biotechnology. 2024; 54(1): 115–25.

Wang LS, Stoner GD. Anthocyanins and their role in cancer prevention. Cancer Letters. 2008; 269(2): 281–90.

Wang WY, Qu HB, Gong XC. Research progress on percolation extraction process of traditional Chinese medicines. Zhongguo Zhong Yao Za Zhi. 2020; 45(5): 1039–46.

Weber F, Boch K, Schieber A. Influence of copigmentation on the stability of spray dried anthocyanins from blackberry. LWT. 2017; 75: 72–7.

Wen P, Hu T-G, Linhardt RJ, Liao S-T, Wu H, Zou Y-X. Mulberry: A review of bioactive compounds and advanced processing technology. Trends in Food Science and Technology. 2019; 83: 138–58.

West ME, Mauer LJ. Color and chemical stability of a variety of anthocyanins and ascorbic acid in solution and powder forms. Journal of Agricultural and Food Chemistry. 2013; 61(17): 4169–79.

Zhang QW, Lin LG, Ye WC. Techniques for extraction and isolation of natural products: A comprehensive review. Chinese Medicine. 2018; 13:20.

Zhao YW, Wang CK, Huang XY, Hu DG. Anthocyanin stability and degradation in plants. Plant Signaling & Behavior. 2021; 16(12): 1987767.

Zou T, Wang D, Guo H, Zhu Y, Luo X, Liu F, Ling W. Optimization of microwave-assisted extraction of anthocyanins from mulberry and identification of anthocyanins in extract using HPLC-ESI-MS. Journal of Food Science. 2012; 77(1): C46–50.

10

HPLC and UHPLC Analysis of Anthocyanins

Mohd. Javed Naim, Kamran Javed Naquvi, and Javed Ahamad

10.1 Introduction

Anthocyanins (ACNs) are glycosylated phenolic compounds/pigments that occur naturally and are soluble in water, making them simple to incorporate into aqueous media. They exist as an extensive group of natural pigments that show various colors such as orange, red, violet, and blue (Riaz et al., 2016; Salehi et al., 2020). The term "anthocyanin" originates from the Greek words "anthos," meaning flower, and "kyanos," meaning blue. ACNs, or ACNs, are a specific kind of flavonoids that may be found in many sections of plants, with a particular abundance in flowers and fruits (Wallace and Giusti, 2013; Sui, 2016). They are described by the presence of a C_6-C_3-C_6 flavan or 2-phenyl benzodihydorpyrane skeleton (Castañeda-Ovando et al., 2009). Nature has more than 700 known ACNs which serve several purposes such as attracting pests to flowers for pollination, attracting grazing animals to fruits for the dispersal of seeds, and protecting plant cells from damage by UV radiation (Warner, 2015). ACNs are often present in several plant families, such as Cruciferae, Caprifoliaceae, Ericaceae, Fabaceae, Rosaceae, Saxifragaceae, and Vitaceae (Mazza, 2018). The concentration of ACNs exhibits substantial variation across various dietary matrices, mostly because of the interplay between production and degradation. Notably, colored grains (purple maize and black rice), berries, and red wine are among the most abundant sources of ACNs (Figure 10.1). The inclusion of ACNs in these food products not only contributes to their visual attractiveness but also improves their nutritional composition (Cappellini et al., 2021; Custodio-Mendoza et al., 2024). The process of ACN production has been well investigated, although our understanding of its breakdown is still restricted (Passeri et al., 2016). ACN metabolism is regulated by genetic, developmental, and environmental variables. Many process-related factors, such as light, pH, oxygen, co-pigmentation, enzymes, ascorbic acid, sulfites, and temperature, may cause ACNs to deteriorate, which is a major downside of employing ACNs (Enaru et al., 2021).

The most prevalent ACNs found in plants are derived from six widely distributed anthocyanidins: pelargonidin, cyanidin, delphinidin, peonidin, petunidin, and malvidin (Figure 10.2) (Kong et al., 2003; Liu et al., 2018). ACNs provide plants with defense mechanisms against a range of living and non-living stressors (Pojer et al., 2013; Ahmed et al., 2014). Currently, neither the USFDA nor EFSA has approved any health claims related to ACNs. Nevertheless, vendors persist in their efforts to communicate the perceived health advantages of ACN supplementation to consumers by interpreting the existing body of literature on the bioactivity of ACNs. They describe various effects on the eyes, blood vessels, lipid profile, liver, antioxidant status, and anti-inflammatory properties. While an

DOI: 10.1201/9781003453260-12

FIGURE 10.1

Different natural sources of anthocyanins (De Pascual-Teresa and Sanchez-Ballesta, 2008; Escribano-Bailón et al., 2006; Wu et al., 2006; Koponen et al., 2007; Horbowicz et al., 2008; Veberic et al., 2009; Montilla et al., 2011; Müller et al., 2012; Fredes et al., 2014; Lao et al., 2017; Guine et al., 2018; Bar-Ya'akov et al., 2019).

official health claim has not been established, other research has investigated the impact of ACNs *via* laboratory experiments and live subjects (Ockermann et al., 2021).

In addition to their antioxidant, anticarcinogenic, anti-inflammatory, cardioprotective, and anti-neurodegenerative properties, ACNs have remarkable bioactive features and provide a host of other health advantages. Additionally, they improve eye health and have a

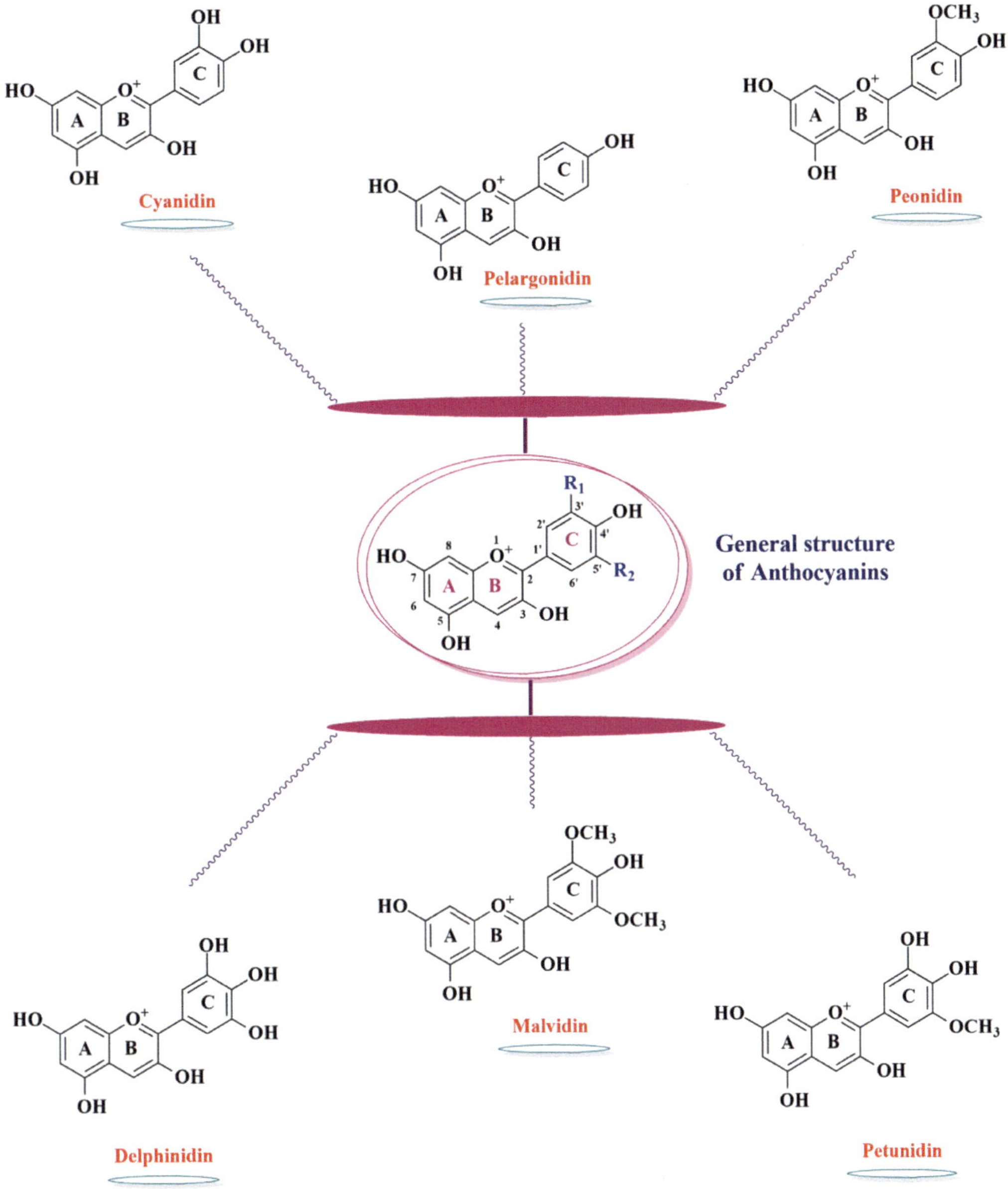

FIGURE 10.2
Chemical structure of important anthocyanins along with their general chemical structure (Salehi et al., 2020; Ockermann et al., 2021; Al-Khayri et al., 2022).

protective impact against diabetes and ocular disorders (Samtiya et al., 2021; Al-Khayri et al., 2022). Additionally, ACNs derived from various berries have shown the ability to safeguard the brain against oxidative damage that comes with aging and to improve cognitive performance. Several food-related applications might benefit from these natural pigments (Speer et al., 2020). According to new research, these compounds may inhibit the replication of several viruses, such as herpes simplex, parainfluenza, syncytial, HIV, rotavirus, and adenovirus (Mohammadi Pour et al., 2019). They are very desirable for pharmacotherapeutic applications because of their low toxicity, a broad variety of pharmacological actions, and supporting experimental and clinical evidence (Figure 10.3).

10.2 Chemistry of Anthocyanins

Anthocyanidins (C3′, C4′, C5′) and flavylium cation backbones (C3, C5, C6, and C7) are the typical sites of hydroxylation in these compounds (Mattioli et al., 2020). The structure of anthocyanidin glycosides (ACNS) typically consists of one glucoside unit; however, some ACNs have more than one glucoside unit attached to them (simple sugar or oligosaccharides), and they have distinctive tricyclic (C_6-C_3-C_6) skeleton pattern (Ockermann et al., 2021). Glucose, arabinose, and galactose are the main monosaccharides, whereas rutinose is a disaccharide that contains 6-O-α-L-rhamnosyl-D-glucose. The most frequent ACN is cyanidin-3-glucoside, while the most prevalent anthocyanidin is cyanidin (Kong et al., 2003; De Pascual-Teresa and Sanchez-Ballesta, 2008; Khoo et al., 2017).

ACNs, due to the presence of their polar hydroxyl group, establish hydrogen bonds with water molecules. According to research by Houghton et al. (2021) and Khoo et al. (2017), the amount of hydroxyl (OH) and methoxyl ($O-CH_3$) groups in an ACN determines its color and its intensity. ACNs with more OH groups tend to be blue, while those with more OCH_3 groups tend to be reddish. There are ACNs in every part of higher plants. The six most common ACNs in nature are based on aglycones, and they are cyanidin (50%), pelargonidin (12%), petunidin (7%), delphinidin (12%), peonidin (12%), and malvidin (7%), with the only difference being the patterns of C-ring substitution (Tena et al., 2020). Anthocyanidins may be glycosylated on various hydroxyl groups of the molecule to generate ACNs, with 3-OH being the most common glycosylation site in nature, resulting in 3-O-β-glucoside. ACNs exist as flavylium cations (oxonium-charged oxygen) at low pH levels, according to biochemical investigations, while uncharged quinones are generated at neutral pH. All ACNs are somewhat stable in basic circumstances (a property that grows in direct proportion to pH), and they may degrade in various ways, losing their color in the process (Figure 10.4) (Mattioli et al., 2020).

10.3 Analysis of Anthocyanins

Natural ACNs have been the focus of several publications in recent literature due to their promising role as healthful pigments. The following are a few examples of their uses: in food (Giusti and Wrolstad, 2003), in plants (Cooper-Driver, 2001), in biosynthesis (Springob et al., 2003), in quantitative analysis using chromatographic and electrophoretic techniques (da Costa et al., 2000), and how they affect plants when stressed (Simmonds, 2003). Considering their bioavailability and steadiness in food is crucial for optimizing their effects and ensuring their safety for consumption. ACNs in various matrices have been identified and characterized using a broad range of analytical techniques. Raman spectroscopy, FT-IR, and near-IR spectroscopy are three crucial analytical tools for the quick and non-destructive assessment of food composition. As a result of their distinct chemical structures, FT-IR and Raman spectra, when coupled with NMR and MS, may help identify food components (ACNs) (Skogerson et al., 2011). On the other hand, if you only need a quick screening, you may utilize these methods alone; complete quantitative analysis

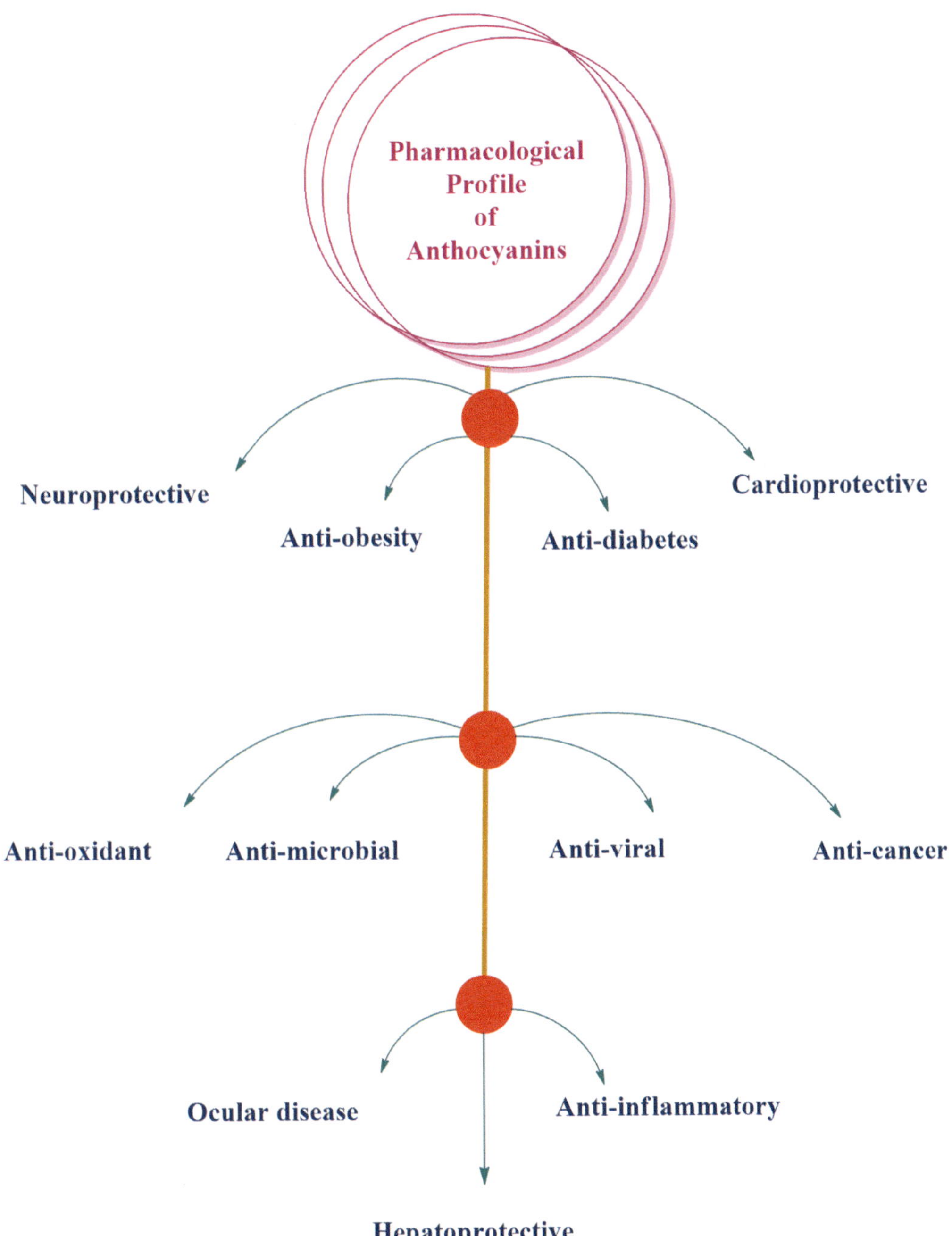

FIGURE 10.3
Important pharmacological activities exhibited by anthocyanins (He and Giusti, 2010; Salehi et al., 2020; Khan et al., 2022).

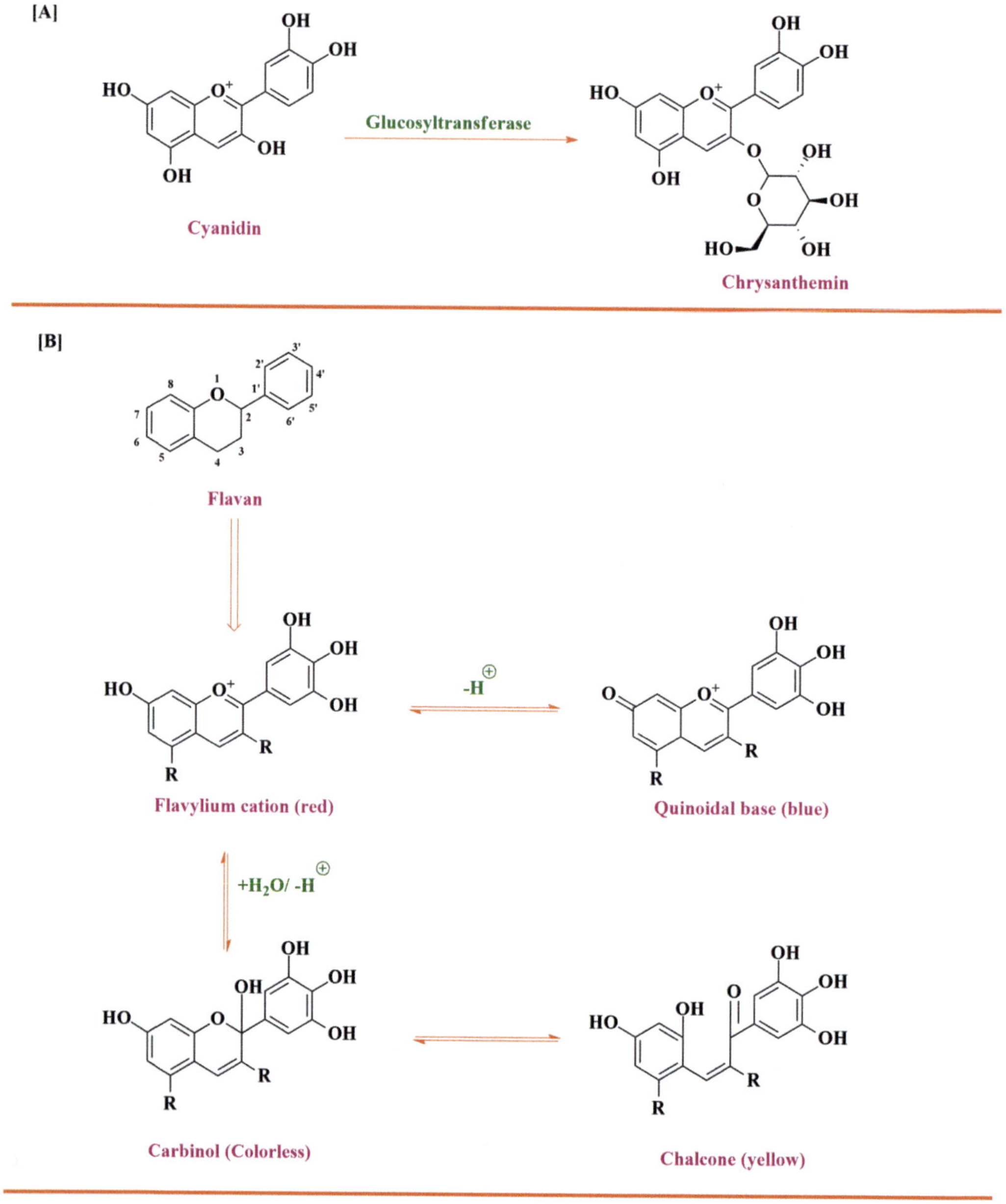

FIGURE 10.4

(A) Conversion of anthocyanidin to anthocyanin via enzymatic action of glucosyltransferase and (B) change in color in anthocyanins due to change in pH aqueous media (Mattioli et al., 2020; Ockermann et al., 2021).

requires a lot of time and effort spent on sample preparation. As described by the training dataset, it identifies out-of-specification conditions by spectroscopic "fingerprint" deviations. Although near-infrared and Raman spectroscopy provide fast analysis, they cannot directly measure ACN content; this subtlety is critical and proves the claim (Amanah et al., 2020).

Liquid chromatography (high-performance or ultra-high-performance) using fluorescence, mass spectrometry (MS), or diode array detectors (DADs) is a typical component of such techniques. Although MS cannot differentiate between positional isomers on its own, it is often used in conjunction with high-performance liquid chromatography (HPLC) to achieve this goal (Garcia-Oliveira et al., 2021). Chromatographic techniques are vital for isolating complicated mixtures according to the physical and chemical characteristics of their constituents, which is essential for identifying specific ACNs in food matrices (Sun et al., 2022). Because of their great sensitivity, selectivity, and resolution, liquid chromatographic (LC) techniques are ideal for analyzing complex mixtures (Lasoń-Rydel et al., 2021; Thuy et al., 2021). The basic idea behind LC is that analytes may be separated from liquid samples by interacting with stationary and mobile phases (Rusli et al., 2022). The mobile phase transports the sample from one step of the system to another; it might be a solvent alone or a mixture of organic and water-based solutions. Based on the separation needs, its composition might be acidified or salt-enriched. Stationary phases, which may be solid or liquid, are used to separate analytes of different compositions. These phases are then packed onto a column or coated onto a support. Analytes with a lower polarity are treated with the normal phase, whereas those with a higher polarity are treated with the reverse phase. Size exclusion is used for size-based separation, whereas ion exchange is employed for charged groups (Sivamaruthi et al., 2020; Leonarski et al., 2023).

While both HPLC and UHPLC use size, electrostatic interactions, and affinity to separate compounds, the two methods vary in the pressures used and the particle sizes that must be considered. With smaller stationary phase particles and elevated pressures (10,000–15,000 psi), UHPLC improves resolution and peak capacity, particularly for tightly eluting peaks (Fibigr et al., 2017; Yabré et al., 2018; Sun et al., 2022). Following similar chromatographic principles, thin-layer chromatography (TLC) uses a stationary phase consisting of an adsorbent material (thin layer) on a plate. This approach is often used for qualitative scrutiny and initial compound separation; it is both quick and simple to use, making it ideal for educational purposes. While UHPLC and HPLC are known for their speed and accuracy, TLC isn't always up to snuff when it comes to resolution and sensitivity (Tuzimski, 2002; Bele and Khale, 2011; Rodriguez-Saona and Allendorf, 2011). When it comes to isolating, detecting, and quantifying varied ACN profiles found in various foods, LC's adaptability, sensitivity, and high resolution provide unmatched capabilities. Understanding nutritional content, disclosing health advantages, and guaranteeing food quality control all rely on the comprehensive study of individual ACNs (Thuy et al., 2021; Avula et al., 2023; Zhang et al., 2023). Optimal gradients, operational systems, and detection techniques are constantly evolving to improve sensitivity, specificity, and structural information (Pereira et al., 2016; Badawy et al., 2022). Photodiode array (PDA) detectors and ultraviolet-visible (UV-Vis) are examples of topical techniques, whereas MS is an example of a quantitative approach. A combination of these advancements makes liquid chromatography for ACN analysis more precise and time-saving (Drabińska et al., 2023).

10.3.1 Role of HPLC in Analysis of Anthocyanins

Using HPLC-DAD-MS analysis, Paun et al. (2022) determined the antioxidant potential, total phenolics, and ACNs in blackberry and black grape skins. Grape skins included seven ACN derivatives, five glycosylated ACN compounds in blackberries, and eight such compounds in grapes. Blackberries mostly contained cyanidin-3-glucoside, but grape skins primarily contained delphinidin-3-O-glucoside. Extracts from blackberries, as compared to grape skins, were richer in polyphenols and ACNs, and they showed a significant improvement in polymeric color whether extracted with methanol or ethanol; however, the TAC (total antioxidant capacity) values were greater in the former case. To measure ACN in different formulations, such as spray-dried aqueous extract, oral powder, and compressible lozenges, Khan et al. (2022) developed an easy, fast, and stability indicating HPLC-UV method using delphinidin-3-O-sambubioside (Dp3S) as a marker component. With higher values of correlation coefficient (0.9998), limit of detection (0.01 μg/mL), and limit of quantification (0.04 μg/mL), the technique was determined to be accurate, precise, and repeatable when validated according to ICH principles (International Council for Harmonisation) as well. Statistical analysis showed that the present procedure is accurate, exact, and repeatable (P < 0.05). When it comes to the detection wavelength and mobile phase pH, the approach holds up well (P < 0.05). Kalogiouri et al. (2022) developed an HPLC-DAD analytical technique to determine ACN levels in three different red Greek wine grape varieties – Kotsifali, Limnio, and Vradiano. Throughout 2020, 16 samples were gathered from various locations around Greece, including the official ampelographic collections and the major areas of each type. Both the limits of detection (LODs – 0.06–0.12 mg/kg) and the limits of quantification (LOQs – 0.20–60 mg/kg) were within the specified ranges. Both the within-day and between-day tests demonstrated sufficient precision with RSD% values below 6.2% and 8.5%, respectively. The major ACNs in blueberries, including delphinidin-3-O-glucoside, cyanidin-3-O-glucoside, petunidin-3-O-glucoside, pelargonidin-3-O-glucoside, peonidin-3-O-glucoside, and malvidin-3-O-glucoside, were identified by Zhou et al. (2021) by optimizing HPLC chromatographic conditions. To begin, the ACN monomers were prepared using semi-preparative HPLC, and then analytical HPLC was used to optimize the separation parameters. Using acetonitrile-water as the mobile phase and gradient elution with a detection wavelength of 520 nm, the results showed that six major ACNs could be successfully separated. The mobile phase also included 0.3% phosphoric acid. The detection limits ranged from 0.010 to 0.035 lg/mL and the quantification limits from 0.03 to 0.117 lg/mL, indicating that the measurement was compatible with the identification of ACNs. There were very significant linear correlations (r > 0.9994) between the six components' concentrations and peak areas as measured by the technique. Using HPLC-DAD-MS analysis, Li et al. (2016) isolated 13 ACNs from 17 samples of various blueberry types in China. Of the overall anthocyanidin content, malvidin, delphinidin, and petunidin made the most significant contributions.

Ştefănuţ et al. (2011) analyzed the antioxidant capacity, total phenolics, ACN composition, and ACN concentration in the extracts of native wild blackberries, mulberries, bilberries, and blackthorns by using HPLC. Chromatograms showed 4 ACN compounds in blackberries, 3 in mulberries, 18 in bilberries, and 5 in blackthorns. Blackberry, mulberry, and bilberry extracts had the highest concentration of cyanidin-3-glucoside, whereas blackthorn extract contained the highest concentration of cyanidin-3-rutinoside. The Folin-Ciocalteu technique was used to measure total phenolics, whereas the pH differential method was used to quantify ACN concentration. To determine the impact of heat treatment on these polyphenolic compounds, Truong et al. (2010) isolated ACNs and anthocyanidins from PFSP (purple-fleshed sweet potatoes). Overall, 17 ACNs were identified by HPLC-DAD/ESI-MS/MS for Stokes Purple and NC 415 varieties with five major compounds: cyanidin-3-caffeoylsophoroside, peonidin-3-caffeoylsophoroside-5-glucoside, cyanidin-3-caffeoyl-p-*hydroxybenzoyl sophoroside*-5-glucoside, peonidin-3-caffeoyl-p-hyfroxybenzoylsophoroside-5-glucoside, and peonidin-caffeoyl-*feruloyl sophoroside*-5-glucoside.

Cyanidin-3-(6″,6‴dicaffeoyl sophoroside)-5-glucoside, cyanidin-3-(6″-caffeoyl-6‴-ferulylsophoroside)-5-glucoside, and a total of 12 pigments were detected in Okinawa variety. The main anthocyanidins in the acid-hydrolyzed extract, cyanidin and peonidin, were successfully isolated and quantified using HPLC in conjunction with external standards and measured using HPLC.

In 2005, Rubinskiene et al. made use of blackcurrants from distinct *Ribes nigrum* breeds analyzed for ACN content and stability using HPLC equipped with UV and MS detectors. This process is carried out at different stages of ripeness. The berries of late blackcurrant varieties, such as Vakariai and Ben Alder, have a greater concentration of ACNs. When berries are reddish (at the beginning of ripening), delphinidin-3-rutinoside is the main component; when berries are black in color (ripe), cyanidin-3-rutinoside is a prominent pigment. Aqueous solutions made from dry colorants are more stable than liquid water and ethanol extracts of coloring compounds, according to studies that examined the impact of light and temperature on the stability of blackcurrant pigments. The ACN with the highest heat stability is cyanidin-3-rutinoside. Of the 25 fruits studied by Wu and Prior (2005), 14 had ACNs; the amounts ranged from 2 in nectarines and peaches to 31 in Concord grapes. They used HPLC-ESI-MS/MS in conjunction with diode array detection to comprehensively identify and characterize the ACNs. Delphinidin, cyanidin, pelargonidin, petunidin, peonidin, and malvidin were detected in all the samples that were examined. An HPLC-UV-visible technique for quantifiable detection of ACN in blood orange juices (varieties: "Moro," "Tarocco," "Sanguinello," and "Sanguinello nocellare") was developed by Mondello et al. (2000). They used a calibration curve for the reference compound cyanidin-3-glucoside. The concentration range of cyanidin-3-glucoside that was achieved was 2–217 parts per million (Table 10.1).

10.3.2 Role of UHPLC in Analysis of Anthocyanins

Jia et al. (2020) utilized the UPLC Quadrupole-Orbitrap MS system to quickly identify and quantify 12 ACNs from berries. Detection and quantification limits were set at 3.86–11.61 mg/L, precision at 0.95–2.38%, repeatability at 0.96–2.08%, stability at 0.86–2.31%, mean recovery at 95.8–103.1%, recovery range at 93.1–107.2%, and RSD at less than 5.21% and the technique was found to be successfully validated. The six ACNs in acai berry and dried blueberry extracts were separated and quantitatively analyzed using a fast UHPLC-UV technique that was developed and validated by Fibigr et al. (2017). The method made use of a core-shell particle column with a pentafluorophenyl stationary phase. In about 5 minutes, the ACNs (peonidin-3-glucoside, delphinidin-3-glucoside, delphinidin-3-rutinoside, cyanidin-3-glucoside, and delphinidin-3-galactoside) were isolated and examined. At a temperature of 50°C, the chromatographic separation was carried out on a Kinetex PFP core-shell column measuring 150 × 2.1 mm and having a particle size of 1.7 μm. Relative standard deviations for within-day and between-day repeatability were from 0.4 to 3.0% and 0.6 to 3.1%, respectively, and recoveries for all ACNs analyzed were 97.8–102.6%. Within the range of 0.11–0.14 μg/mL, the LODs were found.

Shim et al. (2014) employed ultra-high-performance liquid chromatography (UHPLC) in the quick detection of ACN glucosides and free delphinidin in grapes and confirmed its potential via tests of accuracy, precision, and linearity. A reversed phase C18 column with a PDA detector was used for the UHPLC separation. The column dimensions were 2 μm in particle size, 2 mm in injection diameter, and 100 mm in length. For delphinidin aglycone, the UHPLC studies had LODs and quantification ranging from 0.08 to 0.26 mg/kg, and for nine different kinds of ACN glucosides, they varied from 0.07 to 0.65 mg/kg. Delphinidin aglycone and individual ACN glucosides had a precision of less than 9.42% both intra- and inter-day.

TABLE 10.1

Role of HPLC in Analysis of Anthocyanins

S. No.	Anthocyanin	Source	Extraction Method	Procedure	Reference
1	Delphinidin-3-O-sambubioside (3.65–96.99%)	Roselle (*Hibiscus sabdariffa* L.)	HPLC-UV	Mobile phase: Water acidified with 0.2% formic acid and acetonitrile (90:10 *v/v*). Flow rate: 0.8 mL/min Injection volume: 15 µL Column: Reversed phase C18 analytical column (dimension 150 × 4.6 mm × 5 µm) at 45°C.	Khan et al. (2022)
2	Cyanidin-3-glucoside (86.49%) Delphinidin-3-O-glucoside (40.64%)	Blackberries Black grape skin	HPLC-DAD-MS	Mobile phase: A (ultrapure water:formic acid – 99:1 *v/v*) and B (acetonitrile: A – 80:20 *v/v*) Flow rate: 1 mL/min Injection volume: 10 µL Column: Reversed phase C18 analytical column (dimension 4.6 × 150 mm × 5 mm) at 40°C.	Paun et al. (2022)
3	Malvidin-3-O-glucoside (1.25–3.56 µg/mL)	Blueberry extracts of Gardenblue, Britewell, and Baldwin	HPLC-DAD	Mobile phase: A (acetonitrile 10–20%) and B (3% formic acid water) Flow rate: 0.8–1.5 mL/min Injection volume: 8, 10 and 15 µL Column: Reversed phase C18 analytical column (dimension 250 × 4.6 mm × 5 µm) at 20–30°C.	Zhou et al. (2021)
4	Cyanidin (7.7–20.4%)	Red Greek wine grape Cultivars (*Vitis vinifera* L.)	HPLC-DAD	Mobile phase: A (aqueous formic acid 5%) and B (methanol) Flow rate: 1 mL/min Injection volume: 5 µL Column: Reversed phase C18 analytical column (dimension 250 × 4.6 mm × 5 µm) at 40°C.	Kalogiouri et al. (2022)
5	Malvidin (41%), delphinidin (33.1%), and petunidin (17.3%)	Blueberries	HPLC-DAD-MS	Mobile phase: A (water:formic acid:acetonitrile – 80:10:3 *v/v/v*) and B (water:formic acid:acetonitrile – 10:10:50 *v/v/v*) Flow rate: 1 mL/min Injection volume: 2 µL Column: Reversed phase C18 analytical column (dimension 250 × 4.6 mm × 5 µm) at 25°C.	Li et al. (2016)

(Continued)

TABLE 10.1 *(Continued)*

Role of HPLC in Analysis of Anthocyanins

S. No.	Anthocyanin	Source	Extraction Method	Procedure	Reference
6	Cyanidin-3-glucoside Cyanidin-3-rutinoside	Blackberries, mulberries, and bilberries Blackthorn extract	HPLC-DAD-MS	Mobile phase: A (20% methanol + 0.5% TFA) and B (methanol) Flow rate: 1.5 mL/min Injection volume: 20 µL Column: Reversed phase C18 analytical column (dimension 125 × 4 mm × 5 µm) at 40°C.	Ştefănuţ et al. (2011)
7	Anthocyanins and anthocyanidin compounds	Purple-fleshed sweet potatoes	HPLC-DAD/ ESI-MS/MS	Mobile phase: A (formic acid in water 1% v/v) and B (acetonitrile containing 1% formic acid) Flow rate: 0.23 mL/min Injection volume: 10–20 µL Column: Reversed phase C18 analytical column (dimension 250 × 2 mm × 4 µm) maintained at 35°C.	Truong et al. (2010)
8	Cyanidin-3-rutinoside Delphinidin-3-rutinoside	Blackcurrant (black color berries on ripening) Blackcurrant (reddish color berries on the onset of ripening)	HPLC-UV-MS	Mobile phase: A (4% phosphoric acid + water) and B (acetonitrile) Flow rate: 1 mL/min Injection volume: 20 µL Column: Reversed phase C18 analytical column (dimension 125 × 4 mm × 5 µm) at 40°C.	Rubinskiene et al. (2005)
9	Cyanidin-3-glucoside (2–217 ppm)	Blood orange juices	HPLC-UV-Vis	Mobile Phase: A (water:formic aid – 9:1) and B (water:formic acid:acetonitrile – 4:1:5) Flow rate: 1 mL/min Injection volume: 20 µL Column: Restek Pinnacle ODS (dimension 250 × 4.6 mm × 5 µm) at 40°C.	Mondello et al. (2000)
10	Anthocyanins	Fruits and berries	HPLC-ESI-MS/ MS	Mobile phase: A (5% aqueous formic acid) and B (methanol) Flow rate: 1 mL/min Column: Reversed phase C18 analytical column (dimension 250 × 4.6 mm × 5 µm).	Wu and Prior (2005)

Following rapid extraction processes and sample preparation, Dias et al. (2012) developed the first validated UHPLC-PDA approach for detecting major ACNs in *Euterpe oleracea* fruits. The separation took less than 17 minutes on a 1.8 m HSS C18 column using acetonitrile gradient elution and 5% formic acid. The validation technique relied on total error and accuracy metrics. Matrix calibration fared better than non-matrix calibration in terms of accuracy. The qualities of the primary ACNs, cyanidin-3-glucoside, and cyanidin-3-rutinoside, were tested, including trueness (<6.76% relative bias), repeatability (<4.6% RSD), intermediate precision (~5.3% RSD), selectivity, response function, and linearity. The compounds were evaluated at concentrations of 1–48 µg/mL. Dias et al. (2011) developed a rapid UHPLC-DAD-ESI-MS method for the quantification of ACNs from *E. oleracea* fruits harvested at different times of the year in the floodplains of the eastern Amazonian region from Abaetetuba region (Brazil). There was a striking similarity in the ACN profiles of the various extracts. In every stage of maturity, cyanidin-3-glucoside and cyanidin-3-rutinoside were the main components in almost the same amounts, except in the first stage, when the ACN content was too low to be quantified. Nevertheless, cyanidin-3-rutinoside became more prevalent than cyanidin-3-glucoside at the final stage of development. Cyanidin, on the other side, was shown to decrease as samples matured. Stage 4 has the highest ACN concentration (Table 10.2).

TABLE 10.2

Role of UHPLC in the Analysis of Anthocyanins

S. No.	Anthocyanin	Source	Extraction Method	Procedure	Reference
1	Pelargonidin-3-glucoside (2856.1 ± 21.23 µg/g) and cyanidin-3-glucoside (2978 ± 25.67 µg/g)	Berries (*Nitraria tangutorum* B., *Hippophae rhamnoides* L., *Lycium ruthenicum* M., *Lycii fructus*, *Rosa xanthina* L., and *Rubuscor chorifolius* L.)	UHPLC-Quadruple-Orbitrap MS	Mobile phase: 0.9% *v/v* formic acid in water and methanol. Flow rate: 0.4 mL/min. Injection volume: 2 µL. Column: Accucore aQ C18 column (dimension 150 × 2.1 mm × 2.6 µm) at 30°C.	Jia et al. (2020)
2	Cyanidin-3-rutinoside (374.7 ± 3.9 µg/100 g)	Acai berry and dry blueberry extracts	UHPLC-UV	Mobile phase: A (5% formic acid) and B (acetonitrile). Flow rate: 0.55 mL/min. Injection volume: 2 µL. Column: Kinetex PFP core-shell column (dimension 150 × 2.1 mm × 1.7 µm) maintained at 50°C.	Fibigr et al. (2017)
3	Anthocyanin glucosides (9) and delphinidin	Grapes	UHPLC-PDA	Mobile phase: A (water:formic acid – 95:5 *v/v*) and B (acetonitrile:formic acid – 95:5 *v/v*). Flow rate: 0.4 mL/min. Injection volume: 3 µL. Column: Reverse phase C18 column (dimension 100 × 2 mm × 2 µm) maintained at 40°C.	Shim et al. (2014)

(Continued)

TABLE 10.2   (*Continued*)

Role of UHPLC in the Analysis of Anthocyanins

S. No.	Anthocyanin	Source	Extraction Method	Procedure	Reference
4	Cyanidin-3-glucoside and cyanidin-3-rutinoside (1–48 µg/mL)	*Euterpe oleracea* fruits	UHPLC-PDA-ESI-MS/MS	Mobile phase: A (water:formic acid – 95:5) and B (acetonitrile) Flow rate: 400 µL/min Injection volume: 5 µL Column: UPLC HSS C18 column (dimension 100 × 2.1 mm × 1.8 µm).	Dias et al. (2012)
5	Cyanidin-3-glucoside (73 mg/g)	*E. oleracea* fruits	UHPLC-DAD-ESI-MS/MS	Mobile phase: A (5% formic acid in water) and B (acetonitrile) Column: HSS C18 column (dimension 100 × 2.1 mm × 1.8 µm).	Dias et al. (2011)

10.4 Conclusion

In food science, there is a wide variety of analytical approaches to quantifying and characterizing ACNs, including both established and new methods, each with its own set of pros and cons. For an in-depth investigation of ACNs, the gold standard is still HPLC using reverse phase columns made of silica, including C18 and phenyl. ACNs can now be precisely quantified and identified with the use of UHPLC and new packing technologies, such as core-shell columns, which greatly improve the speed, resolution, and efficiency of these analyses. The analysis' goals will determine the detection methods used, which may range from UV detectors to MS and high-resolution MS. UV detectors, such as a PDA or a DAD, are useful tools for identifying different kinds of ACNs while doing profiling processes. On the other hand, MS and HRMS are highly regarded for their exceptional accuracy in differentiating between ACN molecules that have structural similarities, allowing for thorough profiling. Improving current methods, investigating new detection technologies, and creating integrated systems to analyze ACNs comprehensively, accurately, and efficiently are all areas where future research should concentrate, to improve food industry quality assurance procedures.

References

Ahmed NU, Park JI, Jung HJ, Yang TJ, Hur Y, Nou IS. Characterization of dihydroflavonol 4-reductase (DFR) genes and their association with cold and freezing stress in *Brassica rapa*. Gene. 2014;550(1):46–55.

Al-Khayri JM, Asghar W, Akhtar A, Ayub H, Aslam I, Khalid N, Al-Mssallem MQ, Alessa FM, Ghazzawy HS, Attimarad M. Anthocyanin delivery systems: A critical review of recent research findings. Applied Sciences. 2022;12(23):12347.

Amanah HZ, Joshi R, Masithoh RE, Choung MG, Kim KH, Kim G, Cho BK. Nondestructive measurement of anthocyanin in intact soybean seed using Fourier transform near-infrared (FT-NIR) and Fourier transform infrared (FT-IR) spectroscopy. Infrared Physics & Technology. 2020;111:103477.

Avula B, Katragunta K, Osman AG, Ali Z, John Adams S, Chittiboyina AG, Khan IA. Advances in the chemistry, analysis and adulteration of anthocyanin rich-berries and fruits: 2000–2022. Molecules. 2023;28(2):560.

Badawy ME, El-Nouby MA, Kimani PK, Lim LW, Rabea EI. A review of the modern principles and applications of solid-phase extraction techniques in chromatographic analysis. Analytical Sciences. 2022;38(12):1457–87.

Bar-Ya'akov I, Tian L, Amir R, Holland D. Primary metabolites, anthocyanins, and hydrolyzable tannins in the pomegranate fruit. Frontiers in Plant Science. 2019;10:453016.

Bele AA, Khale A. An overview on thin layer chromatography. International Journal of Pharmaceutical Sciences and Research. 2011;2(2):256.

Cappellini F, Marinelli A, Toccaceli M, Tonelli C, Petroni K. Anthocyanins: From mechanisms of regulation in plants to health benefits in foods. Frontiers in Plant Science. 2021;12:748049.

Castañeda-Ovando A, de Lourdes Pacheco-Hernández M, Páez-Hernández ME, Rodríguez JA, Galán-Vidal CA. Chemical studies of anthocyanins: A review. Food Chemistry. 2009;113(4):859–71.

Cooper-Driver GA. Contributions of Jeffrey Harborne and co-workers to the study of anthocyanins. Phytochemistry. 2001;56(3):229–36.

Custodio-Mendoza JA, Aktaş H, Zalewska M, Wyrwisz J, Kurek MA. A review of quantitative and topical analysis of anthocyanins in food. Molecules. 2024;29(8):1735.

da Costa CT, Horton D, Margolis SA. Analysis of anthocyanins in foods by liquid chromatography, liquid chromatography–mass spectrometry and capillary electrophoresis. Journal of Chromatography A. 2000;881(1–2):403–10.

De Pascual-Teresa S, Sanchez-Ballesta MT. Anthocyanins: From plant to health. Phytochemistry Reviews. 2008;7:281–99.

Dias A, Chataigné G, Rozet E, Oliveira A, Silva A, Rogez H, Quetin-Leclercq JA rapid UHPLC-DAD-ESI-MSn method for anthocyanins quantification from *Euterpe oleracea* fruits harvested at different times. In MS in Food and Feed 2011 Jun 9. https://orbi.uliege.be/handle/2268/95197.

Dias AL, Rozet E, Chataigné G, Oliveira AC, Rabelo CA, Hubert P, Rogez H, Quetin-Leclercq J. A rapid validated UHPLC–PDA method for anthocyanins quantification from *Euterpe oleracea* fruits. Journal of Chromatography B. 2012;907:108–16.

Drabińska N, Marcinkowska MA, Wieczorek MN, Jeleń HH. Application of sorbent-based extraction techniques in food analysis. Molecules. 2023;28(24):7985.

Enaru B, Drețcanu G, Pop TD, Stănilă A, Diaconeasa Z. Anthocyanins: Factors affecting their stability and degradation. Antioxidants. 2021;10(12):1967.

Escribano-Bailón MT, Alcalde-Eon C, Muñoz O, Rivas-Gonzalo JC, Santos-Buelga C. Anthocyanins in berries of maqui [*Aristotelia chilensis* (mol.) Stuntz]. Phytochemical Analysis: An International Journal of Plant Chemical and Biochemical Techniques. 2006;17(1):8–14.

Fredes C, Yousef GG, Robert P, Grace MH, Lila MA, Gómez M, Gebauer M, Montenegro G. Anthocyanin profiling of wild maqui berries (*Aristotelia chilensis* [Mol.] Stuntz) from different geographical regions in Chile. Journal of the Science of Food and Agriculture. 2014;94(13):2639–48.

Garcia-Oliveira P, Pereira AG, Fraga-Corral M, Lourenço-Lopes C, Chamorro F, Silva A, Garcia-Perez P, Barroso F, Barros L, Ferreira IC, Simal-Gandara J. Identification, quantification, and method validation of anthocyanins. Chemistry Proceedings. 2021;5(1):43.

Giusti MM, Wrolstad RE. Acylated anthocyanins from edible sources and their applications in food systems. Biochemical Engineering Journal. 2003;14(3):217–25.

Guine RP, Goncalves F, Lerat C, El Idrissi T, Rodrigo E, Correia PM, Goncalves JC. Extraction of phenolic compounds with antioxidant activity from beetroot (*Beta vulgaris* L.). Current Nutrition & Food Science. 2018;14(4):350–7.

He J, Giusti MM. Anthocyanins: Natural colorants with health-promoting properties. Annual Review of Food Science and Technology. 2010;1:163–87.

Horbowicz M, Kosson R, Grzesiuk A, Dębski H. Anthocyanins of fruits and vegetables-their occurrence, analysis and role in human nutrition. Journal of Fruit and Ornamental Plant Research. 2008;68(1):5–22.

Houghton A, Appelhagen I, Martin C. Natural blues: Structure meets function in anthocyanins. Plants. 2021;10(4):726.

Jia Q, Dong Q, Sang Q, Wang M, Zhang H, Zhou Y, Li Y, Xiao T, Hu P, Zhang S. Rapid qualitative and quantitative analyses of anthocyanin composition in berries from the Tibetan Plateau with UPLC-quadruple-Orbitrap MS and their antioxidant activities. European Journal of Mass Spectrometry. 2020;26(4):301–8.

Kalogiouri NP, Karadimou C, Avgidou MS, Petsa E, Papadakis EN, Theocharis S, Mourtzinos I, Menkissoglu-Spiroudi U, Koundouras S. An optimized HPLC-DAD methodology for the determination of anthocyanins in grape skins of red Greek winegrape cultivars (*Vitis vinifera* L.). Molecules. 2022;27(20):7107.

Khan NH, Abdulbaqi IM, Darwis Y, Aminu N, Chan SY. A stability-indicating HPLC-UV method for the quantification of anthocyanin in Roselle (*Hibiscus sabdariffa* L.) spray-dried extract, oral powder, and lozenges. Heliyon. 2022;8(3).

Khoo HE, Azlan A, Tang ST, Lim SM. Anthocyanidins and anthocyanins: Colored pigments as food, pharmaceutical ingredients, and the potential health benefits. Food & Nutrition Research. 2017, 61(1), doi: 10.1080/16546628.2017.1361779.

Kong JM, Chia LS, Goh NK, Chia TF, Brouillard R. Analysis and biological activities of anthocyanins. Phytochemistry. 2003;64(5):923–33.

Koponen JM, Happonen AM, Mattila PH, Törrönen AR. Contents of anthocyanins and ellagitannins in selected foods consumed in Finland. Journal of Agricultural and Food Chemistry. 2007;55(4):1612–9.

Lao F, Sigurdson GT, Giusti MM. Health benefits of purple corn (*Zea mays* L.) phenolic compounds. Comprehensive Reviews in Food Science and Food Safety. 2017;16(2):234–46.

Lasoń-Rydel M, Olejnik TP, Ławińska K, Niculescu MD. Methods of purification of raw polyphenol extract for chromatographic analysis. Technologia i Jakość Wyrobów. 2021;66.

Leonarski E, Cesca K, de Oliveira D, Zielinski AA. A review on enzymatic acylation as a promising opportunity to stabilizing anthocyanins. Critical Reviews in Food Science and Nutrition. 2023;63(24):6777–96.

Li D, Meng X, Li B. Profiling of anthocyanins from blueberries produced in China using HPLC-DAD-MS and exploratory analysis by principal component analysis. Journal of Food Composition and Analysis. 2016;47:1–7.

Liu Y, Tikunov Y, Schouten RE, Marcelis LF, Visser RG, Bovy A. Anthocyanin biosynthesis and degradation mechanisms in Solanaceous vegetables: A review. Frontiers in Chemistry. 2018;6:52.

Mattioli R, Francioso A, Mosca L, Silva P. Anthocyanins: A comprehensive review of their chemical properties and health effects on cardiovascular and neurodegenerative diseases. Molecules. 2020;25(17):3809.

Mazza G. Anthocyanins in Fruits, Vegetables, and Grains. CRC Press; 2018.

Mohammadi Pour P, Fakhri S, Asgary S, Farzaei MH, Echeverría J. The signaling pathways, and therapeutic targets of antiviral agents: Focusing on the antiviral approaches and clinical perspectives of anthocyanins in the management of viral diseases. Frontiers in Pharmacology. 2019;10:485177.

Mondello L, Cotroneo A, Errante G, Dugo G, Dugo P. Determination of anthocyanins in blood orange juices by HPLC analysis. Journal of Pharmaceutical and Biomedical Analysis. 2000;23(1):191–5.

Montilla EC, Arzaba MR, Hillebrand S, Winterhalter P. Anthocyanin composition of black carrot (*Daucus carota* ssp. sativus var. atrorubens Alef.) cultivars antonina, beta sweet, deep purple, and purple haze. Journal of Agricultural and Food Chemistry. 2011;59(7):3385–90.

Müller D, Schantz M, Richling E. High-performance liquid chromatography analysis of anthocyanins in bilberries (*Vaccinium myrtillus* L.), blueberries (*Vaccinium corymbosum* L.), and corresponding juices. Journal of Food Science. 2012;77(4):C340–5.

Ockermann P, Headley L, Lizio R, Hansmann J. A review of the properties of anthocyanins and their influence on factors affecting cardiometabolic and cognitive health. Nutrients. 2021;13(8):2831.

Passeri V, Koes R, Quattrocchio FM. New challenges for the design of high value plant products: Stabilization of anthocyanins in plant vacuoles. Frontiers in Plant Science. 2016;7:181384.

Paun N, Botoran OR, Niculescu VC. Total phenolic, anthocyanins HPLC-DAD-MS determination and antioxidant capacity in black grape skins and blackberries: A comparative study. Applied Sciences. 2022;12(2):936.

Pereira J, Zhou GH, Zhang WG. Effects of rice flour on emulsion stability, organoleptic characteristics and thermal rheology of emulsified sausage. Journal of Food and Nutrition Research. 2016;4(4):216–22.

Pojer E, Mattivi F, Johnson D, Stockley CS. The case for anthocyanin consumption to promote human health: A review. Comprehensive Reviews in Food Science and Food Safety. 2013;12(5):483–508.

Riaz M, Zia-Ul-Haq M, Saad B. Anthocyanins and Human Health: Biomolecular and Therapeutic Aspects. Springer; 2016.

Rodriguez-Saona LE, Allendorf ME. Use of FTIR for rapid authentication and detection of adulteration of food. Annual Review of Food Science and Technology. 2011;2:467–83.

Rubinskiene M, Jasutiene I, Venskutonis PR, Viskelis P. HPLC determination of the composition and stability of blackcurrant anthocyanins. Journal of Chromatographic Science. 2005;43(9):478–82.

Rusli H, Putri RM, Alni A. Recent developments of liquid chromatography stationary phases for compound separation: From proteins to small organic compounds. Molecules. 2022;27(3):907.

Salehi B, Sharifi-Rad J, Cappellini F, Reiner Ž, Zorzan D, Imran M, Sener B, Kilic M, El-Shazly M, Fahmy NM, Al-Sayed E. The therapeutic potential of anthocyanins: Current approaches based on their molecular mechanism of action. Frontiers in Pharmacology. 2020;11:1300.

Samtiya M, Aluko RE, Dhewa T, Moreno-Rojas JM. Potential health benefits of plant food-derived bioactive components: An overview. Foods. 2021;10(4):839.

Shim YS, Kim S, Seo D, Park HJ, Ha J. Rapid method for determination of anthocyanin glucosides and free delphinidin in grapes using UHPLC. Journal of Chromatographic Science. 2014;52(7):629–35.

Simmonds MS. Flavonoid–insect interactions: Recent advances in our knowledge. Phytochemistry. 2003;64(1):21–30.

Sivamaruthi BS, Kesika P, Chaiyasut C. The influence of supplementation of anthocyanins on obesity-associated comorbidities: A concise review. Foods. 2020;9(6):687.

Skogerson K, Wohlgemuth G, Barupal DK, Fiehn O. The volatile compound BinBase mass spectral database. BMC Bioinformatics. 2011;12:1–5.

Speer H, D'Cunha NM, Alexopoulos NI, McKune AJ, Naumovski N. Anthocyanins and human health—A focus on oxidative stress, inflammation and disease. Antioxidants. 2020;9(5):366.

Springob K, Nakajima JI, Yamazaki M, Saito K. Recent advances in the biosynthesis and accumulation of anthocyanins. Natural Product Reports. 2003;20(3):288–303.

Ştefănuţ MN, Căta A, Pop R, Moşoarcă C, Zamfir AD. Anthocyanins HPLC-DAD and MS characterization, total phenolics, and antioxidant activity of some berries extracts. Analytical Letters. 2011;44(18):2843–55.

Sui X. Impact of Food Processing on Anthocyanins. Springer; 2016.

Sun MF, Jiang CL, Kong YS, Luo JL, Yin P, Guo GY. Recent advances in analytical methods for determination of polyphenols in tea: A comprehensive review. Foods. 2022;11(10):1425.

Tena N, Martín J, Asuero AG. State of the art of anthocyanins: Antioxidant activity, sources, bioavailability, and therapeutic effect in human health. Antioxidants. 2020;9(5):451.

Thuy NM, Minh VQ, Ben TC, Thi Nguyen MT, Ha HT, Tai NV. Identification of anthocyanin compounds in butterfly pea flowers (*Clitoria ternatea* L.) by ultra performance liquid chromatography/ultraviolet coupled to mass spectrometry. Molecules. 2021;26(15):4539.

Truong VD, Deighton N, Thompson RT, McFeeters RF, Dean LO, Pecota KV, Yencho GC. Characterization of anthocyanins and anthocyanidins in purple-fleshed sweet potatoes by HPLC-DAD/ESI-MS/MS. Journal of Agricultural and Food Chemistry. 2010;58(1):404–10.

Tuzimski T. Thin-layer chromatography (TLC) as pilot technique for HPLC. Utilization of retention database (RF) vs. eluent composition of pesticides. Chromatographia. 2002;56:379–81.

Veberic R, Jakopic J, Stampar F, Schmitzer V. European elderberry (*Sambucus nigra* L.) rich in sugars, organic acids, anthocyanins and selected polyphenols. Food Chemistry. 2009;114(2):511–5.

Wallace TC, Giusti MM, editors. Anthocyanins in Health and Disease. CRC Press; 2013.

Warner LM, editor. Handbook of Anthocyanins: Food Sources, Chemical Applications, and Health Benefits. Nova Publishers; 2015.

Wu X, Beecher GR, Holden JM, Haytowitz DB, Gebhardt SE, Prior RL. Concentrations of anthocyanins in common foods in the United States and estimation of normal consumption. Journal of Agricultural and Food Chemistry. 2006;54(11):4069–75.

Wu X, Prior RL. Systematic identification and characterization of anthocyanins by HPLC-ESI-MS/MS in common foods in the United States: Fruits and berries. Journal of Agricultural and Food Chemistry. 2005;53(7):2589–99.

Yabré M, Ferey L, Somé IT, Gaudin K. Greening reversed-phase liquid chromatography methods using alternative solvents for pharmaceutical analysis. Molecules. 2018;23(5):1065.

Zhang Z, Yang K, Yan C, Wu M, Wang Y. Revealing the differences in phenolics in different parts of *Taraxacum mongolicum* using UPLC-MS/MS. Phytochemistry Letters. 2023;56:13–8.

Zhou Y, Long S, Xu Q, Yan C, Yang J, Zhou Y. Optimization and application of HPLC for simultaneous separation of six well-known major anthocyanins in blueberry. Preparative Biochemistry & Biotechnology. 2021;51(10):961–70.

11

LC-MS Analysis of Anthocyanins

Kadriye Nur Kasapoglu, Eda Nur Ayar-Sumer, Gamze Hanbeyoglu-Akturk, Evren Demircan, and Beraat Ozcelik

11.1 Introduction

Anthocyanins are bioactive substances classified as colored flavonoids within the phenolic group. These water-soluble pigments naturally occur in a variety of plant-based foods, including fruits, vegetables, flowers, seaweeds, grains, and legumes, imparting colors ranging from red and purple to blue and occasionally black. As members of the flavonoid family, anthocyanins can be classified into various categories based on their specific molecular composition (Constantin & Istrati, 2022). First, the common anthocyanidin derivatives are the basic anthocyanidin compounds are delphinidin, cyanidin, pelargonidin, petunidin, peonidin, and malvidin. Additionally, these primary pigments found in plants can undergo glycosylation, where sugar moieties attach to them, forming delphinidin-3-glucoside, cyanidin-3-glucoside, pelargonidin-3-glucoside, petunidin-3-glucoside, peonidin-3-glucoside, and malvidin-3-glucoside. Furthermore, anthocyanins can undergo acylation, where organic acid groups like caffeic or malonic acid are attached, resulting in more stable compounds such as caffeoylated and malonylated anthocyanins. The acylated forms offer greater stability and color characteristics whereas glycosylated anthocyanins are the most common with higher stability and water solubility (Singh et al., 2020; Tena, Martín & Asuero, 2020).

Due to their antioxidant properties, anthocyanins can neutralize free radicals, thereby protecting cells from oxidative stress. This protective action is associated with mitigating the effects of aging, inflammation, and various diseases, including cancer, cardiovascular ailments, and neurodegenerative conditions (Tena, Martín & Asuero, 2020). However, despite their health-promoting qualities, anthocyanins exhibit low bioavailability. Only 1–2% of consumed anthocyanins are absorbed in their original form, whereas the majority are transformed into other bioactive compounds by enzymes and gut microbiota (Guo et al., 2024).

The extraction of anthocyanins from plant materials can be achieved through the utilization of diverse techniques, which require optimizing the yield and preserving the stability of the extracted compounds. Conventional solvent extraction involves the dissolution of anthocyanins in polar solvents such as water, ethanol, or methanol, which are usually acidified. The subsequent solvent evaporation then concentrates the pigments. Although conventional solvent extraction might be effectively used, it is a time-consuming process that requires careful solvent selection to maintain stability. On the other hand, ultrasound-assisted extraction has gained popularity in recent years, mostly because of its efficiency and shorter extraction times. The ultrasound waves disrupt cell walls, thus improving

DOI: 10.1201/9781003453260-13

solvent penetration and lowering the solvent needed (da Silva Oliveira et al., 2022). Similarly, microwave-assisted extraction employs microwave energy to rapidly heat the plant material, thereby accelerating the extraction of anthocyanins (Nonglait & Gokhale, 2024). Pressurized-liquid extraction involves using high pressure and temperature to extract anthocyanins with high extraction yields and shorter extraction time. Supercritical liquid extraction is an environmentally friendly extraction option that uses supercritical CO_2 as a solvent, frequently with ethanol (Hsieh-Lo, Castillo-Herrera & Mojica, 2020). Another method, enzyme-assisted extraction, uses enzymes like pectinases and cellulases to break down plant cell walls, aiming to release anthocyanins from the food matrix. Pulsed electric field extraction involves the application of short electric field bursts to disrupt cell walls, thus releasing anthocyanins (Singh et al., 2022).

After the extraction, a purification step might be necessary depending on the analysis method. The determination and further identification of anthocyanins is a critical step for them to be accurately studied for their biological and chemical properties (Lucci, Saurina & Núñez, 2017). Because of the complex structure of anthocyanin-rich plants, selective extraction of anthocyanin is a challenge. In order to obtain a purified anthocyanin extract, it is necessary to remove other components such as phenolics, carotenoids, and structural components, including proteins and sugars, from the extract. Column chromatography, membrane separation, high-speed counter-current chromatography, high-performance preparative liquid chromatography (LC), and solid-phase extraction are used for purification, thereby refining the anthocyanin extract (Tan et al., 2022). The extraction method needs to be designed in detail based on the plant type, aimed compounds and their yield and aimed purity. Additionally, column chromatography, countercurrent chromatography, and high-speed counter-current chromatography are employed to eliminate impurities.

The most common methods for analyzing anthocyanins include LC coupled with various detectors. Among all the techniques available, the LC coupled with mass spectrometry (MS) methods is undoubtedly the most widely used analytical technique for anthocyanins. The invention of MS marked a significant breakthrough in analytical science, facilitating enhanced speed and precision in quantitative and qualitative analysis. The introduction of anthocyanins to MS, where ionization of molecules occurs, allows for the measurement of mass-to-charge (*m/z*). This facilitates the identification process by providing information regarding the molecular weight and fragmentation patterns of the molecules (Singh et al., 2020). Furthermore, the HPLC-DAD-MS method has been widely used for anthocyanin analysis (Avula et al., 2023), while reversed-phase (RP) HPLC has also been used by retaining the anthocyanins on the stationary phase and eluting them. For example, Paun et al. (2022) identified several anthocyanins present in grape skins using RP-HPLC-DAD-MS, with delphinidin-3-O-glucoside identified as the most abundant anthocyanin.

One of the most crucial stages in MS-based metabolite analysis is ionization. The characterization of anthocyanins has been achieved through the utilization of electrospray ionization (ESI), matrix-assisted laser desorption ionization (MALDI), and atmospheric pressure chemical ionization (APCI) in conjunction with mass analyzers. Besides, the employment of ESI with LC-MS is the most common for the analysis of anthocyanins in fruit samples (Oliveira et al., 2022). Tandem MS (MS/MS) comprises two MS phases. The initial MS phase is employed for the separation of anthocyanins based on their mass, and the second MS fragments ions thereby creating unique structural information (Ayar-Sümer et al., 2024). For instance, the anthocyanin profile of red cabbage has been analyzed using HPLC-DAD/ESI-MS (Pereira et al., 2024), while the detailed anthocyanin profile of purple sweet potato has been examined using HPLC-DAD/ESI-MS2 (Jiang et al., 2020).

Single quadrupole MS (Q-MS) analysis of anthocyanins is based on *m/z*, but it has limited structural analysis capabilities. However, Triple Quadruple MS (QqQ-MS) is a much more popular MS system for the quantification of anthocyanins, offering superior sensitivity and selectivity through the targeting of ions utilizing the multiple reaction monitoring (MRM) modes. LC-ESI-MS/MS with MRM ion mode has been employed for the analysis of apple anthocyanins, resulting in the identification and quantification of several compounds (Illiano et al., 2022). Additionally, Quadrupole time-of-flight (Q-TOF)-MS provides high-resolution MS with high sensitivity and accuracy. It is widely used for both qualitative and quantitative analysis of anthocyanins (Gonzales et al., 2015).

A comprehensive list of LC-MS methods has been employed for the identification and quantification of anthocyanins in a variety of fruits, vegetables, edible flowers, food matrices, and seaweeds discussed in this chapter. Some of the methods are less commonly employed than others, including ultra-performance LC (UHPLC)-Q-Orbitrap high-resolution MS (HRMS), UPLC VION IMS QTOF MS/MS, UHPLC-TimsTOF Pro-MS, UPLC-APCI-MS/MS, and LC–LTQ-MS, UHPLC-HRMS. These methods are employed for the analysis of anthocyanin-rich fruits and vegetables and a topic that will be elaborated upon in subsequent titles (Bao et al., 2023; Hameed et al., 2023; Ingemann Berentzen et al., 2024; Kodikara et al., 2024; Mirzazadeh et al., 2024; Zhang et al., 2021). The anthocyanin composition of various flowers, including hibiscus, perilla leaves, and edible flowers, was investigated using a range of analytical instruments, including UHPLC-ESI+-Obitrap-MS, UHPLC-QQQ/MS, LC-ESI-ion trap (IT)-TOF-MS, rapid resolution LC (RRLC)-Q/TOF-MS, UPLC–triple-quadrupole (TQ)-MS/MS (Mejía et al., 2023; Shahtouri et al., 2024; Zheng et al., 2020). Each of the methods possesses distinctive advantages based on sensitivity, selectivity, and structural information, which has been discussed in further detail in the following titles. In addition to their presence in plants, anthocyanins are also found in various food matrices, where they are either naturally occurring or intentionally added for enrichment purposes. Consequently, this chapter also considers the utilization of LC-MS for the analysis of anthocyanins in a range of processed food matrices. LC-PDA-ESI-MS/MS, HPLC-PDA-ESI-MS/MS, UHPLC-QE-MS, and LC-MSⁿ have been developed for matrices analysis in addition to LC-MS and LC-MS/MS (Ai et al., 2021; Kostka et al., 2020; Tasinov et al., 2022; Wang et al., 2023a–e). Constantin and Istrati (2022) reviewed advanced techniques for extracting and quantifying anthocyanins. They found that combining LC with LC-MS with ESI or QTOF provides more accurate quantification of anthocyanins in complex food samples.

11.2 LC-MS Analysis of Anthocyanins

11.2.1 Fruits

LC-MS analysis of anthocyanins in fruits is conducted by researchers for several key reasons. One of the primary purposes is to identify the different anthocyanin compounds present. Anthocyanins, which come in various forms like cyanidin and delphinidin, are often bound to sugar molecules. LC-MS allows researchers to accurately distinguish and identify these compounds by analyzing their molecular weight and structure (Singh et al., 2020). This technique has been thoroughly validated and has provided detailed anthocyanin profiles across many fruit types. For instance, Gamage and Choo (2023) examined the impact of diverse extraction techniques (hot water, ultrasound, microwave, and pectinase-assisted extraction) on anthocyanins derived from black goji berries. The individual anthocyanins present in the extracts were identified by LC-MS, with an Agilent 1290 Infinity LC system in conjunction with a Q-TOF mass spectrometer operating in positive ESI mode. A mass range of 600–1100 m/z was scanned, and compounds were separated using a C18 column under specific gradient elution conditions. Eight anthocyanins were identified in each extract, and a comparison of the amounts of these anthocyanins detected in the different extraction methods was also possible with LC-MS. This may facilitate an understanding of the reasons behind the final total anthocyanin content differences among the extracts. In a recent study, Kodikara et al. (2024) employed a novel UHPLC-HRMS metabolomic approach to rapidly and comprehensively analyze the phenolic compounds present in a range of fruit species, including blueberry, raspberry, blackberry, cranberry, and cherry. The HRMS conditions utilized a heated ESI source to create a charged solvent spray. A spray voltage of 3500 V for positive ions and 2200 V for negative ions was applied, while the ion transfer tube and vaporizer were maintained at 325 and 350°C, respectively. Data acquisition included full scan mode (m/z 150–900) at 120,000 resolution, followed by MS2 fragmentation at 30,000 resolution. The study introduced a novel UHPLC-HRMS system that markedly enhances the sensitivity, accuracy, and efficiency of fruit phenolic compound analysis. A significant advancement is the system's capacity to offer highly sensitive and selective detection, enabling the precise identification of 66 phenolic compounds, including anthocyanins, even in complex fruit matrices. Furthermore, the system reduced the analysis time to 20 minutes, representing a substantial advancement over older methods that required considerably longer. Incorporating a biphenyl column enhanced the separation and resolution of phenolic compounds, facilitating unambiguous differentiation of analyte peaks. The mass spectrometer's capacity to fragment parent ions for distinctive identification guaranteed precise quantification of the compounds, establishing this system as a formidable instrument for comprehensive metabolomic analysis.

LC-MS is a highly sensitive method widely used to quantify anthocyanins in fruits. This quantification is crucial because the concentration of anthocyanins can vary significantly depending on factors like the variety and ripeness of the fruit. LC-MS provides precise and reliable data, helping researchers better understand anthocyanin levels under different conditions. This makes the technique essential for both scientific studies and commercial applications (Qu et al., 2021). Ma et al. (2022) quantified 32 distinct types of anthocyanins using UPLC-ESI-MS/MS with an ESI Turbo Ion-Spray interface operating in both positive and negative ion modes. The ESI source parameters included a turbo spray ion source, a source temperature of 550°C, and an ion spray voltage of 5500 V (in positive mode), with a 35-psi curtain gas and high-collision gas settings. The level of anthocyanins in the fruits of four distinct blueberries (Vaccinium sp.) cultivars (Gardenblue, Legacy, Misty, and Brightwell) varied at stages of development. The main findings were that different cultivars had different anthocyanin content. Delphinidins were identified as the primary contributors to the dark blue color of the berries, whereas malvidins and petunidins played a secondary role in the development of color. The study also revealed a notable elevation in anthocyanin concentration as the fruits underwent maturation, manifesting across all four cultivars. In the initial, young fruit stage, the anthocyanin levels were relatively low. As the fruits entered the color-transition stage, at which point the berries changed from green to red or purple, the anthocyanin content increased. The highest levels were recorded in fully mature fruits, which contributed to the deepening of their color. These findings illustrate a clear correlation between fruit maturation and anthocyanin accumulation, with an increase in anthocyanin concentration occurring as the fruits ripened and their color intensified. Additionally, Li et al. (2021) studied the development of fruit color in *Rubus chingii*. This involved investigating anthocyanins and carotenoids using LC-MS/MS, detecting flavanol-anthocyanins, and revealing novel biosynthesis pathways. Anthocyanins were separated using UPLC with an ACQUITY UPLC HSS T3 column and analyzed using an AB Triple TOF 5600plus MS system in negative ion mode (−4.5 kV, 550°C, ±5 ppm, DP: 100 V, CE: 10 V) and positive ion mode (+5.5 kV, 600°C, ±5 ppm, DP: 100 V, CE: 10 V), with MS/ MS acquisition mode (CE: 40 ± 20 V, IRD: 67, IRW: 25). Then, it is found that the fruit color dramatically changes as the fruit ripens in *R. chingii*. The observed color transformation can be attributed to the synthesis of anthocyanins and carotenoids and the crucial genes/ proteins involved in their production, which have been precisely measured using LC-MS/ MS. Consequently, it may elucidate the processes underlying fruit coloration and enhance fruit quality in raspberry breeding program. Kim et al. (2021) focused on optimizing a pectinase-assisted extraction method for mulberry to enhance the anthocyanins extraction yield. An LC-MS analysis was conducted utilizing a UPLC system C18 column and a flow rate of 0.35 mL/min. The mass range scanned was 50–1500 *m/z*, and MS/MS spectra were collected at collision energies of 10–40 eV. These conditions facilitated the identification of metabolites in the mulberry extracts. Table 11.1 lists the MS methods (conditions and validation parameters) for the analysis of anthocyanins in fruits.

LC-MS is facilitating the comprehension of the biosynthesis and transport processes of anthocyanins in fruits. For instance, Qu et al. (2021) explored the browning of litchi pericarp during post-harvest storage, focusing on the balance between anthocyanin biosynthesis and degradation. Researchers used transcriptomic, proteomic, and LC-MS analyses to conduct the investigation. The litchi extracts were subjected to analysis using an AB Sciex QTRAP® 6500 LC-MS/MS. The separation was conducted at 40°C with a flow rate of 0.35 mL/min on an ACQUITY BEHC18 column. Gradient elution was employed, utilizing a solution of 0.1% formic acid in water and methanol at a ratio of 95:5–5:95 (v/v), for 14 minutes. The effluent was connected to an ESI QTRAP-MS in positive ion mode, with a Turbo Ion-Spray interface at 550°C, 5500 V, and 35-psi curtain gas. During storage at 20°C, transcriptomic analysis revealed 4769 genes related to both anthocyanin biosynthesis and degradation, and the proteomic analysis identified 84 proteins with significant roles in anthocyanin biosynthesis and degradation. Besides, LC-MS analysis detected 29 anthocyanins and their derivatives in the litchi pericarp, with cyanidin 3-O-glucoside and cyanidin 3-O-rutinoside being the most dominant. After six days of storage, levels of five anthocyanins, including cyanidin 3-O-galactoside, showed a significant decline, whereas cyanidin 3-O-(6-O-malonyl-beta-D-glucoside) increased. The study concluded that although anthocyanin biosynthesis remained active during the browning process, degradation was the dominant force, leading to the rapid loss of the fruit's red color. These findings offer valuable insights for developing strategies to prevent browning in post-harvest litchi. In addition, Zhang et al. (2022a) investigated the accumulation of anthocyanins in red and white *Prunus tomentosa* cherries using metabolomic and transcriptomic analyses. To better understand the coloration mechanism of *P. tomentosa*, LC-MS/MS analysis was performed on the anthocyanin compounds in mature fruit. The analysis of cherry extracts was conducted using an LC-ESI-MS/MS with a Waters ACQUITY BEH C18 column. The effluent was connected to an API 6500 Q TRAP UPLC/MS/MS, which was operated in both positive and negative ion modes. The key parameters were as follows: a source temperature of 550°C, an ion spray voltage of 5.5 kV, a curtain gas pressure of 35 psi, and a high collision gas pressure. Thus, the study identified 15 anthocyanins in samples, the red cherries exhibited significantly elevated anthocyanin concentrations, with

TABLE 11.1

LC-MS Analysis of Anthocyanins from Fruits

Source of Anthocyanins	Analyte	Analyte Quantity	LC-MS Method	Reference
Ancient apple cultivars (*Malus domestica*)	Petunidin-3-O-glucoside Peonidin-3-O-glucoside	Various concentrations among the samples	LC-MS/MS ESI mode: positive/negative ionization MRM ion mode Ion spray voltage 4.5 kV Source temperature: 380°C	Illiano et al. (2022)
Bilberries (urine and plasma samples) (*Vaccinium myrtillus* L.)	Malvidin-3-O-glucoside Peonidin-3-O-glucoside Petunidin-3-O-glucoside	16.6–34.8 ng/mL 17.5–44.8 ng/mL 17.5–43.0 ng/mL	UHPLC-MS ESI mode: positive ionization LOD: 1.2 to 2.3 ng/mL LOQ 4.1 to 8.1 ng/mL	Kaiser et al. (2020)
Black goji berry (*Lycium ruthenicum*)	Petunidin-3-O-rutinoside(*trans*-p-coumaroyl)-5-O-glucoside Delphinidin-3-O-rutinoside(coumaroyl)-5-O-glucoside Petunidin-3-O-glucoside-5-O-glucoside Delphinidin-3-O-glucoside(coumaroyl)-5-O-glucoside Delphinidin-3-O-rhamnoside-5-O-glucoside Petunidin-3-O-rutinoside(coumaroyl-(4-O-glucoside))-5-O-glucoside Petunidin derivative Malvidin-3-O-rutinoside(coumaroyl)-5-O-glucoside	Petunidin-3-O-rutinoside (*trans*-p-coumaroyl)-5-O-glucoside accounting for more than 80% of total anthocyanins	LC-MS ESI mode: positive ionization Declustering potential 40 eV Collision energy 5 eV Collision cell entrance potential 10 eV Desolvation gas temperature of 300°C Mass Range: 600–1100 *m/z*,	Gamage and Choo (2023)
Blackberry skin (*Rubus fruticosus*)	Cyanidin-3-O-glucoside Delphinidin-3-xyloside Cyanidin-3-O-arabinoside Cyanidin-3-malonyl-glucoside Cyanidin-3-rutinoside	Cyanidin-3-glucoside was the main component (86.49%)	LC-Q-TOF/MS ESI mode: positive ionization Capillary voltage 4200 v Mass range: 100–1500 *m/z*	Paun et al. (2022)
Blackberry (*R. fruticosus*)	Cyanidin-3-glucoside Cyanidin-3-arabinoside Cyanidin-3-rutinoside Delphinidin Delphinidin-3-glucoside Malvidin-3-glucoside	368.4 µg/g DW 134.8 µg/g DW 110.4 µg/g DW 26.74 µg/g DW 18.02 µg/g DW 90.63 µg/g DW	UHPLC-HRMS ESI mode: positive ionization 3500 V for positive ions 2200 V for negative ions Ion transfer tube: 325°C Vaporizer: 350°C Mass range: *m/z* 150–900	Kodikara et al. (2024)

(Continued)

TABLE 11.1 *(Continued)*

LC-MS Analysis of Anthocyanins from Fruits

Source of Anthocyanins	Analyte	Analyte Quantity	LC-MS Method	Reference
Blueberry (*Vaccinium* sp.)	Cyanidin 3-O-(6-O-malonyl-beta-D-glucoside) Cyanidin 3-O-arabinoside Cyanidin 3-O-galactoside Cyanidin 3-O-glucoside Cyanidin 3-O-rutinoside Delphinidin 3,5-O-diglucoside (Delphin) Delphinidin 3-O-(6″-O-malonyl)-beta-D-glucoside Delphinidin 3-O-arabinoside Delphinidin 3-O-galactoside Delphinidin 3-O-glucoside Malvidin 3-O-arabinoside Malvidin 3-O-galactoside Malvidin 3-O-glucoside Malvidin 3-O-rutinoside Pelargonidin 3-O-(6-O-malonyl-beta-D-glucoside) Pelargonidin 3-O-arabinoside Pelargonidin 3-O-galactoside Pelargonidin 3-O-glucoside Pelargonidin 3-O-rutinoside Pelargonidin 3,5-O-diglucoside (Pelargonin) Peonidin 3,5-O-diglucoside Peonidin 3-O-(6-O-malonyl-beta-D-glucoside) Peonidin 3-O-glucoside Peonidin 3-O-arabinoside Peonidin 3-O-galactoside Peonidin 3-O-rutinoside Petunidin 3,5-diglucoside Petunidin 3-O-(6-O-malonyl-beta-D-glucoside) Petunidin 3-O-arabinoside Petunidin 3-O-galactoside Petunidin 3-O-glucoside Petunidin 3-O-rutinoside	Not directly stated in mg/g, relative peak areas available for different stages	UPLC-ESI-MS/MS ESI mode: positive/ negative ionization Ion spray voltage: 5500 V Source temperature: 550°C Mass range: *m/z* 287–639	Ma et al. (2022)

(Continued)

TABLE 11.1 (*Continued*)

LC-MS Analysis of Anthocyanins from Fruits

Source of Anthocyanins	Analyte	Analyte Quantity	LC-MS Method	Reference
Blueberry cultivars (*Vaccinium* sp.)	Malvidin 3-O-glucoside Delphinidin 3-O-glucoside Pelargonidin 3-O-(6-O-malonyl-beta-D-glucoside) Petunidin 3-O-rutinoside Petunidin 3-O-arabinoside Petunidin 3-O-galactoside Petunidin 3,5-diglucoside Petunidin 3-O-(6-O-malonyl-beta-D-glucoside) Petunidin 3-O-glucoside Delphinidin 3,5-O-diglucoside, delphinidin 3-O-(6″-O-malonyl)-beta-D-glucoside Delphinidin 3-O-arabinoside Delphinidin 3-O-galactoside Delphinidin 3-O-glucoside Malvidin 3-O-arabinoside Malvidin 3-O-galactoside Malvidin 3-O-glucoside Malvidin 3-O-rutinoside	N/A	UPLC-ESI-MS/MS ESI mode: positive/ negative ionization Ion spray voltage 5500 V (positive ion) Mass range: 100–1500 *m/z* LOD: 0.02 µg/mL	Ma et al. (2022)
Blueberry (*Vaccinium spp.*)	Cyanidin-3-glucoside Cyanidin-3-arabinoside Delphinidin, Delphinidin-3-glucoside Malvidin-3-glucoside Petunidin Peonidin-3-glucoside	165.5 µg/g DW 23.79 µg/g DW 45.36 µg/g DW 94.44 µg/g DW 252.5 µg/g DW 14.04 µg/g DW 108.4 µg/g DW	UHPLC-HRMS ESI mode: positive ionization Voltage: 3500 V Ion transfer tube: 325°C Vaporizer: 350°C Mass range: *m/z* 150–900	Kodikara et al. (2024)
Cherry (*Prunus* spp.)	Cyanidin-3-rutinoside Cyanidin-3-glucoside Pelargonidin Peonidin Peonidin-3-glucoside	315.6 µg/g DW 66.06 µg/g DW 42.46 µg/g DW 644.2 µg/g DW 225.8 µg/g DW	UHPLC-HRMS ESI mode: positive ionization Voltage: 3500 V Ion transfer tube: 325°C Vaporizer: 350°C HESI, *m/z* 150–900	Kodikara et al. (2024)

(Continued)

TABLE 11.1 *(Continued)*

LC-MS Analysis of Anthocyanins from Fruits

Source of Anthocyanins	Analyte	Analyte Quantity	LC-MS Method	Reference
Cherry sweet (red and yellow) (*Prunus avium* L.)	Cyanidin-3-O-rutinoside Cyanidin-3-O-glucoside Pelargonidin-3-glucoside	3909.32 µg/g FW 13.54 µg/g FW 0.23 µg/g FW	LC-MS/MS ESI mode: positive ionization Source voltage: 4 kV Capillary temperature: 275°C Mass range: *m/z* 100–1000	Wang et al. (2023a)
Cherry (red and white) (*Prunus tomentosa Thunb.*)	Pelargonidin 3-O-arabinoside Pelargonidin 3-O-glucoside Pelargonidin 3-O-(6-O-malonyl-beta-D-glucoside) Pelargonidin 3-O-galactoside Pelargonidin 3-O-rutinoside Peonidin 3-O-galactoside Peonidin 3-O-rutinoside Peonidin 3-O-glucoside Cyanidin 3-O-rutinoside Cyanidin 3-O-glucoside Delphinidin 3-O-rutinoside	N/A	LC-ESI-MS/MS ESI mode: positive ionization Ion spray voltage: 5.5 kV Source temperature: 550°C MRM transitions monitored for each anthocyanin	Zhang et al. (2022a)
Cocoa pod husk (*Theobroma cacao*)	3-Caffeoylpelargodinin 5-glucoside	Not directly stated in mg/g, relative colors available. The concentration differences were indicated by a change in color from blue to red.	HPLC- ESI-QTOF-MS/MS ESI mode: positive ionization Collision energy: 40 eV Source: Dual AJS ESI Vcap: 3000 V Fragmentation voltage: 175 V Mass range: 100–1100 *m/z*	Gallegoet al. (2022)
Cranberry (*Vaccinium macrocarpon*)	Petunidin Peonidin Peonidin-3-glucoside Cyanidin-3-glucoside Delphinidin Delphinidin-3-glucoside	829.7 µg/g DW 1.068 µg/g DW 81.53 µg/g DW 17.44 µg/g DW 10.36 µg/g DW 2.467 µg/g DW	UHPLC-HRMS ESI mode: positive ionization Voltage: 3500 V Ion transfer tube: 325°C Vaporizer: 350°C Mass range: *m/z* 150–900	Kodikara et al. (2024)

(Continued)

TABLE 11.1 *(Continued)*

LC-MS Analysis of Anthocyanins from Fruits

Source of Anthocyanins	Analyte	Analyte Quantity	LC-MS Method	Reference
Grape Skins (*Vitis* sp. *Othello cultivar*)	Malvidin-Delphinidin Malvidin-Cyanidin Delphinidin-3-O-glucoside Cyanidin-3-O-glucoside Cyanidin-3-O-galactoside Delphinidin-3-O-glucoside Delphinidin-3-(6-O-acetylglucoside) Malvidin-3-O-glucoside Cyanidin-3-glucoside-Ethyl-coumaroyl Malvidin-3-glucoside-8- vinyl(epi)catechin Malvidin-3- (6-O-acetylglucoside) pyruvate Malvidin-Petunidin	Delphinidin-3-O-glucoside was the major (40.64%)	RP-HPLC-DAD-MS ESI mode: Positive ionization Capillary voltage 4200 V Mass range: 100–1500 *m/z*	Paun et al. (2022)
Kadsura coccinea fruits	Cyanidin 3,5-O-diglucoside Cyanidin 3-O-(6-O-malonyl-beta-D-glucoside) Cyanidin 3-O-arabinoside Cyanidin 3-O-glucoside Cyanidin 3-O-rutinoside Delphinidin 3,5-O-diglucoside (Delphin) Delphinidin 3-O-glucoside Delphinidin 3-O-rutinoside Malvidin 3-O-glucoside Pelargonidin 3-O-glucoside Pelargonidin 3-O-rutinoside Peonidin 3,5-O-diglucoside Peonidin 3-O-galactoside Peonidin 3-O-glucoside Peonidin 3-O-rutinoside Petunidin 3-O-glucoside Petunidin 3-O-rutinoside	0.47–0.55 µg/g 0.01 µg/g 0.02 µg/g 30.90–33.10 µg/g 55.90–59.00 µg/g 0.52–0.53 µg/g 11.30–11.40 µg/g 1.56–1.83 µg/g 0.07 µg/g 0.15–0.18 µg/g 1.47–1.61 µg/g 0.02 µg/g 0.02 µg/g 0.46–0.53 µg/g 3.79–4.41 µg/g 0.03 µg/g 0.07–0.08 µg/g	UPLC/MS/MS ESI mode: Positive ionization Ion spray voltage: 5500 V Source temperature: 550°C	Huang (2021)
Lingonberry (*Vaccinium vitis-idaea* L.)	Cyanidin-3-O-glucoside Cyanidin-3-O-galactoside Cyanidin-3-O-arabinoside	N/A	UHPLC-QqQ-MS/MS ESI mode: Positive/negative ionization Capillary voltage: 4 kV for positive ion mode 3.5 kV for negative ion mode	Xu et al. (2024a)

(Continued)

TABLE 11.1 *(Continued)*

LC-MS Analysis of Anthocyanins from Fruits

Source of Anthocyanins	Analyte	Analyte Quantity	LC-MS Method	Reference
Litchi (*Litchi chinensis* Sonn.)	Cyanidin 3,5-O-diglucoside Cyanidin 3-O-(6-O-malonyl-beta-D-glucoside) Cyanidin 3-O-arabinoside Cyanidin 3-O-galactoside Cyanidin 3-O-glucoside Cyanidin 3-O-rutinoside Cyanidin 3-O-sambubioside Cyanidin 3-O-sophoroside Cyanidin-3-O-xyloside Cyanidin-3-rutinoside 5-glucoside Delphinidin 3-O-glucoside Delphinidin 3-O-rutinoside Delphinidin-3-rutinoside 5-glucoside Pelargonidin 3-O-(6-O-malonyl-beta-D-glucoside)	N/A	HPLC-ESI-MS/MS ESI mode: Positive ionization Ion spray voltage (IS): 5500 V Source temperature: 550°C	Qu et al. (2021)
Litchi aril (flesh) and pericarp (*L. chinensis*)	Cyanidin-3-O-glucoside Cyanidin-3-O-rutinoside	Cyanidin-3-O-rutinoside: 21.95 µg/g DW in arils	UPLC-ESI-MS/MS ESI mode: positive ionization Curtain gas: 35 psi Ion spray voltage: 5500 V Temperature: 550°C	Wang et al. (2023b)
Mulberry (*Morus alba* L.)	Cyanidin-3-O-glucoside Delphinidin hexoside Pelargonidin 3-glucoside, Delphinidin 3-(6″-p-coumaroyl glucoside) Cyanidin 3-rutinoside	The concentration differences were indicated by a change in color from blue to red	UPLC-Q-TOF MS ESI mode: positive ionization Capillary voltage: 3.0 kV Desolvation temperature 400°C Mass range: 50–1200 *m/z*	Kim et al. (2021)
Passion fruit (*Passiflora edulis* Sims)	Cyanidin-3-O-glucoside Cyanidin-3-O-(6″-p-coumaroyl)-glucoside Delphinidin-3-O-(6″-p-coumaroyl)-glucoside Delphinidin-3-O-glucoside Cyanidin-3-O-(6″-p-coumaroyl)-glucoside (repeated in a different form) Cyanidin-3-O-hexoside-rhamnoside Delphinidin-di-C,C-hexosyl-O-rhamnoside	737.69 mg/100 g 789.90 mg/100 g 774.82 mg/100 g 85.72 mg/100 g 98.44 mg/100 g 500.43 mg/100 g 47.96 mg/100 g	LC-QTOF/ESI-MS ESI mode: positive ionization Capillary voltage: 2500V Cone voltage: 30V Temperature: 300°C Mass range: *m/z* 285–759	Siniawska and Wojdyło (2023)

(Continued)

TABLE 11.1 *(Continued)*

LC-MS Analysis of Anthocyanins from Fruits

Source of Anthocyanins	Analyte	Analyte Quantity	LC-MS Method	Reference
Pistacia fruit (*Pistacia lentiscus* L.)	Delphinidin dihexoside 1 Delphinidin dihexoside 2 Delphinidin 3-O-galactoside Delphinidin 3-O-glucoside Cyanidin 3-O-galactoside Delphinidin pentoside 1 Cyanidin 3-O-glucoside Cyanidin pentoside Delphinidin pentoside 2	122 mg/100 g DW 248 mg/100 g DW 513 mg/100 g DW 646 mg/100 g DW 319 mg/100 g DW 71 mg/100 g DW 201 mg/100 g DW 103 mg/100 g DW 37 mg/100 g DW	UHPLC-QqQ-MS/MS ESI mode: positive ionization Sheath gas: 60 a.u Voltage: 1500 V Vaporizer Temperature: 450°C Capillary Temperature: 150°C LOD: 0.08 mg/100 g LOQ: 0.23 mg/100 g	Aissat et al. (2022)
Red pistachio drupes (*Pistacia vera* L.)	Cyanidin 3-O-β-d-galactopyranoside Cyanidin pentoside	N/A	HPLC-DAD-ESI-MS^n ESI mode: positive ionization Capillary: ±2287 V CID: 1.2 V Nebulizer temperature: 365°C Mass range: *m/z* 100–2000	Ersan et al. (2016)
Cherry (*P. tomentosa* Thunb.)	Pelargonidin 3-O-glucoside Cyanidin 3-O-rutinoside Pelargonidin 3-O-rutinoside	N/A	LC-ESI-MS/MS Mass range: 100–1500 *m/z* LOD: 0.02 µg/mL	Zhang et al. (2022a)
Raspberry (*Rubus idaeus*)	Cyanidin-3-rutinoside Cyanidin-3-glucoside Pelargonidin Peonidin-3-glucoside Delphinidin Delphinidin-3-glucoside	310.3 µg/g DW 65.63 µg/g DW 30.45 µg/g DW 117.5 µg/g DW 5.204 µg/g DW 2.18 µg/g DW	UHPLC-HRMS ESI mode: positive ionization Voltage: 3500 V Ion transfer tube: 325°C Vaporizer: 350°C Mass range: *m/z* 150–900	Kodikara et al. (2024)
Chinese raspberry (*Rubus chingii*)	Cyanidin-3-(6″-cinnamoyl)glucoside (Epi)Afzelechin(4α->8) pelargonidin 3,5-O-diglucoside (Epi)Catechin(4α->8)pelargonidin 3-O-glucopyranoside (Epi)Afzelechin(4α->8)pelargonidin 3-O-sophoroside (Epi)Afzelechin(4α->8)pelargonidin 3-O-glucopyranosid Pelargonidin 3-O-glucoside	10–115 µg/g FW	UPLC-MS/MS ESI mode: positive ionization Voltage: +5.5 kV Temperature 600°C Mass Range: 100–2000 *m/z*	Li et al. (2021)

(Continued)

TABLE 11.1 *(Continued)*

LC-MS Analysis of Anthocyanins from Fruits

Source of Anthocyanins	Analyte	Analyte Quantity	LC-MS Method	Reference
Sumac fruits (*Rhus coriaria* L.)	Delphinidin-3,5-diglucoside Delphinidin-3-galactoside Cyanidin-3-glucoside Petunidin-3-glucoside Pelargonidin-3-glucoside	15.51–20,889.81 mg/kg DE	HPLC-MS/MS ESI mode: positive ionization Capillary voltage: 4000 V Ionization source temperature: 350°C The gas flow: 12 L/min Nebulizer pressure: 55 psi	Mazzara et al. (2023)
Sumac fruits (*R. coriaria* L.)	Petunidin-3-O-galactoside Petunidin Cyanidin	N/A	HPLC-DAD-ESI-MS/MS ESI mode: positive ionization Temperature: 325°C Capillary voltage: 1.5 kV	Viola et al. (2024)
Staghorn sumac fruits (*Rhus hirta* L.)	7-O-methyl-cyanidin-3-O-(2″galloyl)-galactoside 7-O-methyl-cyanidin-3-O-galactoside) 7-O-methyl-delphinidin-3-O-(2″galloyl)-galactoside Cyanidin-3-O-(2″galloyl)-galactoside Peonidin	N/A	UPLC–MS ESI mode: positive ionization Capillary voltage: 4 kV Cone voltage: 50–70 V Source temperature: 150°C Gas temperature: 400°C MS scans: 400–1000 amu	Peng et al. (2016)
Staghorn sumac fruits (*Rhus hirta* L.)	Peonidin-3-O-galloyl-hexose Cyanidin-3-glucoside 2 novel anthocyanidins: (449 Da)-hexose-3-O-galloyl-hexose (419 Da)-hexose-3-O-galloyl-hexose	270 ± 6.53 mg/g DW	LC-ESI-MS ESI mode: positive ionization Capillary voltage: 32.5 kV Needle voltage 5 kV Temperature 350°C	Wu et al. (2013)

pelargonidin and cyanidin derivatives representing the most prevalent derivatives. The study concluded that the observed color difference is attributable to anthocyanin content and gene regulation, offering insights for breeding. In a further study, Wang et al. (2023b) investigated the deficiency of anthocyanin biosynthesis in litchi aril through the utilization of UPLC-ESI-MS/MS and transcriptomic analysis. The synthesis of anthocyanins is a complex process that involves the action of multiple enzymes and is typically regulated by transcription factors. In litchi, while the pericarp (skin) accumulates anthocyanins such as cyanidin-3-O-glucoside and cyanidin-3-O-rutinoside, the aril (flesh) lacks visible anthocyanins despite having the early precursors for anthocyanin biosynthesis. LC-MS/MS was employed to profile the metabolites involved in the anthocyanin biosynthesis pathway in both arils and pericarps. The presence of intermediates such as phenylalanine and cyanidin confirmed the functionality of the initial steps of anthocyanin biosynthesis in the aril. LC-MS/MS was instrumental in identifying the point of obstruction in the anthocyanin biosynthesis pathway, specifically at the glycosylation stage where cyanidin is converted to stable anthocyanins.

Another important use of LC-MS is in studying the bioavailability and health effects of anthocyanins. Known for their antioxidant properties and potential health benefits, anthocyanins are of great interest to researchers. LC-MS helps track how these compounds are metabolized in the body, giving valuable insights into their bioavailability and effectiveness. This is especially useful in research focused on the nutritional and therapeutic benefits of fruits rich in anthocyanins (Avula et al., 2023). In a recent study, Xu et al. (2024a) employed high-resolution UHPLC-QqQ-MS/MS to profile anthocyanins in lingonberry and to evaluate their potential biological activity. Kaiser et al. (2020) developed and validated a method for the analysis of anthocyanin metabolites in plasma using UHPLC-MS. The MS system was a Thermo Scientific LTQ XL Iontrap MSn with ESI in positive mode. The spray voltage was 4 kV, the capillary temperature 325°C, with capillary/tube lens voltages of 9 V/40 V for anthocyanidins and 15 V/60 V for anthocyanins. The method demonstrated acceptable sensitivity for the detection of metabolites, providing insights into anthocyanin bioavailability.

LC-MS also serves as a key tool for indicating fruit ripening and post-harvest changes. Anthocyanin levels can fluctuate during storage and ripening, and LC-MS helps to monitor these changes. This information is critical for optimizing storage conditions and improving the color and nutritional quality of fruit products, which is essential for maintaining fruit quality during transportation and storage (Qu et al., 2021). Gallego et al. (2022) analyzed the ripening process of cacao pod husks to avoid late-stage metabolic changes, assisting in the determination of the optimal harvest time and the standardization of practices. The deployment of chemical markers and visual indicators may prove instrumental in the standardization of the process. Therefore, the biochemical disturbance was analyzed in cacao pod husks, resulting in the detection of five potential biochemical markers, one of which was 3-caffeoylpelargodinin 5-glucoside. The biochemicals, including anthocyanin, were identified through the use of an LC-MS coupled to a Q-TOF 6545 mass spectrometer, equipped with an ESI source. The analysis was conducted in positive ionization mode, with a mass scan range from 100 to 1100 *m/z*. Additionally, MS/MS analysis was conducted with a collision energy of 40 eV, to facilitate metabolite fragmentation and identification. The findings helped to avoid late-stage metabolic changes that harm bean quality, helping to determine the optimal harvest time and standardize practices. The study of Wang et al. (2023a) concentrated on the biosynthesis of anthocyanins and carotenoids in red and yellow sweet cherry cultivars at different stages of ripening and in the post-harvest period. The objective was to examine the discrepancies in pigment accumulation, particularly anthocyanins and carotenoids, and the associated gene regulation throughout the fruit ripening process. The study emphasized the pivotal role of anthocyanins in the formation of red cherry color and the regulatory genes that are responsible for pigment biosynthesis during ripening. The utilization of TripleTOF 5600 LC-MS/MS enabled the precise quantification necessary to monitor these alterations, thereby providing insights into fruit ripening, post-harvest management and potential applications in the agricultural and food industries for the improvement of fruit quality through the implementation of controlled ripening and color development. In another study, Chen et al. (2023a) examined the impact of non-thermal treatments, encompassing ultrasound, pulsed light, and magnetic energy, on fruit quality, anthocyanin accumulation, and gene expressions associated with anthocyanin biosynthesis in blood oranges during low-temperature storage. Furthermore, the researchers employed LC-MS with AB Triple TOF 5600plus to identify and quantify anthocyanins. Fourteen individual anthocyanins were identified and quantified through HPLC using LC-MS, which played a crucial role in ensuring precise identification and quantification of the anthocyanins.

LC-MS metabolic profiling of different varieties of *Pistacia vera* L. nuts (family *Anacardiaceae*) from Spain revealed cyanidin-3-O-galactoside, cyanidin-3-O-glucoside, and peonidin-3-O-hexoside (Ojeda-Amador et al., 2019). On the other hand, Turkish red pistachio hulls were found to contain cyanidin 3-O-β-d-galactopyranoside and cyanidin pentoside. Aissat et al. (2022) examined the phenolic composition of *Pistacia lentiscus* L. fruits at varying stages of ripening, with a particular emphasis on anthocyanins, utilizing UHPLC-MS/MS (6430 QqQ MS, with MRM in either positive or negative mode) and LC-NMR methodologies. As the fruits ripened, there was a notable increase in anthocyanin content, which became the dominant phenolic class in the fully ripe black fruits, with concentrations reaching 1273–1373 mg/100 g DW. The most abundant anthocyanin in the ripe stages was delphinidin derivative, which constituted 88% of the total anthocyanin content. In contrast, the prevalence of flavonols and flavanols, which were more abundant in the initial unripe stages, diminished as the fruits matured. The UHPLC-MS/MS system enabled the precise identification and quantification of these anthocyanins, revealing the presence of nine different anthocyanins, including cyanidin and delphinidin derivatives, with high sensitivity and accuracy. The study emphasized the potential of *P. lentiscus* L. fruits as a rich source of anthocyanins, with applications in nutraceuticals and the food industry, particularly as natural colorants, while also underscoring the importance of ripening in enhancing anthocyanin content.

Beyond its scientific use, LC-MS plays a crucial role in the food and agriculture industries, especially for quality control. By measuring the anthocyanin content in fruits, LC-MS helps in breeding programs designed to produce fruit varieties with higher anthocyanin levels. This is particularly valuable for boosting the nutritional value and market appeal of commercial fruit products (Singh et al., 2020). Kostka et al. (2020) generated pomegranate extracts from commercially available pomegranate juice, not from concentrate, and analyzed focusing on its anthocyanin and copigment fractions. An HPLC system coupled with an HCT Ultra Ion Trap mass, which featured an ESI source was used. Hence, the HPLC-PDA-ESI-MS/MS analysis identified and showed that the combination of anthocyanins, copigments, and other polymeric compounds results in a synergistic effect. Furthermore, the extract was found to notably decrease induced oxidative stress in human HepG2 cells. The research concludes that pomegranate extract (thus juice) shows promising potential for use in functional foods and nutraceuticals designed to alleviate oxidative stress. Besides, another study employed LC-QTOF/ESI-MS to analyze polyphenolic compounds in purple passion fruit, identifying seven anthocyanins. The fruit's antioxidant, anti-inflammatory, and anti-diabetic activities were evaluated, thereby providing insights into its potential as a functional food ingredient (Siniawska & Wojdyło, 2023). *Aronia melanocarpa* L. is another fruit rich in polyphenolic compounds and anthocyanins. The juice samples of aronia were found to vary considerably depending on factors such as the harvesting period and the processing techniques employed. LC QTOF/ESI-MS was employed in both negative and positive ionization modes in full scan mode with voltage ramping from 0.3 to 2 V. Collision-induced fragmentation was conducted using argon as the collision gas. The optimal LC-MS parameters were as follows: capillary voltage of 2500 V, cone voltage of 30 V, capillary temperature of 300°C, and source heater temperature of 100°C. Tasinov et al. (2022) used LC-PDA-ESI-MS/MS to identify and quantify bioactive compounds in juices, ensuring they meet health claims. MS analysis was performed using an LTQ Orbitrap with an ESI source. Key parameters were a 4 kV source voltage, 20 units of sheath gas, 10 units of auxiliary gas, 2 units of sweep gas, and a capillary temperature of 275°C. Full scan mode was used at a 30,000 resolution (*m/z* 400), with MS/MS events at 15,000 resolution. The study found that factors like harvesting time, processing methods (e.g., cold pressing and pasteurization), and growing conditions (soil, climate, region) influence nutrient content. LC-MS played an important role in accurately identifying and quantifying bioactive compounds such as anthocyanins, thereby ensuring product consistency. The study underscored the importance of utilizing LC-PDA-ESI-MS/MS for quality control, thereby ensuring that chokeberry juices retained their beneficial properties and complied with the standards set by the food and health industries.

11.2.2 Root Vegetables

Table 11.2 summarizes LC-MS analysis of anthocyanins from vegetal sources. Wan et al. (2024) compared the anthocyanin profiles of alcohol extracted white, orange, and purple sweet potato flesh using an LC-MS/MS system. Chromatographic analysis was conducted on a column (1.8 μm, 2.1 mm × 100 mm), and MS was performed with an ion spray voltage of +5.5/−4.5 kV, a temperature of 550°C, and ion source gases at 50 psi and 60 psi. The curtain gas was set to 35 psi, with de-clustering potentials of ±60 V, MS1 collision energy of ±10 V, MS2 collision energy of ±30 V, and a collision energy spread of 15 V. The MS1 and MS2 TOF masses ranged from 50 to 1250 Da. A total of nine significantly different anthocyanins were identified, including three paeonidins. The purple variety exhibited 74.78 times higher levels of peonidin-3-rhamnoside compared to the white variety and 13.31 times higher than the orange variety, while the orange variety had 5.62 times more than white. Peonidin 3-O-beta-galactopyranoside levels in purple varieties were 16.98 times higher than in white and 5.76 times higher than in orange, though differences between white and orange were not significant. Additionally, peonidin 3-(6″-p-coumarylglucoside) was 2.51 times more abundant in orange than in white varieties. The most notable difference was observed in delphinidin 3,5-diglucoside, which was significantly higher in the purple variety than in the other ones.

Jiang et al. (2020) determined the free anthocyanin content in purple sweet potatoes using an Ultimate 3000 HPLC system (Thermo Fisher Scientific Inc., USA) coupled with a Thermo Scientific LTQ XL mass spectrometer and a Phenomenex C18 column (250 mm × 4.6 mm, 5 μm). The system employed gradient elution for separation, and the mass spectrometer, equipped with an ESI source, performed scans in positive ion mode (*m/z* 100–2000) to identify molecular and fragment ions. The capillary voltage was set at 26 V, with nitrogen used as the drying and nebulizing gas at a flow rate of 20.0 L/min and a temperature of 270°C. Through HPLC-MS analysis, 13 distinct anthocyanins were identified, with

TABLE 11.2

LC-MS Analysis of Anthocyanins from Veggies, Grains and Legumes

Source of Anthocyanins	Analyte	Analyte Quantity	LC-MS Method	Reference
Black bean (*Phaseolus vulgaris* L.)	Delphinidin-3-O-glucoside Petunidin-3-O-glucoside Malvidin-3-O-glucoside	4.67 mg/g DW 3.23 mg/g DW 1.08 mg/g DW	LC-ESI-MS ESI mode: positive ionization Interface voltage 4.5 kV RF-beam voltage 60 V Spray interface temperature 350°C Heat block temperature 200°C	Kuasnei et al. (2022)
Chilean bean (*P. vulgaris* L.)	Delphinidin 3-O-hexoside Petunidin 3-O-hexoside Malvidin-3-O-glucoside Delphinidin 3-(6″-malonylhexoside)	17.16 mg/100 g of dry seeds	HPLC-DAD-Q-TOF-MS/MS ESI mode: positive ionization Ionization voltage + 4500 V Nebulizer temperature 200°C Full scan mode Mass range: m/z 50–1500	Nina et al. (2023)
Mung bean (*Vigna radiata* L.)	Peonidin-3-O-arabinoside Peonidin-3-O-galactoside Delphinidin-3-O-(6-O-malonyl)-glucoside Peonidin-3-O-glucoside Petunidin-3-O-rutinoside Cyanidin-3-O-glucoside Delphinidin-3-O-rutinoside Petunidin-3-O-(6-O-malonyl)-glucoside Cyanidin-3-O-galactoside Petunidin-3-O-glucoside Delphinidin-3-O-galactoside Delphinidin-3,5-O-diglucoside Delphinidin-3-O-arabinoside Cyanidin-3-O-rutinoside Cyanidin-3-O-(6-O-malonyl)-glucoside Delphinidin-3-O-glucoside Pelargonidin-3-O-glucoside Cyanidin-3-O-arabinoside Pelargonidin-3-O-rutinoside	N/A	LC-MS/MS ESI mode: positive ionization Voltage 5500V Multiple reaction monitoring mode Ion source temperature 550°C	Ma et al. (2023)

(Continued)

TABLE 11.2 *(Continued)*

LC-MS Analysis of Anthocyanins from Veggies, Grains and Legumes

Source of Anthocyanins	Analyte	Analyte Quantity	LC-MS Method	Reference
Black rice bran (*Oryza sativa* L.)	Cyanidin-3-O-glucoside	1.63 mg/g DW	HPLC-MS ESI mode: positive	Leonarski et al. (2023)
Red rice bran (*O. sativa* L.)	Cyanidin-3-arabinoside Petunidin-3-galactoside Cyanidin-3-galactoside Delphinidin-3-arabinoside Delphinidin-3-xyloside Cyanidin Cyanidin-3-glucoside Peonidin-3-galactoside Peonidin-3-glucoside Delphinidin Delphinidin-3,5-diglucoside Malvidin Peonidin Petunidin Malvidin-3-galactoside Malvidin-3-glucoside Delphinidin-3-sambubioside	N/A	UPLC-ESI-QTOF-MS ESI mode: positive ionization Source temperature 150°C Desolvation temperature 350°C Mass range *m/z* 50–1000	Chen et al. (2022)
Sweet potato (*Ipomoea batatas* L.)	Peonidin-3-rhamnoside Peonidin 3-beta-galactopyranoside Peonidin 3-(6″-p-coumarylglucoside) Delphinidin 3,5-diglucoside	N/A	LC-MS/MS ESI mode: positive ionization Spray voltage +5.5 KV Temperature 550°C De-clustering potential ±60 V MS1 collision energy ±10 V MS2 collision energy ±30 V Collision energy spread 15 V Mass range 50–1250 Da	Wan et al. (2024)

(Continued)

TABLE 11.2 *(Continued)*

LC-MS Analysis of Anthocyanins from Veggies, Grains and Legumes

Source of Anthocyanins	Analyte	Analyte Quantity	LC-MS Method	Reference
Purple sweet potato (*I. batatas* L.)	Cyanidin-3-sophoroside-5-glucoside Peonidin-3-sophoroside-5-glucoside Cyanidin-3-*p*-hydroxybenzoyl sophoroside-5-glucoside Peonidin-3-*p*-hydroxybenzoyl sophoroside-5-glucoside Cyanidin-3-(6″-feruloyl sophoroside)-5-glucoside Peonidin-3-(6″-feruloyl sophoroside)-5-glucoside Cyanidin-3-caffeoyl-phydroxybenzoyl sophoroside-5-glucoside Cyanidin-3-(6″-caffeoyl sophoroside)-5-glucoside Cyanidin-3-(6″-caffeoyl-6‴-feruloyl sophoroside)-5-glucoside Peonidin-3-caffeoyl-phydroxybenzoyl sophoroside-5-glucoside Peonidin-3-caffeoyl sophoroside-5-glucoside Peonidin-3-(6″-caffeoyl-6‴-feruloyl sophoroside)-5-glucoside Peonidin-3-(6″- *p*-hydroxybenzoyl-6‴-feruloyl sophoroside)-5-glucoside	N/A	HPLC-DAD/ESI-MS2 ESI mode: positive ionization Nebulizer temperature 270°C Mass range *m/z* 100 and 2000	Jiang et al. (2020)
Red-fleshed radish (*Raphanus sativus* L.)	Pelargonidin 3-O-beta-D-glucoside Cyanidin 3,5-O-diglucoside Cyanidin 3-O-rutinoside Pelargonidin Cyanidin Peonidin 3-feruloylsophoroside-5-glucoside Cyanidin 3-O-galactoside	70.29% 9.09% 7.97% 6.60% 5.29% 0.42% 0.33%	LC–ESI–MS/MS ESI mode: positive ionization Source temperature 500°C Ion spray voltage 5500 V	Liu et al. (2023)
Radish (*R. sativus* L.)	Delphinidin-3-O-(6″-O-alpha-rhamnopyranosyl-beta-glucopyranoside) Delphinidin-3-O-rutinoside-5-O-glucoside Peonidin 3-O-glucoside-5-O-glucoside	N/A	UHPLC–Q–TOF–MS ESI mode: positive ionization Capillary voltage 3.5 kV Nozzle voltage 500 V Fragmentor voltage 110 V Full scan mode Mass range 50 and 1000	Wang et al. (2023c)

(Continued)

TABLE 11.2 *(Continued)*

LC-MS Analysis of Anthocyanins from Veggies, Grains and Legumes

Source of Anthocyanins	Analyte	Analyte Quantity	LC-MS Method	Reference
Red-skin onion bulbs (*Allium cepa* L.)	Petunidin 3-rutinoside Delphinidin 3-rutinoside Cyanidin-3-rutinoside Delphinidin 3-glucoside Cyanidin 3-glucoside Pelargonidin 3-glucoside	N/A	LC–LTQ-MS ESI mode: positive ion Mass range 100–2000 *m/z*	Mirzazadeh et al. (2024)
Carrots (*Daucus carota* L. cultivar "Guanghong")	Peonidin-3-O-arabinoside Peonidin-3-O-glucoside Cyanidin-3-O-glucoside Pelargonidin- 3-O-glucoside	0.01–5 µg/mL 0.001–5 µg/mL 0.002–5 µg/mL 50–5000 µg/mL	UPLC-APCI-MS/MS ESI mode: positive ionization Entrance potential 10 V Collision cell exit potential 13 V Multiple reaction monitoring mode (MRM) Ion source temperature 350°C	Bao et al. (2023)
Black carrot (*Daucus carota* L.)	Cya-xyl-glc-gal Cya-xyl-gal Cya-xyl-glc-gal sinapoly Cya-xyl-glc-gal feruloyl Cya-xyl-glc-gal p-coumaroyl	N/A	UHPLC-TimsTOF Pro-MS ESI mode: positive ionization Ion mobility separation and PASEF	Ingemann Berentzen et al. (2024)
Red cabbage (*Brassica oleracea* L.)	Cyanidin 3-sophoroside 5-glucoside Cyanidin 3-(sinapoyl)-diglucoside-5-glucoside isomer Cyanidin 3-(caffeoyl)(p-coumaroyl)-diglucoside-5- glucoside Cyanidin 3-(sinapoyl)-triglucoside-5-glucoside Cyanidin 3-(feruloyl)(feruloyl)-triglucoside-5-glucoside Cyanidin 3-(feruloyl)(sinapoyl)-triglucoside-5- glucoside Cyanidin 3-(signapoyl) (sinapoyl)-triglucoside-5- glucoside g Cyanidin 3-(p-coumaroyl)-diglucoside-5-glucoside Cyanidin 3-(sinapoyl)-diglucoside-5-glucoside isomer Cyanidin 3-(feruloyl)-diglucoside-5-glucoside Cyanidin 3-(feruloyl)-triglucosides-5-glucoside Cyanidin 3-(feruloyl)(feruloyl)-diglucoside-5-glucoside Cyanidin 3-(feruloyl) (sinapoyl)-diglucoside-5- glucoside Cyanidin 3-(sinapoyl)(sinapoyl)-diglucoside-5- glucoside	307 ± 3 mg/100 g 60 ± 1 mg/100 g 27 ± 2 mg/100 g 17.6 ± 0.6 mg/100 g 154 ± 1 mg/100 g 58 ± 1 mg/100 g 46.6 ± 0.7 mg/100 15.2 ± 0.3 mg/100 g 700 ± 6 mg/100 g 262 ± 2 mg/100 g 10 ± 3 mg/100 g 110 ± 2 mg/100 g 140 ± 12 mg/100 g 420 ± 10 mg/100 g	UHPLC-ESI-QTOF-MS/MS ESI mode: positive ionization Capillary voltage 4.5 kV Full scan mode Temperature 220°C Mass range *m/z* 50–1500	Dong et al. (2024)

(Continued)

TABLE 11.2 *(Continued)*

LC-MS Analysis of Anthocyanins from Veggies, Grains and Legumes

Source of Anthocyanins	Analyte	Analyte Quantity	LC-MS Method	Reference
Red cabbage (*B. oleracea* L.)	Delphinidin-3-sambubioside Cyanidin-3-sambubioside Cyanidin-3-diglucoside-5-glucoside Cyanidin-3-(sinapoyl)-diglucoside-5-glucoside Cyanidin-3-(caffeoly)(p-coumaroyl)-diglucosides-5-glucoside Cyanidin-3-(feruloyl)-triglucoside-5-glucoside Cyanidin-3-(sinapoyl)-triglucoside-5-glucoside Cyanidin 3-(feruloyl)(sinapoyl)triglucoside-5-glucoside Cyanidin 3-(sinapoyl)(p-coumaroyl)triglucoside-5-glucoside Cyanidin 3-(sinapoyl)(sinapoyl)triglucoside-5-glucoside Cyanidin-3-(p-coumaroyl)-diglucoside-5-glucoside Cyanidin-3-diglucoside-5-(sinapoyl)-glucoside Cyanidin-3-(feruloyl)-diglucoside-5-glucoside Cyanidin-3-(feruloyl)(feruloyl)-diglucoside-5-glucoside Cyanidin-3-(feruloyl)(sinapoyl)-diglucoside-5-glucoside Cyanidin-3-(sinapoyl)-(sinapoyl)-diglucoside-5-glucoside	N/A	HPLC-DAD/ESI-MS ESI mode: positive ionization Capillary voltage: 49 V Tube lens voltage: 250 V Capillary temperature: 275°C Full scan mass (selected ion monitoring, SIM) Relative collision energies: 30 and 60 V Mass range m/z 300 and 2000	Pereira et al. (2024)
Red lettuce (*Lactuca sativa* L.)	Delphinidin 3-(6-p-coumaroylgalactoside)	N/A	UPLC VION IMS QTOF MS/MS ESI mode: positive ionization Capillary voltage 2 kV Collision energy 20–55 eV Source temperature 115°C Desolvation temperature 450°C Mass range 50–1000 m/z	Hameed et al. (2023)
Okra pod skins (*Abelmoschus esculentus* L.)	Delphinidin 3-osambubioside Cyanidin 3-osambubioside	7.00 ± 0.22 mg/g DW 4.05 ± 0.12 mg/g DW	UHPLC-Q-Orbitrap HRMS	Zhang et al. (2021)

cyanidin-3-sophoroside-5-glucoside and peonidin-3-sophoroside-5-glucoside identified as the main acylated anthocyanins, either monoacyl or diacyl forms. These anthocyanins contained one or two p-hydroxybenzoic, caffeic, or ferulic acid groups. Most anthocyanins in purple sweet potatoes were conjugated forms of peonidin or cyanidin, primarily glycosylated with glucose and sophorose.

Anthocyanins in red-fleshed radish were analyzed using an LC–ESI–MS/MS system consisting of a Shim-pack UFLC SHIMADZU CBM30A HPLC system and an Applied Biosystems 6500 quadrupole-linear ion trap (Q TRAP) mass spectrometer (Liu et al., 2023). The separation was achieved on a Waters ACQUITY UPLC HSS T3 C18 column (1.8 µm, 2.1 mm × 100 mm), and the eluate was directed into the ESI-triple Q TRAP-MS. Mass data were acquired in positive ion mode and processed with Analyst 1.6.3 software (AB Sciex), using turbo spray ionization at 500°C and a spray voltage of 5500 V. Gas settings were 55, 60, and 25 psi for ion source gases I and II and curtain gas, respectively, with high collision gas pressure. Instrument tuning and calibration were done using polypropylene glycol solutions in QQQ and LIT modes. MRM experiments were performed with nitrogen as the collision gas at 5 psi. Each period's MRM transitions were optimized for specific metabolites, and metabolite identification was based on primary and secondary spectral data compared with the Metware Biotechnology Co., Ltd. database and public sources. Quantification of metabolites was conducted through MRM and partial least squares discriminant analysis (PLS-DA). Seven anthocyanins were identified, with pelargonidin 3-O-beta-D-glucoside and pelargonidin being the primary contributors to the color of the red-fleshed radish, constituting 70.29% and 6.6% of the total anthocyanin content, respectively. Additionally, four cyanidin derivatives were detected, making up 22.68% of the overall anthocyanin content.

In another study, ultrasound-assisted extraction was employed to extract anthocyanins from radishes (Wang et al., 2023c) and their identification was performed using UHPLC-Q-TOF-MS. The UHPLC analysis utilized an Agilent 1290 Infinity II LC system with an Eclipse-plus C18 column (150 × 3.0 mm id, 1.8 µm) and a gradient elution. MS was conducted with an Agilent 6545 Q-TOF mass spectrometer in both positive and negative ion modes. Full-scan mode covered the *m/z* range of 50–1000, with two spectra recorded per second. The source parameters included a capillary voltage of 3.5 kV, a nozzle voltage of 500 V in positive ion mode and 1490 V in negative ion mode, a fragmentor voltage of 110 V, sprayer pressure of 45 psi, and sheath gas flow rate of 8 L/min. Collision-induced dissociation voltages were set at 10, 20, and 40 V. The analysis revealed three distinct anthocyanins: delphinidin-3-O-(6″-O-alpha-rhamnopyranosyl-beta-glucopyranoside), delphinidin-3-O-rutinoside-5-O-glucoside, and peonidin 3-O-glucoside-5-O-glucoside.

Mirzazadeh et al. (2024) used LC-MS to identify anthocyanins in red onion skin. Various extraction methods were employed, including solvent extraction, ultrasound-assisted extraction, subcritical water extraction, microwave-assisted extraction, pulsed electric field extraction, supercritical fluid extraction, and high hydrostatic pressure-assisted extraction. High hydrostatic pressure extraction yielded the greatest efficiency for anthocyanin extraction. LC-MS (Agilent 1200-6410, USA) with a C18 column, a pump connected to a photo diode array (PDA), and a linear trap quadrupole (LTQ) mass spectrometer was used in gradient mode for compound isolation and identification. Mass scanning was conducted in the range of 100–2000 *m/z* in positive ion mode. The identified anthocyanins included petunidin 3-rutinoside, delphinidin 3-rutinoside, cyanidin-3-rutinoside, delphinidin 3-glucoside, cyanidin 3-glucoside, and pelargonidin 3-glucoside.

Extracted Guanghong carrot anthocyanins were identified by using UPLC-APCI-MS/MS system (UPLC, SCIEX ExionLC™ AD, Framingham, MA, USA; MS, Applied Biosystems

6500 Q TRAP, Foster City, CA, USA) (Bao et al., 2023). A YMC Carotenoid C30 column (3 μm, 100 mm × 2.0 mm) was used with a gradient program for the LC, and mass data were acquired using a 6500 Q TRAP mass spectrometer (Applied Biosystems, Foster City, CA, USA) equipped with an atmospheric pressure chemical ionization (APCI) interface in positive ion mode. Detection was conducted in MRM mode. The ion source temperature was set to 350°C, while the entrance potential (EP) and collision cell exit potential (CXP) were set at 10 V and 13 V, respectively. The anthocyanins identified and their concentrations were as follows: peonidin-3-O-arabinoside ranged from 0.01 to 5 μg/mL, peonidin-3-O-glucoside from 0.001 to 5 μg/mL, cyanidin-3-O-glucoside from 0.002 to 5 μg/mL, and pelargonidin-3-O-glucoside ranged from 50 to 5000 μg/mL. In the study of Ingemann Berentzen et al. (2024), black carrot anthocyanins were extracted and analyzed using a TimsTOF Pro mass spectrometer (Bruker, MA, USA), which was coupled with an ELUTE UHPLC system outfitted with an Intensity Solo 2, C18 column (2 × 100 mm) (Bruker, MA, USA). The compounds were separated via gradient elution, and data collection was performed using the vendor's 4D metabolomics method, incorporating ion mobility separation and PASEF in positive ion mode. The researchers identified five different anthocyanins: cya-xyl-glc-gal, cya-xyl-gal, cya-xyl-glc-gal sinapoyl, cya-xyl-glc-gal feruloyl, and cya-xyl-glc-gal p-coumaroyl.

11.2.3 Other Veggies

Anthocyanins from red cabbage were extracted using ultrasound-assisted methods and analyzed for identification (Dong et al., 2024). The analysis was performed using a Waters Acquity UHPLC HSS T3 RP18 column (100 × 2.1 mm i.d., 1.8 μm, 100 Å) with an UltraLine UHPLC In-Line Filter (RESTEK, Bellefonte, PA, USA). The extracted anthocyanins were screened using a Bruker Elute UHPLC coupled with a Bruker Compact QTOF mass spectrometer equipped with an ESI probe (Bruker, Bremen, Germany). Data were acquired in full scan mode (*m/z* range 50–1500) in positive ion mode, with a capillary voltage of 4.5 kV. Nitrogen was used as the nebulizer (2.0 bar) and drying gas (8.0 L/min) at a temperature of 220°C. MS/MS spectra were recorded in Auto MS/MS mode, with collision energy set between 20 and 50 eV for the top three precursor ions in the *m/z* 300–1500 range. Instrument calibration was performed with a 10-mM sodium formate solution introduced to the ion source via a 20-μL loop before each run. Compounds were identified by comparing fragmentation patterns and high-accuracy mass spectra, cross-referenced with the MetaboScape 2023 database (Bruker, Bremen, Germany). The majority of the identified anthocyanins were derivatives of cyanidin.

Pereira et al. (2024) identified anthocyanins in red cabbage using LC coupled with MS (LC-DAD/ESI-MS). The analysis was conducted with a Finnigan Surveyor series liquid chromatograph, equipped with a RP C18 column (Agilent), measuring 250 × 4.6 mm. Mass detection was performed using a Finnigan LCQ DECA XP MAX quadrupole ion trap mass detector (Finnigan Corp., San José, CA, USA), featuring an Atmospheric Pressure Ionization (API) source and an ESI interface. Spectra were recorded in positive ion mode, covering a range between *m/z* 300 and 2000. Fourteen anthocyanins were identified in the red cabbage extract, the majority of which were cyanidin derivatives.

The anthocyanins in red lettuce extracts were analyzed using the Acquity UPLC VION IMS QTOF-MS/MS system (Waters Corp., Milford, MA, USA) (Hameed et al., 2023). A Waters Acquity UPLC HSS T3 column (100 mm × 2.1 mm, 1.7 μm) was utilized for the analysis with a 3-μL sample injection. The column temperature was maintained at 40°C. The mobile phase consisted of water and acetonitrile with a flow rate of 0.4 mL/min, applied in a gradient. For the first 0–3 minutes, a 100% gradient was used. From 3 to 5 minutes, a linear gradient of 90–100% water was maintained. Between 5 and 11.5 minutes, the gradient ranged from 65 to 90%, followed by 1–65% from 11.5 to 14 minutes. For 14–17 minutes, 1% water was used, with a 1–100% transition between 17 and 17.1 minutes. The column was then re-equilibrated for 5 minutes. Capillary voltage was set to 2 kV, with a source temperature of 115°C and desolvation temperature of 450°C, and the desolvation gas flow rate was 800 L/h. The collision energy varied from 20 to 55 eV, and spectra were recorded in negative ion mode across a scan range of m/z 50 to 1000. Delphinidin 3-(6-p-coumaroyl-galactoside) was the reported anthocyanin from red lettuce.

Ultrasound-assisted extracts of okra pod skins anthocyanins were identified by UHPLC-Q-Orbitrap HRMS (Zhang et al., 2021). Anthocyanins were separated using a Vanquish Flex UHPLC system (Thermo Scientific, Waltham, MA, USA) paired with a Waters XBridge C18 column (1.7 µm, 2.1 mm × 150 mm). UV detection was performed at 535 nm for anthocyanins and at 360 nm for flavonols. Gradient elution was applied. For metabolite identification, a Q-Exactive mass spectrometer (Thermo Scientific) was connected to the UHPLC system. The identified anthocyanins were reported as Delphinidin 3-O-sambubioside and Cyanidin 3-O-sambubioside.

Wang et al. (2024a) investigated the thermal stability of anthocyanins within a model dough system. To conduct this study, anthocyanins extracted from eggplant peels were incorporated into dough composed of microcrystalline cellulose, wheat starch, and isolated soy protein which was then exposed to thermal processing. Following the heat treatment, anthocyanins were re-extracted from the dough and analyzed using UPLC-MS/MS method on a Thermo Accela 600 UPLC system (Thermo Scientific, Bremen, Germany) with an Inertsil ODS-3 C18 column (4.6 mm × 250 mm, 5 µm, GL Sciences, Nakano, Tokyo, Japan). Detection was performed with diode array monitoring at 280 and 520 nm. The gradient elution system was employed, and mass detection was carried out using a linear ion trap quadrupole-Orbitrap mass spectrometer equipped with a heated ESI probe (Thermo Fisher Scientific, San Jose, CA, USA) operating in both positive and negative ion modes. The parameters included a capillary voltage of −40 V, capillary temperature of 320°C, source voltage of 4 kV, and tube lens voltage of 110 V, with sheath and auxiliary gas flow rates (nitrogen) set at 42 and 11 L/min, respectively. MS spectra were acquired in data-dependent mode across a mass range of 100–1500 m/z. The researchers identified three anthocyanins: delphinidin-3-O-rutinoside-5-O-glucoside, delphinidin-3-O-rutinopyranosyl-glucoside, and delphinidin-3-O-rutinoside.

11.2.4 Grains

Leonarski et al. (2023), compared the extraction of anthocyanins from rice bran using pressurized liquid extraction and heating-stirring extraction methods. Individual anthocyanins were identified using a Shimadzu HPLC-MS chromatography system (model LCMS-2020, Kyoto, Japan), which included an LC-20AD binary pump, photodiode array detector (PDA), SIL-20 AC HT autosampler, central controller, and single-quadrupole MS detector with an ESI interface. The system utilized a Kromasil® C18 column (100 Å, 300 mm × 4.6 mm i.d.) under gradient conditions, with eluates detected at 25°C. Ionization was performed using nitrogen with a drying gas flow of 15 L/min and a nebulizing gas flow of 1.5 L/min. Cyanidin-3-O-glucoside was identified (m/z 449) and quantified at 1.63 mg/g for pressurized liquid extraction and 1.07 mg/g for heating-stirring extraction.

Chen et al. (2022), explored the anthocyanin profile of Wugong Mount purple red rice bran pigment extracts. The anthocyanins were analyzed using an Agilent 1290 UPLC system coupled with an Agilent 6538 Q-TOF mass spectrometer (Agilent Technologies, Santa Clara, CA, USA). Separation was achieved using a Zorbax Eclipse Plus C18 column (2.1 × 100 mm, 1.8 μm). The optimal MS conditions involved analysis in positive ion mode, with a scan range of m/z 50–1000, a source temperature of 150°C, a desolvation temperature of 350°C, a cone gas flow of 10 L/h, and a desolvation gas flow of 900 L/h. The identified anthocyanins included cyanidin-3-arabinoside, petunidin-3-galactoside, cyanidin-3-galactoside, delphinidin-3-arabinoside, delphinidin-3-xyloside, cyanidin, cyanidin-3-glucoside, peonidin-3-galactoside, peonidin-3-glucoside, delphinidin, delphinidin-3,5-diglucoside, malvidin, peonidin, petunidin, malvidin-3-galactoside, malvidin-3-glucoside, and delphinidin-3-sambubioside.

11.2.5 Legumes/Beans

Kuasnei et al. (2022) extracted anthocyanins from black bean hulls using subcritical water extraction. They compared the use of deep eutectic solvents with choline chloride-citric acid mixture as extraction solvents and found that eutectic water demonstrated greater efficiency. The deep eutectic solvents were prepared by combining choline chloride with various organic acids and alcohols, including citric acid, propanediol, malic acid, lactic acid, and a mixture of lactic acid and propanediol, all at a concentration of 1%. Extracts were analyzed via LC-MS to determine the anthocyanin profile and quantify individual components. The analysis was conducted using a Shimadzu liquid chromatograph (Kyoto, Japan) equipped with a diode array detector (SPD-M20A) and an ESI mass spectrometer (MS-ESI, LCMS-2020). A C18 column (NST, Santos, Brazil), with dimensions of 4.6 mm × 250 mm, 5.0 μm, was used at 40°C with gradient elution. Anthocyanins were identified by confirming their masses (m/z) in the mass spectrometer. The mass spectrometer parameters included spray interface temperature of 350°C and heat block temperature of 200°C. The interface voltage was set to 4.5 kV, and the RF-beam voltage was 60 V. The three primary anthocyanins detected, based on their respective ions, were delphinidin-3-O-glucoside [M+] 465.1, petunidin-3-O-glucoside [M+] 479.12, and malvidin-3-O-glucoside [M+] 493.13, which are the predominant anthocyanins present in black beans.

Nina et al. (2023) extracted anthocyanins from Chilean bean varieties using ultrasound assisted extraction method. An HPLC-DAD-Q-TOF-MSn analysis was conducted using a Bruker Elute UHPLC/HPLC-DAD system, coupled with a Q-TOF Compact mass spectrometer (Bruker Daltonics, Bremen, Germany). Separation was achieved using a Kromasil® C-18 column (250 mm × 4.6 mm, 5 μm particle size, Kromasil, Akzo Nobel, Bohus, Sweden) alongside a C-18 pre-column (Nova-Pak Waters, 22 × 3.9 mm, 4 μm particle size, Waters Corp., Milford, Connecticut, USA). Data were collected in full scan mode across a mass-to-charge ratio (m/z) range of 50–1500 in both positive and negative ionization modes. Nitrogen was employed as the nebulizer gas at 200°C, and MS detection was performed using base peak chromatograms. The MS conditions were set at +4500 V for positive ionization and −3500 V for negative ionization. Collision-induced dissociation spectra were generated using a fragmentation amplitude of 1.00 V with helium as the collision gas. Auto MS/MS mode (four precursor ions per cycle) was employed with a scan range of 50–1500 m/z in centroid mode, with a scan time of 0.2 seconds. The main anthocyanins in black beans: delphinidin 3-O-hexoside was identified by the [M + H]+ ion at m/z 465 and the neutral loss of 162 amu leading to the aglycone at *m/z* 303; petunidin 3-O-hexoside was identified by the [M + H]+ at *m/z* 479 and the MS2 ion at *m/z* 317; malvidin-3-O-glucoside was characterized by the [M + H]+ ion at *m/z* 331; delphinidin 3-(6''-malonylhexoside) presented a neutral loss of 248 amu, an additional 86 amu to the neutral loss of the hexoside (162 amu), indicating the presence of a malonyl group.

Ma et al. (2023) examined black mung bean extracts using a LC-tandem MS system (Applied Biosystems 6500 QTRAP). The chromatographic setup involved a C18 column (ACQUITY BEH, 1.7 µm, 2.1 mm × 100 mm). For MS, the conditions were as follows: the electrospray ion source temperature was set to 550°C, with a MS voltage of 5500 V and a curtain gas pressure of 35 psi. The MS data obtained were qualitatively assessed using a local database. Subsequently, the anthocyanin components were quantitatively evaluated using the MRM mode of triple quadrupole MS. Across the three stages, LC-MS/MS identified anthocyanin derivatives including five types of cyanidin, six types of delphinidin, two types of pelargonidin, three types of peonidin, and three types of petunidin.

11.2.6 Edible Flowers

Edible flowers have been used traditionally for their health promoting effects depending on a wealth of phytochemical content, and even for culinary purposes in some areas (Kandylis, 2022). The wide research of anthocyanin rich edible plants is not only due to their potential in preventing several diseases but also owing to their potential as appealing colorants (Teixeira et al., 2023). The red, thick, fleshy calyces are the most common products of commerce from the hibiscus plant (also known as roselle, family Malvaceae) have been well-investigated by means of their rich anthocyanin content (Li, Simon & Wu, 2024). *Hibiscus sabdariffa* has been extensively used as an antihypertensive tea worldwide (Hopkins et al., 2013). Delphinidin and its glycosides were demonstrated in calyces of *H. sabdariffa* L. upon MS/MS analysis in the negative mode (Salem et al., 2020). Hibiscin, or delphinidin 3-sambubioside (C26H27O16), was confirmed due to the deprotonated molecules [M − H]− at *m/z* 595 and fragment ions at *m/z* 521, *m/z* 485, *m/z* 419, *m/z* 334, *m/z* 300, *m/z* 249, *m/z* 217, and *m/z* 189. Additionally, delphinidin (C15H9O7) aglycone (*m/z* =301), delphinidin 3-neohesperidoside (C27H30O16) (*m/z* =609), and delphinidin 3-O-galactoside (C21H19O12) (*m/z* =463).

When three roselle cultivars were compared, metabolome analysis using UPLC-Q-TOF/MS showed that the total anthocyanin content in red calyx was the highest, followed by that in stripped calyx, whereas green calyx had the lowest content (Li et al., 2022). The flower color of *Hibiscus syriacus* (Qiansiban) transitions from fuchsia to pink–purple and finally to pale purple, thereby enhancing the ornamental value. In order to elucidate the pigmentation mechanism, Li et al. (2024) analyzed Qiansiban and reported that malvacin-3-O-glucoside being the predominant compound in the flowers at different stages, followed by petunidin-3-O-glucoside. Except peonidin, the amounts of the detected five anthocyanin aglycones with a sugar moiety indicated gradual declines throughout the flowering process leading the color of the petals to convert from purplish red to purple. Mejía et al. (2023) evaluated flowers of 16 *Hibiscus rosa-sinensis* cultivars using an Orbitrap high-resolution mass detector (Exactive Plus, Thermo-Fisher Scientific, Sunnyvale, CA, USA). Chromatographic separation was carried out with a Hypersil GOLD aQ column (Thermo Scientific, Sunnyvale, CA, USA) of 100 mm, L, × 2.1 mm, id, × 1.9 µm particle size. Cyanidin-sophoroside ([M]+ at *m/z* 611.16066 with product ions at *m/z* values of 287, 147, 137, and 121) was identified in extracts of lilac, pink, orange, and red flowers, but was not detected in extracts of white or yellow flowers. In the white flowers, the level of flavan-3-ols (catechin and epicatechin) were 120 times lower than that found in the red cultivars which was attributed to the biosynthetic pathway of anthocyanins.

In another work, researchers found a significant positive correlation between delphinidin-3-O-(6-O-Malonyl-β-D-glucoside) a* coordinate measured with a colorimeter in the red-purple cultivars of H. syriacus (Chen et al., 2023a). Hoang et al. (2023) purified anthocyanins using macroporous D101 resin isolated from *H. sabdariffa* L. flowers, where cyanidin-3-O-sambubioside was dominant. Anthocyanin (purity of 72.18%) possessing high antioxidant capacity and inhibiting both Gram (+) and Gram (−) bacteria had a very good color stability at low temperatures and in mildly alkaline media. On the other hand, when aqueous *H. sabdariffa* extract was subjected to in vitro digestion, the contents of cyanidin-3-O-sambubioside and delphinidin-3-O-sambubioside drastically dropped from 2.91 ± 0.03 and 8.53 ± 0.08 µg/g (w/w) cyanidin-3-O-glucoside (CG) in the raw extract, respectively, to 0.12 ± 0.01 0.12 ± 0.01 µg/g (w/w) CG at the end of duodenal digestion (Majdoub et al., 2021). Song et al. (2024) reported that nanoencapsulation within chitosan derivatives or β-lactoglobulin could improve the bioaccessibility of cyanidin-3-sambubioside the most among the anthocyanin monomers present in roselle extract. A list of anthocyanins analysis through LC-MS from edible flowers can be seen in Table 11.3.

In order to avoid quality loss and extend the shelf life of Roselle products available, LC-MS analysis is a helpful tool for the measurement of anthocyanins (Gweyi-Onyango, Osei-Kwarteng & Mahunu, 2021). Food processing such as fermentation has a crucial impact on the resulting sensory quality, including color, aroma, taste, and mouthfeel. Using an untargeted UHPLC-QE-MS approach, delphinidin 3,5-diglucoside, cyanidin 3-(6″-(E)-caffeylsambubioside), delphinidin 3-gentiobioside, pelargonidin 3-lathyroside, pelargonidin 3-(4‴-p-coumarylrutinoside)-5-glucoside, and petunidin were identified in roselle wine for the first time (Ai et al., 2021). When different commercially available hibiscus products were compared, significant variations were discovered in the contents of anthocyanins following the order of calyces > extracts > beverages (Li, Simon & Wu, 2024). Recently, several research groups developed functional hibiscus drinks by taking the advantage of its characteristic flavor and therapeutic effects by analyzing the anthocyanin metabolites (Arce-Reynoso et al., 2023; Villalobos-Vega et al., 2023).

Saffron (family *Iridaceae*) plant belongs to the genus Crocus that produces the most expensive spice in the world. The aerial part of the plant consists of leaves and flowers, where stigma −the commercial saffron, constitutes only 7% of the flower (Bakshi et al., 2022). Crocin, picrocrocin, and safranal are the main constituents responsible for its color, taste, and aroma, respectively. Saffron is also important in pharmaceutics, cosmetics, perfumery, and textile dye-producing industries (Kothari, Thakur & Kumar, 2021). The quality of saffron depends on the concentration of secondary metabolites, thus saffron from different origins have been studied from Iran, Spain, and Türkiye (Amanpour et al., 2024), China (Li et al., 2024), and Italy (De Angelis et al., 2024). Due to its high cost, saffron adulteration is common by synthetic additives or safflower and turmeric (Salehi et al., 2022). Thus, LC-MS analyses of saffron containing samples contribute to identify adulterants, contaminant as well as its authentication (Avila-Sosa et al., 2022; Eghbali, Farhadi & Askari, 2023). Anthocyanins were quantified for the first time in saffron tepals by Goupy et al. (2013), 4804 ± 233 µg/g DW, using an UPLC-DAD/ESI-MSn. Senizza and co-authors (2019) observed that anthocyanins were strongly accumulated into the adulterated saffron products. Delphinidin 3,5-di-O-glucoside accounts for 80% of the total anthocyanins followed by petunidin 3,5-di-O-glucoside, delphinidin 3-O-glucoside, malvidin 3,5-di-O-glucoside, and petunidin 3-O-glucoside (Ruggieri et al., 2023). The production of 1 kg of saffron spice is estimated to generate around 350 kg of floral by-products, thereby the

TABLE 11.3

LC-MS Analysis of Anthocyanins from Edible Flowers

Source of Anthocyanins	Analyte	Analyte Quantity	LC-MS Method	Reference
Calyces of Hibiscus (*Hibiscus sabdariffa* L.)	Delphinidin 3-sambubioside (Hibiscin) Delphinidin 3-neohesperidoside Delphinidin 3-O-galactoside Delphinidin Dimethyl-delphinidin-glucosyl acetate	N/A	UPLC-MS/MS ESI mode: negative ionization Capillary voltage: 3 kV Alternating full-scan and all-ion fragmentation-scan modes Capillary temperature: 150°C Drying gas: 350°C Mass range: 100–1500 m/z	Salem et al. (2020)
Flowers of 16 Hibiscus cultivars (*Hibiscus rosa-sinensis*)	Cyanidin-sophoroside 1 Cyanidin-sambubioside1 Cyanidin-3-glucoside	21–1.0 mg/g of extract	UHPLC-ESI+-Obitrap-MS ESI mode: positive ionization Vspray: 3.5 kV Cell energies: 10, 20, 30, 40 eV Full scan mode Nebulizer temperature: 350°C Capillary temperature: 320°C Mass range: m/z 80–1000	Mejía et al. (2023)
Hibiscus market samples (calyces, beverages, and extracts) (*H. sabdariffa* L.)	Delphinidin-3-sambubioside Cyanindin-3-sambubioside	< 0.001–2.372% of total anthocyanins	UHPLC-QQQ/MS Voltage: 4 V MRM scan mode Drying gas: 300oC Nebulizer: 250oC LOD:98.63, LOQ:197.27 ng/mL (delphinidin derivative) LOD:26.56, LOQ:53.13 ng/mL(cyanidin derivative)	Li (2023)

(Continued)

TABLE 11.3 *(Continued)*

LC-MS Analysis of Anthocyanins from Edible Flowers

Source of Anthocyanins	Analyte	Analyte Quantity	LC-MS Method	Reference
Different calyx colors of roselle (red, green, stripped) (*H. sabdariffa* L.)	Cyanidin-3-O-glucoside Cyanidin-3-O-(6-O-malonyl-beta-D-glucoside) Cyanidin-3-O-sambubioside-5-O-glucoside Cyanidin-3-O-xyloside Cyanidin-3-O-rutinoside-5-O-glucoside Cyanidin-3-O-galactoside Cyanidin-3-O-arabinoside Cyanidin-3-O-sophoroside Cyanidin-3-O-rutinoside Cyanidin-3,5,3'-O-triglucoside Cyanidin-3-O-sambubioside Cyanidin-3,5-O-diglucoside Delphinidin-3-O-rutinoside Delphinidin-3-O-(6-O-malonyl-beta-D-glucoside) Delphinidin-3-O-rutinoside-5-O-glucoside Delphinidin-3-O-sambubioside-5-O-glucoside Delphinidin-3-O-rhamnoside Delphinidin-3-O-glucoside Delphinidin-3-O-arabinoside Delphinidin-3-O-sambubioside Delphinidin-3-O-sophoroside Malvidin-3-O-glucoside Pelargonidin-3-O-galactoside Pelargonidin-3-O-sambubioside Pelargonidin-3-O-glucoside Pelargonidin-3-O-sophoroside Peonidin-3-O-sambubioside-5-O-glucoside Peonidin-3,5-O-diglucoside Peonidin-3-O-sambubioside Peonidin-3-O-glucoside Peonidin Petunidin-3-O-galactoside Petunidin-3-O-glucoside Petunidin-3-O-arabinoside Petunidin-3,5-O-diglucoside	Red calyx had the highest contents of anthocyanin derivatives such as delphinidin-3-O-sambubioside (955.11 µg/g) and cyanidin-3-O-sambubioside (531.37 µg/g). Stripped calyx: (851.97 and 330.06 µg/g, respectively).	UPLC-Q-TOF/MS ESI mode: ionization Drying gas: 350°C HV capillary: –4100 V End plate offset: –500 V Capillary exit: 134.3 V Skimmer: 40 V	Li et al. (2022)

(Continued)

TABLE 11.3 *(Continued)*

LC-MS Analysis of Anthocyanins from Edible Flowers

Source of Anthocyanins	Analyte	Analyte Quantity	LC-MS Method	Reference
Hibiscus flowers (*Hibiscus syriacus* L.)	Delphinidin-3-O-glucoside Cyanidin-3-O-glucoside Petunidin-3-O-glucoside Pelargonidin-3-O-glucoside Malvacin-3-O-glucoside	N/A	HPLC-MS ESI mode: positive ionization Dryer temperature: 400°C Ion source temperature: 120°C Capillary voltage: 3.5 kV Cone hole voltage: 40 V Mass range: 100–1200 m/z	Li et al. (2024)
Hibiscus flowers from different strains (*H. syriacus* L.)	Pelargonidin-3,5-Odiglucoside Pelargonidin-3-Ogalactoside Pelargonidin-3-Oglucoside Pelargonidin-3-Oarabinoside Pelargonidin-3-Orutinoside Pelargonidin-3-O-(6″-Omalonyl-β-D-glucoside) Cyanidin-3,5-Odiglucoside Cyanidin-3-Ogalactoside Cyanidin-3-O-glucoside Cyanidin-3-Oarabinoside Cyanidin-3-O-rutinoside Cyanidin-3-O-(6″-Omalonyl-β-D-glucoside) Delphinidin-3,5-Odiglucoside Delphinidin-3-Ogalactoside Delphinidin-3-Oglucoside Delphinidin-3-Oarabinoside Delphinidin-3-Orutinoside Delphinidin-3-O-(6″-Omalonyl-β-D-glucoside) Peonidin-3,5-Odiglucoside Peonidin-3-Ogalactoside Peonidin-3-O-glucoside Peonidin-3-Oarabinoside Peonidin-3-O-(6″-Omalonyl-β-D-glucoside) Petunidin-3-Ogalactoside Petunidin-3-Oarabinoside Petunidin-3-O-rutinoside Petunidin-3-O-(6″-Omalonyl-β-D-glucoside)	Various concentrations among the samples	UPLC-ESI-MS/MS ESI mode: positive ionization Source temperature: 550°C Ion spray voltage: 5500 V	Zhang et al. (2022b)

(Continued)

TABLE 11.3 *(Continued)*

LC-MS Analysis of Anthocyanins from Edible Flowers

Source of Anthocyanins	Analyte	Analyte Quantity	LC-MS Method	Reference
	Malvidin-3,5-Odiglucoside Malvidin-3-Ogalactoside Malvidin-3-O-glucoside Malvidin-3-Oarabinoside Malvidin-3-O-rutinoside Malvidin-3-O-(6″-Omalonyl-β-D-glucoside)			
Saffron stigma from Iran, Spain, Türkiye	Petunidin-3-O-glucoside Delphin-3,5-di-O-glucoside Delphinidin-3-O-glucoside Cyanidin-3-O-glucoside	0.31–105 mg/kg	LC-DAD-ESI-MS/MS ESI mode: positive ionization Collision energy: −35 V Full MS scans Desolvation temperature: 250 °C Heat block temperature: 400 °C Mass range: 100 and 1000 m/z	Amanpour et al. (2024)
Saffron Floral By-Products (*Crocus sativus* L.)	Delphinidin 3,5-di-O-glucoside Petunidin 3,5-di-O-glucoside Delphinidin 3-O-glucoside Petunidin 3-O-glucoside	0.6–3.2 ppm	LC-ESI-QTOF MS/MS ESI mode: positive ionization Capillary voltage: 3500 V Nozzle voltage: 500 V Full scan Drying gas: 300°C Sheath gas: 250°C Mass range: 40–1300 m/z	Masala et al. (2024)
Saffron petals (*C. sativus* L.)	Delphinidin 3,5-di-O-glucoside Petunidin 3,5-di-O-glucoside Cyanidin-3-O-glucoside Cyanidin-3,5-O-diglucoside Cyanidin-3-O-(6-O-acetyl)-glucoside	N/A	LC-ESI-IT-TOF-MS ESI mode: positive and negative ionization Capillary voltage: 3 kV Cone voltage: 30 V Extractor voltage: 3.3 V Source temperature: 100°C Desolvation temperature: 210°C Full scan over 70–700 m/z (MS1) and 70–500 m/z (MS2)	Shahtouri et al. (2024)

(Continued)

TABLE 11.3 *(Continued)*

LC-MS Analysis of Anthocyanins from Edible Flowers

Source of Anthocyanins	Analyte	Analyte Quantity	LC-MS Method	Reference
Pistacia chinensis leaves	Cyanidin-3-O-(6-O-malonyl-beta-D-glucoside)	All in a range of 0.001–121.1 ng/g	UPLC-MS/MS 0.01–50,000 ng/mL	Song et al. (2022)
	Cyanidin-3-O-(coumaryl)-glucoside			
	Cyanidin-3-O-arabinoside			
	Cyanidin-3-O-galactoside			
	Cyanidin-3-O-glucoside			
	Cyanidin-3-O-rutinoside			
	Cyanidin-3-rutinoside 5-glucoside			
	Cyanidin-3-O-sophoroside			
	Cyanidin-3-O-xyloside			
	Cyanidin-3,5,3'-O-triglucoside			
	Delphinidin-3-O-(6-O-acetyl)-glucoside			
	Delphinidin-3-O-(6-O-malonyl)-glucoside-3'-glucoside			
	Delphinidin-3-O-arabinoside			
	Delphinidin-3-O-galactoside			
	Delphinidin-3-O-rhamnoside			
	Delphinidin-3-O-sophoroside			
	Malvidin-3-O-glucoside			
	Pelargonidin-3-O-(coumaryl)-glucoside			
	Pelargonidin-3-O-arabinoside			
	Pelargonidin-3-O-glucoside			
	Peonidin-3-O-(6-O-p-coumaryl)-glucoside			
	Peonidin-3-O-arabinoside			
	Peonidin-3-O-galactoside			
	Peonidin-3-O-glucoside			
	Peonidin-3-O-sophoroside			
	Petunidin-3-O-galactoside			
	Petunidin-3-O-glucoside			

(Continued)

TABLE 11.3 *(Continued)*

LC-MS Analysis of Anthocyanins from Edible Flowers

Source of Anthocyanins	Analyte	Analyte Quantity	LC-MS Method	Reference
Perillae Folium leaf (*Perilla frutescens var. crispa*)	Caffeylcyanin Cis-shisonin Caffeylmalonylcyanin Shisonin Cyanidin-3-O-feruloylglucoside-5-O-glucoside Manonyl-cis-shisonin Malonylshisonin Cyanidin-3-O-feruloylglucoside-5-O-malonylglucoside Peonidin-3-O-coumaroylglucoside-5-Omalonylglucoside	N/A	RRLC-Q/TOF-MS UPLC–TQ-MS/MS ESI mode: positive ionization Temperature: 500°C Ion spray voltage: +5500 V Declustering potential (DP): 80 eV Collision energy (CE):15 eV CID: 35/−35 ± 15 eV Mass range: m/z 100–2000	Zheng et al. (2020)
Perilla leaves (*P. frutescens var. acuta*)	Cis-shisonin Cyanidin 3-O-caffeoylglucoside-5-O-glucoside Cyanidin 3-O-caffeoylglucoside-5-O-malonylglucoside Malonyl-cis-shisonin Shisonin Cyanidin 3-O-feruloylglucoside-5-O-glucoside Malonyl-shisonin	N/A	UPLC-ESI-IT-TOF-MS^n ESI mode: positive ionization Interface voltage: 4.5 kV Detector voltage: 1.60 kV Temperature: 200°C The relative collision induced dissociation energy adjusted between 20% and 30% Mass range: m/z 150–1000 Da	He, Yao and Chang (2015)
Lemon bottlebrush flowers (*Callistemon citrinus*)	Cyanidin-3,5-O-diglucoside (cyanin) Peonidin-3,5-O-diglucoside (peonin) Cyanidin-3-O-glucoside Cyanidin-coumaroylglucoside-pyruvic acid.	295.58 ± 2.226 mg CyG/100 g DE (Cyanin) 182.03 ± 1.90 mg CyG/100 g DE (Peonin)	RP-HPLC-DAD-ESI-MS/MS ESI mode: positive ionization Temperature: 350°C CID: 1.0 V Collision gas: 1.46 × 10−5 bar, helium	Laganà et al. (2020)

(Continued)

TABLE 11.3 *(Continued)*

LC-MS Analysis of Anthocyanins from Edible Flowers

Source of Anthocyanins	Analyte	Analyte Quantity	LC-MS Method	Reference
6 Halophytes (*Aptenia cordifolia var. variegata, Glottiphyllum linguiforme, Carpobrotus edulis, Ferocactus glaucescens, F. pottsii* and *F. Herrerae*)	Delphinidin malonyl hexoside Malvidin acetyl hexoside Petunidin coumaroyl hexoside Cyanidin pentosyl hexoside Petunidin Cyanidin Pelargonidin pentosyl hexoside Delphinidin Delphinidin dihexoside	N/A	LC-HR-ESI-MS ESI mode: positive & negative ionization Spray voltage: 4.5 kV Capillary temperature: 320°C Mass range: *m/z* 100–2000	Hamed et al. (2020)
Blue flowers of six grape hyacinth cultivars (*Muscari* spp.)	Delphinidin 3-O-glucoside Petunidin 3-O-glucoside Unknown compound Malvidin 3-O-glucoside Unknown compound Pelargonidin-3-O-caffeoylsophoroside-5-O-arabinoside Cyanidin Cyanidin-3-O-caffeoylrutinoside Cyanidin-3-O-(p-coumaroyl)glucoside-5-O-malonylglucoside Pelargonidin-3-O-sinapylglucoside-5-O-glucoside Pelargonidin-3-O-sinapylglucoside-5-O-arabinoside Pelargonidin-3-O-ferulylglucoside-5-O-arabinoside	N/A	UPLC-QTOF-MS ESI mode: positive ionization Capillary voltage: 3000 V Cone voltage: 45 V Extract cone voltage: 4 V Ion source temperature: 100°C Drying temperature: 350°C Mass range: 100–1000 *m/z*	Lou et al. (2017)

utilization of these has a considerable economical and bioactive potential (Cerdá-Bernad et al., 2022; Wu et al., 2023). In the study of Masala et al. (2024), delphinidin and petunidin derivatives have also been detected in saffron flower by-products. (HR) LC-ESI-QTOF MS/MS analysis revealed the presence of delphinidin 3,5-di-O-glucoside and delphinidin 3-O-glucoside due to the [M+H]+ at m/z 627.1564 with a fragment at 465.1039 and [M+H]+ at m/z 465.1015 with a fragment at 303.0092, respectively. Similarly, petunidin 3,5-di-O-glucoside and petunidin 3-O-glucoside were identified based on the [M+H]+ at m/z 641.1717 with a fragment at 465.1044 and [M+H]+ at m/z 479.1108 with a fragment at 317.2589, respectively. Recently, Shahtouri et al. (2024) synthesized zirconium-based nanocomposites for anthocyanin purification. Used in saffron petal waste, the adsorbents exhibited an excellent selectivity leading to specific and sharper peaks. They identified new cyanidin derivatives: cyanidin-3,5-O-diglucoside and cyanidin-3-O-(6-O-acetyl)-glucoside.

To date, different parts of saffron plant were attempted to be used in several food and pharmaceutical formulations such as cookies, pasta, and probiotic beverage in free (Bakshi et al., 2022) or encapsulated form (Akramiani, Afsahi & Akhavan, 2024). Moreover, Wang et al. (2023d) investigated the metabolomics of color change of saffron floral residues during storage. Another group of researchers used saffron petals for fermentation for the first time, where 23 kinds of anthocyanin markers were found after fermentation based on untargeted metabolomics (Wang et al., 2024b).

Coming from *Rhus coriaria* L., sumac (family *Anacardiaceae*) is a medicinal plant that is native to the Mediterranean area and widely used in Turkish and Middle Eastern cuisine, especially in Lebanon, Iran, and Iraq (Zannou et al., 2025). The bark and leaves of the plant are astringent, antiseptic, alterative, and tonic properties while the berries have refrigerant, diuretic, emmenagogue, diaphoretic, and cephalic effects (Khoshkharam et al., 2022). The fruits can also serve as a coloring agent source in the food industry thanks to anthocyanin pigments. The anthocyanin content was reported to be ranging between 3.57 and 66.14 mg cyanidin-3-glucoside equivalents per gram of DW (Fereidoonfar et al., 2019). Mazzara et al. (2023) characterized sumac fruits from five Sicilian accessions by means of phytochemical composition using HPLC-MS/MS followed by principal component analysis. The dominant anthocyanins were cyanidin 3-glucoside and delphinidin-3,5-diglucoside. Except for delphinidin-3,5-diglucoside, which was detected in higher amounts in San Biagio Platani sumac extract, all the other anthocyanins were prevalent in Giarratana sumac extract.

A number of studies demonstrated the addition of sumac extract or powder as natural food preservatives to bread (Dziki et al., 2021; Xu et al., 2024b), soybean oil (Rahmati, Bazargani-Gilani & Aghajani, 2022), and beef burgers (Grassi et al., 2024). LC-MS analysis of anthocyanins from different processed food matrices is shown in Table 11.4. Wang and co-authors (2023e) investigated the use of the Staghorn sumac fruit powder in cooked beef meatballs and Cheddar cheese. LC-MSn profiling disclosed the contribution of nonanthocyanin phenolics in addition to anthocyanins to the antimicrobial and antioxidant activities observed. The prevalence of 7-O-methyl anthocyanins and their galloylated derivatives in Staghorn sumac (*Rhus typhina* L.) was shown by UPLC–ESI-MS and NMR spectroscopic methods previously, which had been highlighted as being unusual by Kirby et al. (2013).

Color being one of the essential aspects in a consumer's sensory perception of a food product leads to carrying out research on new and underexplored sources of anthocyanins. In that sense, *Perilla frutescens* L. belonging to the family Labiatae is widely consumed as either dietary vegetable or medicinal herb especially in China, Japan, Korea, Vietnam, and other regions in East Asia (Fan et al., 2022). Previously, cyanidin 3-O-feruloylglucoside-5-O-glucoside, shisonin, and cyanidin 3-O-caffeoylglucoside-5-O-glucoside-plus their malonyl derivatives were identified in perilla leaves (He, Yao & Chang, 2015). Zheng et al.

TABLE 11.4

LC-MS Analysis of Anthocyanins from Different Processed Food Matrices

Source of Anthocyanins	Food Matrix	Analyte	LC-MS Method	Reference
Black chokeberry (*Aronia melanocarpa* L.)	Chokeberry fruit juice	Cyanidin-3-O-galactoside Cyanidin-3-O-arabinoside Cyanidin-3-O-glucoside Cyanidin-3-O-xyloside	LC-PDA-ESI-MS/MS ESI mode: positive ionization Source voltage: 4 kV Capillary temperature: 275°C Mass range: *m/z* 100– 1000	Tasinov et al. (2022)
Eggplant peels (*Solanum melongena* L.)	Dough analoque; Microcrystalline cellulose Wheat starch Soy protein isolate	Delphinidin-3-O-rutinoside-5-O-glucoside Delphinidin-3-Orutinopyranosyl-Glucoside Delphinidin-3-O-rutinoside	UPLC-MS/MS Capillary voltage: 40 V Source voltage: 4 kV Tube lens voltage: 110 V Capillary temperature: 320°C Mass range: 100–1500 *m/z*	Wang et al. (2024a)
Functional hibiscus beverage and commercial hibiscus beverage (*Hibiscus sabdariffa* L.)	Citric acid Mint Stevia	Delphinidin-3-sambubioside Cyanidin-3-Glucoside Delphinidin	HPLC-DAD-ESI-MS ESI mode: Negative & positive ionization Capillary voltage: 3500 V Drying gas temperature: 350°C Mass range: 100–1000 *m/z*	Rodríguez-Romero et al. (2023)
Roselle wine (*H. sabdariffa* L.)	Sucrose Dry yeast Roselle calyces Potassium metabisulfite	Delphinidin 3-gentiobioside Cyanidin 3-(6″-(E)-caffeylsambubioside) Delphinidin 3,5-diglucoside Petunidin Pelargonidin 3-lathyroside Pelargonidin 3-(4‴-p-coumarylrutinoside)-5-glucoside Delphinidin 3-gentiobioside Cyanidin 3-(6″-(E)-caffeylsambubioside) Delphinidin 3,5-diglucoside Pelargonidin 3-lathyroside Petunidin Pelargonidin 3-(4‴-p-coumarylrutinoside)-5-glucoside Pinotin A (Malvidin 3-glucoside-4-vinylcatechol) Delphinidin 3-(6″-O-4-malyl-glucosyl)-5-glucoside	UHPLC-QE-MS ESI mode: Negative & positive ionization Spray voltage: 3500 V Capillary temperature: 320°C Probe heater temperature: 350°C Mass range: (*m/z*) 100–1000	Ai et al. (2021)

(Continued)

TABLE 11.4 *(Continued)*

LC-MS Analysis of Anthocyanins from Different Processed Food Matrices

Source of Anthocyanins	Food Matrix	Analyte	LC-MS Method	Reference
Mulberry (*Morus alba*)	Wine residues	Cyanidin-3-O-glucoside Cyanidin-3-O-rutinoside	UPLC-MS ESI mode: positive ionization Capillary voltage: 3.0 kV Source temperature: 100°C Temperature: 52°C Mass Range: 50–1500 *m/z*	Zhang et al. (2020)
Pomegranate (*Punica granatum L.*)	Commercially available pomegranate juice not from concentrate	Delphinidin-3,5-diglucoside Cyanidin-3,5-diglucoside Delphinidin-3-glucoside Cyanidin-3-glucoside Pelargonidin-3-glucoside	HPLC-PDA-ESI-MS/MS ESI mode: positive ionization Voltage: 3000 V Nebulizer (nitrogen): 50 psi Drying gas: 365°C	Kostka et al. (2020)
Saffron floral by-products beverage (*C. sativus* L.)	Food thickeners (xanthan gum, guar gum, β-glucan, pectin) Citric acid Freeze-dried saffron floral by-products	Delphinidin 3,5-di-O-glucoside Petunidin 3,5-di-O-glucoside Delphinidin 3-O-glucoside Petunidin 3-O-glucoside	HPLC-DAD-ESI/MS^n ESI mode: positive ionization Capillary voltage: 4 kV CID fragmentation: 0.3–2 V Capillary temperature: 350°C	Cerdá-Bernad et al. (2024)
Saffron petals (*C. sativus* L.)	Yeast Sodium bicarbonate Sucrose Potassium metabisulfite	Sixty-three anthocyanins were screened during fermentation; 13 cyanidins, 5 delphinidins, 10 malvidins, 11 pelargonidins, 10 peonidins, 5 petunians, 2 procyanidins, and 7 flavonoids	UPLC-MS/MS ESI mode: positive ionization Voltage: 5500 V ESI temperature: 550°C In Q-Trap 6500+, each ion pair is scanned against an optimized Declustering Potential and Collision Energy	Wang et al. (2024b)
Sumac fruits (*Rhus typhina* L.)	Cheddar cheese and ground beef	Petunidin Peonidin Unknown aglycone (443 Da)	LC-MS^n ESI mode: positive ionization Capillary voltage: 32.5 kV ESI needle voltage: 5 kV Temperature: 350°C	Wang et al. (2023e)

(2020) conducted screening of anthocyanin compounds in two color varieties of perillae folium using rapid resolution LC coupled with quadruple-time-of-flight MS (RRLC-Q/TOF-MS). The authors identified a new peonidin-derived anthocyanin, namely, peonidin-3-Ocoumaroylglucoside-5-O-malonylglucoside in the purple variety.

El Bishbishy, Gad and Aborehab (2020) recorded pelargonidin-3-O-hexoside, malvidin-3-O-glucoside and malvidin-3,5-O-diglucoside via UHPLC-ESI-MS analysis in leaves of economically important Egyptian *Pistacia* plants, namely: *P. chinensis*, *P. lentiscus*, and *P. khinjuk*. In addition, the red leaves of *Pistacia chinensis* L. was studied through a detailed UPLC-MS/MS profiling by Song et al. (2022) who reported the presence of at least 27 anthocyanins. Non-methylated methylated anthocyanins accounted for 99.80% of the total anthocyanin content while six of the 27 identified anthocyanins were acylated.

Laganà and co-workers (2020) reported on RP-HPLC-DAD-ESI-MS/MS identification of anthocyanin's profile of lemon bottlebrush flowers for the first time. Chromatographic separation was carried out using a Luna Omega PS C18 column (150 × 2.1 mm, 5 μm; Phenomenex). In addition to cyanidin-3-O-glucoside, the most abundant anthocyanins in plant kingdom, the following ones were detected in positive ionization mode: cyanidin 3,5-O-diglucoside by the molecular ion with m/z 611 (product ions at m/z 449, m/z 287); peonidin-3,5-O-diglucoside by the molecular ion with m/z 625 (product ions at m/z 463, m/z 301); and cyanidin-coumaroylglucoside-pyruvic acid by the molecular ion with m/z 661 (product ions at m/z 595, m/z 482).

Halophyte species, salt-resistant plants that mostly grow in deserts and adapted to harsh environmental conditions, have been also investigated by means of their anthocyanin metabolites as these plants contain high polyphenol levels and possess strong biological activity (Falleh et al., 2011). For instance, *Aptenia cordifolia* and *Carpobrotus edulis* were used in African traditional medicine. Thus, Hamed et al. (2021) studied the chemopreventive effect of six halophyte plants belonging to *Aizoaceae* and *Cactaceae* families through LC/MS—bioassay guided approach. LC-HR-ESI-MS metabolite profiling displayed the presence of mostly delphinidin derivatives in addition to petunidin, malvidin, pelargonidin, and cyanidin.

The sources of blue color in nature are limited to: phycocyanin from spirulina, iridoid from genipap fruit, and anthocyanin from butterfly pea flower according to previous research (Buchweitz, 2024). Belonging to family *Fabaceae*, butterfly pea flower (*Clitoria ternatea* L.) has been traditionally used as a food colorant in Southeast Asia. Different flower colors including light blue, dark blue, white, and mauve can be observed from several *C. ternatea* plants. The specific anthocyanin in butterfly pea flower is known as "ternatin" which is delphinidin triacylated by p-coumaroyl groups, resulting in the blue color (Jeyaraj, Lim & Choo, 2021). Handayani et al. (2024) investigated the use of an ethanol solvent combination acidified with HCl extract of flowers as a colorimetric indicator pigment for monitoring seafood products. LC-MS/Q-TOF analysis identified delphinidin, and its glycosides, contributing to a reddish-blue to purple color of the flower.

Having distinctive dark blue colored flowers in spring, grape hyacinth (*Muscari spp.*) is one of the economically important species that can benefit to food, pharmaceutical, and dyeing industries (Khaleghi & Khadivi, 2022). The flower buds and bulbs of these plants also take their place in Mediterranean cuisine (Lim, 2014). A total of 14 anthocyanins were identified in the tepals of eight grape hyacinth cultivars for the first time via a Waters UPLC-Q-TOF-MS (Milford, MA, USA) system (Lou et al., 2017). From MS results, the presence of pelargonidin-3-O-caffeoylsophoroside-5-O-arabinoside with a molecular ion at m/z 889 and three mass fragment ions, one at m/z 757, due to the elimination of one molecule of arabinoside, one at m/z 595, indicating the loss of caffeoyl-group, and one at *m/z* 271 for the loss of sophoroside. Moreover, the characteristic fragment ion at *m/z* 287, cyanidin-3-O-caffeoylrutinoside was identified due to mass ions *m/z* 757 and 595 (loss of caffeoyl-group), respectively. Similarly, cyanidin-3-O-(p-coumaroyl)-glucoside-5-O-malonylglucoside was detected exhibiting a molecular cation at *m/z* 887 and a fragment ion at *m/z* 757 for the loss of malonyl-group. Three pelargonidin (*m/z* 271) derivatives were assigned based on the losses of (i) sinapylglucoside (*m/z* 595) and sophoroside (*m/z* 433) confirming pelargonidin-3-O-sinapylglucoside-5-O-glucoside (*m/z* 801); (ii) sinapyl-group (*m/z* 565) and arabinoside (*m/z* 433) confirming pelargonidin-3-O-sinapylgluco-side-5-O-arabinoside (*m/z* 771); and (iii) ferulyl-group (*m/z* 565) and arabinoside (*m/z* 433) confirming pelargonidin-3-O-sinapylglucoside-5-O-arabinoside (*m/z* 741). Although delphinidin-3-O-glucoside was found, no acylated delphinidin derivative was detected in the studied cultivars.

11.2.7 Seaweeds

Natural pigments are also found in marine sources. Seaweeds, named also macroalgae, comprise a few thousand species that each group is characterized by specific combinations of photosynthetic pigments such as green (*Chlorophyta*), brown (*Phaeophyta*), and red (*Rhodophyta*) (Peng et al., 2015; Siddik et al., 2023). They have also emerged as a rich source of bioactive compounds with several health-promoting properties, thereby attracting interest of pharmaceutical, food, cosmetic, and textile industries, as reviewed recently by Sadeghi et al. (2024). LC-MS technique has been extensively used to identify the anthocyanin composition of algae species from different origins. The secondary metabolite analysis of Indian red seaweed *Jania rubens* was achieved using LC-TOF MS/MS (Micromass, Waters, Milford, MA, USA) by Dixit and Reddy (2017). Their study unveiled the presence of two anthocyanins: malonylshisonin and 4'''-demalonylsalvianin at *m/z* 825.19, that have the potential for the cosmetic industry.

Fresh seaweeds have been consumed as a valuable food source, especially in Southeast Asia and coastal communities. Thus, a comprehensive phenolic profiling of Australian red seaweeds was performed by Ebrahimi et al. (2024). Cyanidin 3-O-glucoside was found in two species (*Areschougia congesta* and *Hypnea* sp.). However, 6''-acetyl-glucoside ([M–H]– *m/z* 490.1110 and a product ion at *m/z* 287) and 6''-p-coumaroyl-glucoside ([M+H]+ *m/z* 580.1789 with fragmentation ions at *m/z* 271 and 433) derivatives of cyanidin was observed in *Hypnea* sp. and *Plocamium preissianum*, respectively. Pelargonidin 3-O-rutinoside ([M+H]+ *m/z* 580.1789 with fragmentation ions at *m/z* 271 and 433) was only detected in *Crouania attenuate*. Duan et al. (2023) characterized the phenolic profile of four freeze-dried brown seaweeds using LC–ESI-QTOF–MS/MS, and only detected delphinidin 3-O-glucosyl-glucoside by the [M + H]+ at *m/z* 628 with MS2 ions at *m/z* 465 and 303 in *Fucus* sp. while delphinidin 3-O-glucoside by the [M + H]+ at *m/z* 466 with fragment ion at *m/z* 303 in *Ecklonia* sp.

Khan, Mohamed and Islam (2024) investigated three freeze-dried South African seaweeds, namely *Codium fragile*, *Radicilingua thysanorhizans* (red), and *Mazzaella capensis* using a Shimadzu LC-MS system. Applying the following operational parameters – scan polarity is positive, event duration is 1.00 s, detector voltage is +1.00 kV, threshold is 0, start *m/z* is 50.00, finish *m/z* is 1700.00, and scan speed is 1.667 u/sec; delphinidin 3-O-sambubioside was isolated from cold-water extract whereas malvidin 3-O-glucoside was found in ethanolic extract. The seaweeds exhibited anti-diabetic and anti-obesogenic potential in addition to antioxidant, making Codium fragile the most promising among them. Examples of LC-MS analysis of seaweed anthocyanins is given in Table 11.5.

TABLE 11.5

LC-MS Analysis of Anthocyanins from Seaweeds (Macroalgae)

Source of Anthocyanins (Scientific Name)	Analyte	Analyte Quantity	LC-MS Method	Reference
Red seaweed (*Jania rubens*)	Malonylshisonin 4‴-demalonylsalvianin	N/A	LC-TOF-MS/MS ESI mode: positive ionization Capillary: 2.5 kV Cone voltage: 25 V Source temperature: 110°C Desolvation temperature: 200°C Mass range: 0–1000 *m/z*	Dixit and Reddy (2017)
Australian red seaweeds (*Crouania attenuata, Hypnea* sp., *Areschougia congesta, Hypnea* sp., *Plocamium preissianum*)	Pelargonidin 3-O-rutinoside Cyanidin 3-O-(6″-acetyl-glucoside) Cyanidin 3-O-glucoside Cyanidin 3-O-(6″-p-coumaroyl-glucoside)	N/A	LC-ESI-QTOF-MS/MS ESI mode: positive ionization Capillary voltage: 3.5 kV Nozzle voltage: 500 V Nebulizer temperature: 300°C Collision energies: 10, 15, 30 eV Mass range: 50–1300 amu	Ebrahimi et al. (2024)
Australian shoreline seaweeds (*Phyllospora comosa, Ecklonia radiata, Durvillaea* sp., *Sargassum* sp., *Cystophora* sp.)	Delphinidin 3-O-glucoside Pelargonidin Cyanidin 3,5-O-diglucoside Peonidin 3-O-diglucoside-5-O-glucoside	N/A	LC-ESI-QTOF-MS/MS Drying gas: 300°C Capillary voltage: 3.5 kV Nozzle voltage: 500 V Collision energy: 10, 15, 30 eV Mass range: 50–1300 amu	Subbiah et al. (2023a)
Brown seaweeds (*Fucus* sp., *Ecklonia* sp.)	Delphinidin 3-O-glucosyl-glucoside Delphinidin 3-O-glucoside	N/A	LC-ESI-QTOF-MS/MS ESI mode: positive ionization Capillary voltage: 3.5 kV Nozzle voltage: 500 V Nebulization gas: 300°C Sheath gas: 250°C Collision energy 10, 15, 30 eV Mass scan: *m/z* 50–1300	Duan et al. (2023)

Subbiah et al. (2023a) analyzed the phenolic composition of six Australian seaweeds obtained through two different extraction procedures. An Agilent HPLC connected via ESI to the Agilent 6530 Accurate-Mass Quadrupole Time-of-Flight (Q-TOF) LC/MS was used where nitrogen gas has been used as a nebulizer. The separation was carried out using a Synergi Hydro-Reverse Phase 80 Å, LC column 250 × 4.6 mm, 4 μm with temperature 25°C and sample temperature at 10°C. Notably, the authors observed anthocyanins including delphinidin 3-O-glucoside ([M − H]− at *m/z* 465.1016), pelargonidin aglycone ([M − H]− at *m/z* 271.0618), cyanidin 3,5-O-diglucoside [M + H]+ at *m/z* 611.1632), and peonidin 3-O-diglucoside-5-O-glucoside) [M − H]− at *m/z* 787.2331) in conventional extract, yet not the one obtained by ultrasound-assisted method.

Fresh seaweeds contain a large amount of water (75–85%) and easily deteriorate within a few days after harvest, therefore are usually dried prior to industrial processing, which remarkably affects their nutritional and biological properties (Uribe et al., 2018). When the impact of drying conditions on phytochemical constituents of seaweeds, the same research group compared the effect of different drying methods (freeze, oven, vacuum) on metabolites of five brown seaweeds to gain insight regarding the resultant shelf-life (Subbiah et al., 2023b). Besides two cyanidin derivatives, petunidin 3-O-(6″-acetyl-glucoside) ([M+H]+ m/z 522.1373) was identified by the fragmentation peak at m/z 317 due to the loss of glucose and acetyl moiety from the precursor ion, which seems to be detected for the first time in brown seaweeds.

11.3 Conclusion

In conclusion, LC-MS analysis of anthocyanins provides detailed insights into their molecular composition, concentration, and potential health benefits. This makes LC-MS an indispensable tool in the fields of food science, nutrition, and agricultural research. The analysis of anthocyanins in fruits requires methods that are both sensitive and precise, given the complex nature of these compounds and the presence of numerous isomers. UPLC-MS/MS coupled with QTOF and ESI is a popular method for both qualitative and quantitative analyses of anthocyanins in numerous sources in nature due to its high sensitivity, resolution, and ability to provide structural information. While each technique has its advantages, this configuration offers a good combination of speed, accuracy, and detailed insight; making it a suitable choice for anthocyanin metabolome studies.

References

Ai, J., Wu, Q., Battino, M., Bai, W., & Tian, L. (2021). Using untargeted metabolomics to profile the changes in roselle (*Hibiscus sabdariffa* L.) anthocyanins during wine fermentation. Food Chemistry, 364, 130425.

Aissat, A. K., Chaher-Bazizi, N., Richard, T., Kilani-Atmani, D., Pedrot, E., Renouf, E., … & Fonayet, J. V. (2022). Analysis of individual anthocyanins, flavanols, flavonols and other polyphenols in *Pistacia lentiscus* L. fruits during ripening. Journal of Food Composition and Analysis, 106, 104286.

Akramiani, S., Afsahi, M. M., & Akhavan, H. R. (2024). Optimization of encapsulation and spouted bed drying of saffron petal extract and evaluation of its thermal stability in natural pomegranate juice. Journal of Food Measurement and Characterization, 1–11.

Amanpour, A., Soltani, M., Lipan, L., Garcia-Garví, J. M., Hernández-García, F., Carbonell-Barrachina, Á. A., & Nadal, E. S.. (2024). Comparative study on nutraceutical and sensorial characteristics of saffron (*Crocus sativus* L.) cultivated in Iran, Spain, and Türkiye. Journal of the Science of Food and Agriculture, 104(12), 7580–7591.

Arce-Reynoso, A., Mateos, R., Mendivil, E. J., Zamora-Gasga, V. M., & Sáyago-Ayerdi, S. G. (2023). Bioavailability of bioactive compounds in *Hibiscus sabdariffa* beverage as a potential anti-inflammatory. Food Research International, 174, 113581.

Avila-Sosa, R., Nevárez-Moorillón, G. V., Ochoa-Velasco, C. E., Navarro-Cruz, A. R., Hernández-Carranza, P., & Cid-Pérez, T. S. (2022). Detection of saffron's main bioactive compounds and their relationship with commercial quality. Foods, 11(20), 3245.

Avula, B., Katragunta, K., Osman, A. G., Ali, Z., John Adams, S., Chittiboyina, A. G., & Khan, I. A. (2023). Advances in the chemistry, analysis and adulteration of anthocyanin rich-berries and fruits: 2000–2022. Molecules, 28(2), 560.

Ayar-Sümer, E. N., Verheust, Y., Özçelik, B., & Raes, K. (2024). Impact of lactic acid bacteria fermentation based on biotransformation of phenolic compounds and antioxidant capacity of mushrooms. Foods, 13(11), 1616.

Bakshi, R. A., Sodhi, N. S., Wani, I. A., Khan, Z. S., Dhillon, B., & Gani, A. (2022). Bioactive constituents of saffron plant: Extraction, encapsulation and their food and pharmaceutical applications. Applied Food Research, 2(1), 100076.

Bao, S., Li, X., Lan, T., Wang, J., Hu, Y., Sun, X., & Ma, T. (2023). Effects of different cooking treatments on the sensory qualities and pigmented phytochemicals of carrots. Food Chemistry, 405(PB), 135015. https://doi.org/10.1016/j.foodchem.2022.135015

Buchweitz, M. (2024). Natural solutions for blue colors in food. In Handbook on natural pigments in food and beverages (pp. 437–464). Woodhead Publishing.

Cerdá-Bernad, D., Clemente-Villalba, J., Valero-Cases, E., Pastor, J. J., & Frutos, M. J. (2022). Novel insight into the volatile profile and antioxidant properties of *Crocus sativus* L. flowers. Antioxidants, 11(9), 1650.

Cerdá-Bernad, D., D'costa, A. S., Moreno, D. A., Bordenave, N., & Frutos, M. J. (2024). Functional model beverages of saffron floral by-products: Polyphenolic composition, inhibition of digestive enzymes, and rheological characterization. Foods, 13(10), 1440.

Chen, J., Ye, H., Wang, J., & Zhang, L. (2023a). Relationship between anthocyanin composition and floral color of *Hibiscus syriacus*. Horticulturae, 9(1), 48.

Chen, J., Zhang, Y., Liu, F., Chen, J., Sun, Y., Ye, X., … & Cheng, H. (2023b). Ultrasound treatment improves fruit quality of postharvest blood oranges (*Citrus sinensis* L. Osbeck): Anthocyanin enrichment and its biosynthesis. Journal of Agricultural and Food Chemistry, 71(38), 14013–14026.

Chen, T., Lu, H., Shen, M., Yu, Q., Chen, Y., Wen, H., & Xie, J. (2022). Phytochemical composition, antioxidant activities and immunomodulatory effects of pigment extracts from Wugong Mountain purple red rice bran. Food Research International, 157(235), 111493. https://doi.org/10.1016/j.foodres.2022.111493

Constantin, O. E., & Istrati, D. I. (2022). Extraction, quantification and characterization techniques for anthocyanin compounds in various food matrices—A review. Horticulturae, 8(11), 1084.

da Silva Oliveira, J. P., de Oliveira, R. T., Guedes, A. L., da Costa Oliveira, M., & Macedo, A. F. (2022). Metabolomic studies of anthocyanins in fruits by means of a liquid chromatography coupled to mass spectrometry workflow. Current Plant Biology, 32, 100260.

De Angelis, E., Al-Ayoubi, O., Pilolli, R., Monaci, L., & Bejjani, A. (2024). Time-of-flight secondary ion mass spectrometry coupled with unsupervised methods for advanced saffron authenticity screening. Foods, 13(13), 2033.

Dixit, D., & Reddy, C. R. K. (2017). Non-targeted secondary metabolite profile study for deciphering the cosmeceutical potential of red marine macro alga *Jania rubens*—An LCMS-based approach. Cosmetics, 4(4), 45.

Dong, W., Yang, X., Zhang, N., Chen, P., Sun, J., Harnly, J. M., & Zhang, M. (2024). Study of UV–Vis molar absorptivity variation and quantitation of anthocyanins using molar relative response factor. Food Chemistry, 444(February), 138653. https://doi.org/10.1016/j.foodchem.2024.138653

Duan, X., Subbiah, V., Xie, C., Agar, O. T., Barrow, C. J., Dunshea, F. R., & Suleria, H. A. (2023). Evaluation of the antioxidant potential of brown seaweeds extracted by different solvents and characterization of their phenolic compounds by LC–ESI-QTOF–MS/MS. Journal of Food Science, 88(9), 3737–3757.

Dziki, D., Cacak-Pietrzak, G., Hassoon, W. H., Gawlik-Dziki, U., Sułek, A., Różyło, R., & Sugier, D. (2021). The fruits of sumac (*Rhus coriaria* L.) as a functional additive and salt replacement to wheat bread. LWT, 136, 110346.

Ebrahimi, F., Subbiah, V., Agar, O. T., Bringloe, T. T., Legione, A. R., & Suleria, H. A. (2024). LC-ESI-QTOF-MS/MS characterization of phenolic compounds from Victorian shorebound red seaweeds and their antioxidant capacity. Algal Research, 82, 103609.

Eghbali, S., Farhadi, F., & Askari, V. R. (2023). An overview of analytical methods employed for quality assessment of *Crocus sativus* (saffron). Food Chemistry: X, 20, 100992.

El Bishbishy, M. H., Gad, H. A., & Aborehab, N. M. (2020). Chemometric discrimination of three Pistacia species via their metabolic profiling and their possible in vitro effects on memory functions. Journal of Pharmaceutical and Biomedical Analysis, 177, 112840.

Ersan, S., Güçlü Üstündağ, O., Carle, R., & Schweiggert, R. M. (2016). Identification of phenolic compounds in red and green pistachio (*Pistacia vera* L.) hulls (exo-and mesocarp) by HPLC-DAD-ESI-(HR)-MS n. Journal of Agricultural and Food Chemistry, 64(26), 5334–5344.

Falleh, H., Oueslati, S., Guyot, S., Dali, A. B., Magné, C., Abdelly, C., & Ksouri, R. (2011). LC/ESI-MS/MS characterisation of procyanidins and propelargonidins responsible for the strong antioxidant activity of the edible halophyte *Mesembryanthemum edule* L. Food Chemistry, 127(4), 1732–1738.

Fan, Y., Cao, X., Zhang, M., Wei, S., Zhu, Y., Ouyang, H., & He, J. (2022). Quantitative comparison and chemical profile analysis of different medicinal parts of *Perilla frutescens* (L.) Britt. from different varieties and harvest periods. Journal of Agricultural and Food Chemistry, 70(28), 8838–8853.

Fereidoonfar, H., Salehi-Arjmand, H., Khadivi, A., Akramian, M., & Safdari, L. (2019). Chemical variation and antioxidant capacity of sumac (*Rhus coriaria* L.). Industrial Crops and Products, 139, 111518.

Gallego, A. M., Zambrano, R. A., Zuluaga, M., Rodríguez, A. V. C., Cortés, M. S. C., Vergel, A. P. R., & Valencia, J. W. A. (2022). Analysis of fruit ripening in Theobroma cacao pod husk based on untargeted metabolomics. Phytochemistry, 203, 113412.

Gamage, G. C. V., & Choo, W. S. Effect of hot water, ultrasound, microwave, and pectinase-assisted extraction of anthocyanins from black goji berry for food application. Heliyon 9 (3)(2023).

Gonzales, G. B., Raes, K., Vanhoutte, H., Coelus, S., Smagghe, G., & Van Camp, J. (2015). Liquid chromatography–mass spectrometry coupled with multivariate analysis for the characterization and discrimination of extractable and nonextractable polyphenols and glucosinolates from red cabbage and Brussels sprout waste streams. Journal of Chromatography A, 1402, 60–70.

Goupy, P., Vian, M. A., Chemat, F., & Caris-Veyrat, C. (2013). Identification and quantification of flavonols, anthocyanins and lutein diesters in tepals of *Crocus sativus* by ultra performance liquid chromatography coupled to diode array and ion trap mass spectrometry detections. Industrial Crops and Products, 44, 496–510.

Grassi, G., Di Gregorio, P., Rando, A., & Perna, A. M. (2024). Quality and sensorial evaluation of beef burgers added with Sicilian sumac (*Rhus coriaria* L). Heliyon, 10(5).

Guo, W., Mehrparvar, S., Hou, W., Pan, J., Aghbashlo, M., Tabatabaei, M., & Rajaei, A. (2024). Unveiling the impact of high-pressure processing on anthocyanin-protein/polysaccharide interactions: A comprehensive review. International Journal of Biological Macromolecules, 132042.

Gweyi-Onyango, J. P., Osei-Kwarteng, M., & Mahunu, G. K. (2021). Measurement and maintenance of *Hibiscus sabdariffa* quality. In Roselle (*Hibiscus sabdariffa*) (pp. 47–67). Academic Press.

Hamed, A. R., El-Hawary, S. S., Ibrahim, R. M., Abdelmohsen, U. R., & El-Halawany, A. M. (2021). Identification of chemopreventive components from halophytes belonging to Aizoaceae and Cactaceae through LC/MS—Bioassay guided approach. Journal of Chromatographic Science, 59(7), 618–626.

Hameed, M. K., Umar, W., Razzaq, A., Wei, S., Niu, Q., Huang, D., & Chang, L. (2023). Quantification of total polyphenols, antioxidants, anthocyanin and secondary metabolites by UPLC VION IMS QTOF MS/MS analysis in green and red lettuce cultivars. Scientia Horticulturae, 315(January), 111994. https://doi.org/10.1016/j.scienta.2023.111994

Handayani, L., Aprilia, S., Arahman, N., & Bilad, M. R. (2024). Identification of the anthocyanin profile from butterfly pea (*Clitoria ternatea* L.) flowers under varying extraction conditions: Evaluating its potential as a natural blue food colorant and its application as a colorimetric indicator. South African Journal of Chemical Engineering, 49, 151–161.

He, Y. K., Yao, Y. Y., & Chang, Y. N. (2015). Characterization of anthocyanins in Perilla frutescens var. acuta extract by advanced UPLC-ESI-IT-TOF-MSn method and their anticancer bioactivity. Molecules, 20(5), 9155–9169.

Hoang, T. N. N., Nguyen, N. P. M., Dong, T. A. D., & Le, T. H. A. (2023). Anthocyanin isolation from *Hibiscus sabdariffa* L. flowers by extraction, macroporous D101 resin purification, and biological evaluation. Journal of Agriculture and Food Research, 14, 100848.

Hopkins, A. L., Lamm, M. G., Funk, J. L., & Ritenbaugh, C. (2013). *Hibiscus sabdariffa* L. in the treatment of hypertension and hyperlipidemia: A comprehensive review of animal and human studies. Fitoterapia, 85, 84–94.

Hsieh-Lo, M., Castillo-Herrera, G., & Mojica, L. (2020). Black bean anthocyanin-rich extract from supercritical and pressurized extraction increased in vitro antidiabetic potential, while having similar storage stability. Foods, 9(5), 655.

Huang, Y., Zhou, S., Zhao, G., & Ye, F. (2021). Destabilisation and stabilisation of anthocyanins in purple-fleshed sweet potatoes: A review. Trends in Food Science & Technology, 116, 1141–1154.

Illiano, A., Pinto, G., Carrera, M. A., Palmese, A., Di Novella, R., Casoria, P., & Amoresano, A. (2022). LC–MS/MS-based quantification method of polyphenols for valorization of ancient apple cultivars from Cilento. ACS Food Science & Technology, 2(4), 647–654.

Ingemann Berentzen, E., Hauer Møller, A., Danielsen, M., Jensen, M., Joernsgaard, B., & Kastrup Dalsgaard, T. (2024). Stability of individual anthocyanins from black carrots stored in light and darkness – Impact of acylation. Food Research International, 186(January). https://doi.org/10.1016/j.foodres.2024.114382

Jeyaraj, E. J., Lim, Y. Y., & Choo, W. S. (2021). Extraction methods of butterfly pea (*Clitoria ternatea*) flower and biological activities of its phytochemicals. Journal of Food Science and Technology, 58(6), 2054–2067.

Jiang, T., Shuai, X., Li, J., Yang, N., Deng, L., Li, S., He, Y., Guo, H., Li, Y., & He, J. (2020). Protein-bound anthocyanin compounds of purple sweet potato ameliorate hyperglycemia by regulating hepatic glucose metabolism in high-fat Diet/Streptozotocin-induced diabetic mice. Journal of Agricultural and Food Chemistry, *68*(6), 1596–1608. https://doi.org/10.1021/acs.jafc.9b06916

Kaiser, M., Müller-Ehl, L., Passon, M., & Schieber, A. (2020). Development and validation of methods for the determination of anthocyanins in physiological fluids via UHPLC-MSn. Molecules, 25(3), 518.

Kandylis, P. (2022). Phytochemicals and antioxidant properties of edible flowers. Applied Sciences, 12(19), 9937.

Khaleghi, A., & Khadivi, A. (2022). Genetic diversity of wild grape hyacinth (*Muscari neglectum* Guss. ex Ten.) germplasm with ornamental potential in the Central region of Iran. South African Journal of Botany, 148, 307–314.

Khan, M. Z., Mohamed, A. I., & Islam, M. S. (2024). The antioxidant, anti-obesogenic, and anti-diabetic potential of selected South African marine algal species. Scientific African, 25, e02284.

Khoshkharam, M., Shahrajabian, M. H., Singh, R. B., Sun, W., Magadlela, A., Khatibi, M., & Cheng, Q. (2022). Sumac: A functional food and herbal remedy in traditional herbal medicine in the asia. In Functional foods and nutraceuticals in metabolic and non-communicable diseases (pp. 261–266). Academic Press.

Kim, M., Nam, D. G., Choe, J. S., Hwang, K. A., & Choi, A. J. (2021). Optimization of pectinase-assisted extraction condition of mulberry (*Morus alba* L.) fruit using response surface methodology and its effect on anthocyanin synthesis pathway-related metabolites. Journal of Food Science, 86(9), 3926–3938.

Kirby, C. W., Wu, T., Tsao, R., & McCallum, J. L. (2013). Isolation and structural characterization of unusual pyranoanthocyanins and related anthocyanins from Staghorn sumac (*Rhus typhina* L.) via UPLC–ESI-MS, 1H, 13C, and 2D NMR spectroscopy. Phytochemistry, 94, 284–293.

Kodikara, C., Netticadan, T., Bandara, N., Wijekoon, C., & Sura, S. (2024). A new UHPLC-HRMS metabolomics approach for the rapid and comprehensive analysis of phenolic compounds in blueberry, raspberry, blackberry, cranberry and cherry fruits. Food Chemistry, 445, 138778.

Kostka, T., Ostberg-Potthoff, J. J., Briviba, K., Matsugo, S., Winterhalter, P., & Esatbeyoglu, T. (2020). Pomegranate (*Punica granatum* L.) extract and its anthocyanin and copigment fractions—Free radical scavenging activity and influence on cellular oxidative stress. Foods, 9(11), 1617.

Kothari, D., Thakur, R., & Kumar, R. (2021). Saffron (*Crocus sativus* L.): Gold of the spices—A comprehensive review. Horticulture, Environment, and Biotechnology, 62(5), 661–677.

Kuasnei, M., Wojeicchowski, J. P., Santos, N. H., Pinto, V. Z., Ferreira, S. R. S., & Zielinski, A. A. F. (2022). Modifiers based on deep eutectic mixtures: A case study for the extraction of anthocyanins from black bean hulls using high pressure fluid technology. Journal of Supercritical Fluids, 191(June). https://doi.org/10.1016/j.supflu.2022.105761

Laganà, G., Barreca, D., Smeriglio, A., Germanò, M. P., D'Angelo, V., Calderaro, A., ... & Trombetta, D. (2020). Evaluation of anthocyanin profile, antioxidant, cytoprotective, and anti-angiogenic properties of *Callistemon citrinus* flowers. Plants, 9(8), 1045.

Leonarski, E., Kuasnei, M., Moraes, P. A. D., Cesca, K., de Oliveira, D., & Zielinski, A. A. F. (2023). Pressurized liquid extraction as an eco-friendly approach to recover anthocyanin from black rice bran. Innovative Food Science and Emerging Technologies, 86(May). https://doi.org/10.1016/j.ifset.2023.103372

Li, D., Zuo, W., Ma, S., Li, R., & Ye, Z. (2024). Metabolomics characterization of two saffron from Iran and China using GC–MS and LC–MS methods. Journal of Analytical Science and Technology, 15(1), 7.

Li, J., Li, Y., Li, M., Lin, L., Qi, J., Xu, J., ... & Tao, A. (2022). Novel insights into anthocyanin synthesis in the calyx of roselle using integrated transcriptomic and metabolomic analyses. International Journal of Molecular Sciences, 23(22), 13908.

Li, N., Simon, J. E., & Wu, Q. (2024). Determination of anthocyanins, organic acids, and phenolic acids in hibiscus market products using LC/UV/MS. Journal of Food Science, 89(2), 1098–1113.

Li, X., Wang, Y., Jin, L., Chen, Z., Jiang, J., & Jackson, A. (2021). Development of fruit color in *Rubus chingii* Hu (Chinese raspberry): A story about novel offshoots of anthocyanin and carotenoid biosynthesis. Plant Science, 311, 110996.

Li, Z., Liu, D., Wang, D., Sun, M., Zhang, G., Wu, Y., ... & Cheng, B. (2024). Study on the causes of changes in colour during *Hibiscus syriacus* flowering based on transcriptome and metabolome analyses. BMC Plant Biology, 24(1), 431.

Lim, T. K. (2014). Muscari neglectum. In: Edible medicinal and non-medicinal plants. Springer, Dordrecht. https://doi.org/10.1007/978-94-007-7395-0_7

Liu, T., Liu, T., Zhang, X., Song, J., Qiu, Y., Yang, W., Jia, H., Wang, H., & Li, X. (2023). Combined widely targeted metabolomics and transcriptomics analysis reveals differentially accumulated metabolites and the underlying molecular bases in fleshy taproots of distinct radish genotypes. Plant Physiology and Biochemistry, 195(January), 351–361. https://doi.org/10.1016/j.plaphy.2023.01.028

Lou, Q., Wang, L., Liu, H., & Liu, Y. (2017). Anthocyanin profiles in flowers of grape hyacinth. Molecules, 22(5), 688.

Lucci, P., Saurina, J., & Núñez, O. (2017). Trends in LC-MS and LC-HRMS analysis and characterization of polyphenols in food. TrAC Trends in Analytical Chemistry, 88, 1–24.

Ma, C., Feng, Y., Zhou, S., Zhang, J., Guo, B., Xiong, Y., Wu, S., Li, Y., Li, Y., & Li, C. (2023). Metabolomics and transcriptomics provide insights into the molecular mechanisms of anthocyanin accumulation in the seed coat of differently colored mung bean (*Vigna radiata* L.). Plant Physiology and Biochemistry, 200(May), 107739. https://doi.org/10.1016/j.plaphy.2023.107739

Ma, C., Meng, L., Wang, R., Fan, Y., & Wang, R. (2022). Dynamics of anthocyanin profiles of the fruits of four blueberry (*Vaccinium* sp.) cultivars during different growth stages. International Journal of Food Properties, 25(1), 1302–1316.

Majdoub, Y. O. E., Ginestra, G., Mandalari, G., Dugo, P., Mondello, L., & Cacciola, F. (2021). The digestibility of *Hibiscus sabdariffa* L. polyphenols using an in vitro human digestion model and evaluation of their antimicrobial activity. Nutrients, 13(7), 2360.

Masala, V., Jokić, S., Aladić, K., Molnar, M., & Tuberoso, C. I. G. (2024). Exploring phenolic compounds extraction from saffron (*C. sativus*) floral by-products using ultrasound-assisted extraction, deep eutectic solvent extraction, and subcritical water extraction. Molecules, 29(11), 2600.

Mazzara, E., Caprodossi, A., Mustafa, A. M., Maggi, F., & Caprioli, G. (2023). Phytochemical investigation of sumac (*Rhus coriaria* L.) fruits from different sicilian accessions. Foods, 12(23), 4359.

Mejía, J. J., Sierra, L. J., Ceballos, J. G., Martínez, J. R., & Stashenko, E. E. (2023). Color, antioxidant capacity and flavonoid composition in *Hibiscus rosa*-sinensis cultivars. Molecules, 28(4), 1779.

Mirzazadeh, N., Bagheri, H., Mirzazadeh, M., Soleimanimehr, S., Rasi, F., & Akhavan-Mahdavi, S. (2024). Comparison of different green extraction methods used for the extraction of anthocyanin from red onion skin. Food Science and Nutrition, February, 1–11. https://doi.org/10.1002/fsn3.4354

Nina, N., Theoduloz, C., Paillán, H., Jiménez-Aspee, F., Márquez, K., Schuster, K., Becker, L., Oellig, C., Frank, J., & Schmeda-Hirschmann, G. (2023). Chemical profile and bioactivity of Chilean bean landraces (*Phaseolus vulgaris* L.). Journal of Functional Foods, 104(January). https://doi.org/10.1016/j.jff.2023.105513

Nonglait, D. L., & Gokhale, J. S. (2024). Review insights on the demand for natural pigments and their recovery by emerging microwave-assisted extraction (MAE). Food and Bioprocess Technology, 17(7), 1681–1705.

Ojeda-Amador, R. M., Salvador, M. D., Fregapane, G., & Gómez-Alonso, S. (2019). Comprehensive study of the phenolic compound profile and antioxidant activity of eight pistachio cultivars and their residual cakes and virgin oils. Journal of Agricultural and Food Chemistry, 67(13), 3583–3594.

Peng, Y., Hu, J., Yang, B., Lin, X. P., Zhou, X. F., Yang, X. W., & Liu, Y. (2015). Chemical composition of seaweeds. In Seaweed sustainability (pp. 79–124). Academic Press.

Peng, Y., Zhang, H., Liu, R., Mine, Y., McCallum, J., Kirby, C., & Tsao, R. (2016). Antioxidant and anti-inflammatory activities of pyranoanthocyanins and other polyphenols from staghorn sumac (*Rhus hirta* L.) in Caco-2 cell models. Journal of Functional Foods, 20, 139–147.

Pereira, A. R., Fernandes, V. C., Delerue-Matos, C., de Freitas, V., Mateus, N., & Oliveira, J. (2024). Exploring acylated anthocyanin-based extracts as a natural alternative to synthetic food dyes: Stability and application insights. Food Chemistry, 461(August), 140945. https://doi.org/10.1016/j.foodchem.2024.140945

Qu, S., Li, M., Wang, G., Yu, W., & Zhu, S. (2021). Transcriptomic, proteomic and LC-MS analyses reveal anthocyanin biosynthesis during litchi pericarp browning. Scientia Horticulturae, 289, 110443.

Rahmati, S., Bazargani-Gilani, B., & Aghajani, N. (2022). Effect of extraction methods on the efficiency of sumac (*Rhus coriaria* L.) fruit extract in soybean oil quality during accelerated conditions. Food Science & Nutrition, 10(10), 3302–3313.

Rodríguez-Romero, J. D. J., Arce-Reynoso, A., Parra-Torres, C. G., Zamora-Gasga, V. M., Mendivil, E. J., & Sáyago-Ayerdi, S. G. (2023). In vitro gastrointestinal digestion affects the bioaccessibility of bioactive compounds in *Hibiscus sabdariffa* beverages. Molecules, 28(4), 1824.

Ruggieri, F., Maggi, M. A., Rossi, M., & Consonni, R. (2023). Comprehensive extraction and chemical characterization of bioactive compounds in tepals of *Crocus sativus* L. Molecules, 28(16), 5976.

Sadeghi, A., Rajabiyan, A., Nabizade, N., Meygolinezhad, N., & Ahmady, A. Z. (2024). Seaweed-derived phenolic compounds as diverse bioactive molecules: A review on identification, application, extraction and purification strategies. International Journal of Biological Macromolecules, 131147.

Salehi, A., Shariatifar, N., Pirhadi, M., & Zeinali, T. (2022). An overview on different detection methods of saffron (*Crocus sativus* L.) Adulterants. Journal of Food Measurement and Characterization, 16(6), 4996–5006.

Salem, M. A., Michel, H. E., Ezzat, M. I., Okba, M. M., El-Desoky, A. M., Mohamed, S. O., & Ezzat, S. M. (2020). Optimization of an extraction solvent for angiotensin-converting enzyme inhibitors from *Hibiscus sabdariffa* L. based on its UPLC-MS/MS metabolic profiling. Molecules, 25(10), 2307.

Senizza, B., Rocchetti, G., Ghisoni, S., Busconi, M., Pascual, M. D. L. M., Fernandez, J. A., ... & Trevisan, M. (2019). Identification of phenolic markers for saffron authenticity and origin: An untargeted metabolomics approach. Food Research International, 126, 108584.

Shahtouri, M. G., Fooladi, E., Feizy, J., Jahani, M., & Marrazza, G. (2024). Valorization of saffron production waste: Isolation of anthocyanins from saffron petal extract by zirconium-based adsorbents. LWT, 201, 116233.

Siddik, M. A., Francis, P., Rohani, M. F., Azam, M. S., Mock, T. S., & Francis, D. S. (2023). Seaweed and seaweed-based functional metabolites as potential modulators of growth, immune and antioxidant responses, and gut microbiota in fish. Antioxidants, 12(12), 2066.

Singh, M. C., Kelso, C., Price, W. E., & Probst, Y. (2020). Validated liquid chromatography separation methods for identification and quantification of anthocyanins in fruit and vegetables: A systematic review. Food Research International, 138, 109754.

Singh, M. C., Probst, Y., Price, W. E., & Kelso, C. (2022). Relative comparisons of extraction methods and solvent composition for Australian blueberry anthocyanins. Journal of Food Composition and Analysis, 105, 104232.

Siniawska, M., & Wojdyło, A. (2023). Polyphenol profiling by LC QTOF/ESI-MS and biological activity of purple passion fruit Epicarp extract. Molecules, 28(18), 6711.

Song, W., Yuan, Q., Xie, Y., Wang, Y., Deng, D., & Guo, H. (2024). Formulation and characterization of nanocapsules loaded with roselle anthocyanins extract and enhancement of anthocyanins bioaccessibility. Food Chemistry, 459, 140446.

Song, X., Zhang, J., Chang, X., Xian, L., & Liu, Y. (2022). Characterization of the anthocyanin biosynthesis pathway at the metabolic level in the red leaves of *Pistacia chinensis*. Scientia Horticulturae, 302, 111158.

Subbiah, V., Ebrahimi, F., Agar, O. T., Dunshea, F. R., Barrow, C. J., & Suleria, H. A. (2023a). Comparative study on the effect of phenolics and their antioxidant potential of freeze-dried Australian beach-cast seaweed species upon different extraction methodologies. Pharmaceuticals, 16(5), 773.

Subbiah, V., Duan, X., Agar, O. T., Dunshea, F. R., Barrow, C. J., & Suleria, H. A. (2023b). Comparative study on the effect of different drying techniques on phenolic compounds in Australian beach-cast brown seaweeds. Algal Research, 72, 103140.

Tan, J., Han, Y., Han, B., Qi, X., Cai, X., Ge, S., & Xue, H. (2022). Extraction and purification of anthocyanins: A review. Journal of Agriculture and Food Research, 8, 100306.

Tasinov, O., Dincheva, I., Badjakov, I., Grupcheva, C., & Galunska, B. (2022). Comparative phytochemical analysis of *Aronia melanocarpa* L. fruit juices on Bulgarian market. Plants, 11(13), 1655.

Teixeira, M., Tao, W., Fernandes, A., Faria, A., Ferreira, I. M., He, J., … & Oliveira, H.. (2023). Anthocyanin-rich edible flowers, current understanding of a potential new trend in dietary patterns. Trends in Food Science & Technology, 138, 708–725.

Tena, N., Martín, J., & Asuero, A. G. (2020). State of the art of anthocyanins: Antioxidant activity, sources, bioavailability, and therapeutic effect in human health. Antioxidants, 9(5), 451.

Uribe, E., Vega-Gálvez, A., Vargas, N., Pasten, A., Rodríguez, K., & Ah-Hen, K. S. (2018). Phytochemical components and amino acid profile of brown seaweed *Durvillaea antarctica* as affected by air drying temperature. Journal of Food Science and Technology, 55, 4792–4801.

Villalobos-Vega, M. J., Rodríguez-Rodríguez, G., Armijo-Montes, O., Jiménez-Bonilla, P., & Álvarez-Valverde, V. (2023). Optimization of the extraction of antioxidant compounds from roselle hibiscus calyxes (*Hibiscus sabdariffa*), as a source of nutraceutical beverages. Molecules, 28(6), 2628.

Viola, E., Mannino, G., Serio, G., La Rosa, L., Garofalo, G., Schicchi, R., … & Gaglio, R. (2024). Phytochemical profiling and investigation of antioxidant, anti-proliferative, and antibacterial properties in spontaneously grown Sicilian sumac (*Rhus coriaria* L.) fruits. Food Bioscience, 61, 104704.

Wan, X., Wu, J., Wang, X., Cui, L., & Xiao, Q. (2024). Accumulation patterns of flavonoids and phenolic acids in different colored sweet potato flesh revealed based on untargeted metabolomics. Food Chemistry: X, 23(March), 101551. https://doi.org/10.1016/j.fochx.2024.101551

Wang, D., Chen, L., Yang, Y., Abbas, F., Qin, Y., Lu, H., … & Hu, G. (2023b). Integrated metabolome and transcriptome analysis reveals the cause of anthocyanin biosynthesis deficiency in litchi aril. Physiologia Plantarum, 175(1), e13860.

Wang, J., Wei, Q., Wang, W., Hu, H., Yan, Y., Wang, Y., Li, Y., Jiang, Y., Wu, G., Hu, T., & Bao, C. (2023c). Understanding the nutraceutical diversity through a comparative analysis of the taproot metabolomes of different edible radish types via UHPLC–Q–TOF–MS. Food Chemistry, 403(June 2022). https://doi.org/10.1016/j.foodchem.2022.134469

Wang, L., Qin, Y., Wang, Y., & Zhou, Y. (2024b). Changes of anthocyanin and amino acid metabolites in saffron petals (*Crocus sativus* L.) during fermentation based on untargeted metabolomics. LWT, 192, 115724.

Wang, Q., Jing, L., Xu, Y., Zheng, W., & Zhang, W. (2023a). Transcriptomic analysis of anthocyanin and carotenoid biosynthesis in red and yellow fruits of sweet cherry (*Prunus avium* L.) during ripening. Horticulturae, 9(4), 516.

Wang, R., Chen, N., Li, J., Qian, D., Huang, X., & Yang, B. (2023d). Ultra-performance liquid chromatography-quadrupole time-of-flight mass spectrometry-based metabolomics to clarify the mechanism of color change of saffron floral bio-residues. Journal of Food Science, 88(2), 732–743.

Wang, S., Nie, S., Gan, R. Y., & Zhu, F. (2023e). Properties of cheese and ground beef in the presence of staghorn sumac. eFood, 4(2), e74.

Wang, Z., Zhang, Y., Tu, Z., Yu, C., Liu, R., Deng, Z., & Luo, T. (2024a). The degradation and antioxidant capacity of anthocyanins from eggplant peels in the context of complex food system under thermal processing. Food Bioscience, 59(March), 103914. https://doi.org/10.1016/j.fbio.2024.103914

Wu, T., McCallum, J. L., Wang, S., Liu, R., Zhu, H., & Tsao, R. (2013). Evaluation of antioxidant activities and chemical characterisation of staghorn sumac fruit (*Rhus hirta* L.). Food Chemistry, 138(2–3), 1333–1340.

Wu, Y., Gong, Y., Sun, J., Zhang, Y., Luo, Z., Nishanbaev, S. Z., … & Benito, M. J. (2023). Bioactive components and biological activities of *Crocus sativus* L. byproducts: A comprehensive review. Journal of Agricultural and Food Chemistry, 71(49), 19189–19206.

Xu, J., Li, H., Yang, H., Wang, T., Chang, Y., Nie, C., & Fu, Y. (2024a). Lingonberry (*Vaccinium vitis-idaea* L.) fruits: Potential characterization of flavor and functional profiles during ripening based on UHPLC-QqQ-MS/MS. Journal of Food Composition and Analysis, 130, 106143.

Xu, X., Ding, L., Fu, Y., Wang, Y., & Cai, S. (2024b). Polyphenol-rich extract of Chinese sumac (Rhus chinensis mill.) Fruits and its main component decrease in vitro starch digestibility of bread: Exploring the potential mechanisms from the perspective of molecular interactions. LWT, 195, 115856.

Zannou, O., Oussou, K. F., Chabi, I. B., Alamou, F., Awad, N. M., Miassi, Y. E., … & Ibrahim, S. A. (2025). Phytochemical and nutritional properties of sumac (*Rhus coriaria*): A potential ingredient for developing functional foods. Journal of Future Foods, 5(1), 21–35.

Zhang, A., Yang, H., Ji, S., Tian, C., Chen, N., Gong, H., & Li, J. (2022a). Metabolome and transcriptome analyses of anthocyanin accumulation mechanisms reveal metabolite variations and key candidate genes involved in the pigmentation of *Prunus tomentosa* Thunb. cherry fruit. Frontiers in Plant Science, 13, 938908.

Zhang, L., Fan, G., Khan, M. A., Yan, Z., & Beta, T. (2020). Ultrasonic-assisted enzymatic extraction and identification of anthocyanin components from mulberry wine residues. Food Chemistry, 323, 126714.

Zhang, P., Li, Y., Chong, S., Yan, S., Yu, R., Chen, R., … & Zhang, X. (2022b). Identification and quantitative analysis of anthocyanins composition and their stability from different strains of *Hibiscus syriacus* L. flowers. Industrial Crops and Products, 177, 114457.

Zhang, Y., Zhang, T., Zhao, Q., Xie, X., Li, Y., Chen, Q., Cheng, F., Tian, J., Gu, H., & Huang, J. (2021). Comparative transcriptome analysis of the accumulation of anthocyanins revealed the underlying metabolic and molecular mechanisms of purple pod coloration in okra (*Abelmoschus esculentus* L.). Foods, 10(9), 1–19. https://doi.org/10.3390/foods10092180

Zheng, Y. F., Li, D. Y., Sun, J., Cheng, J. M., Chai, C., Zhang, L., & Peng, G. P. (2020). Comprehensive comparison of two color varieties of perillae folium using rapid resolution liquid chromatography coupled with quadruple-time-of-flight mass spectrometry (RRLC-Q/TOF-MS)-based metabolic profile and in vivo/in vitro anti-oxidative activity. Journal of Agricultural and Food Chemistry, 68(49), 14684–14697.

12

UHPLC/MS Analysis of Anthocyanins

Leo M.L. Nollet

12.1 UHPLC and UPLC of Anthocyanins

12.1.1 UHPLC and UPLC

Ultra-performance liquid chromatography (UPLC) is a technique similar to that of high-performance liquid chromatography (HPLC) when it comes to separating different components of a sample, as well as identifying, quantifying, and separating the components of a mixture. The major difference in UPLC is the particle size in the column which is less than 2 µm providing better separation than in HPLC where the size of these is limited to 5 µm. These smaller particles require higher pump pressures (100 MPa vs. 40 MPa) which makes this method very efficient with rapid analysis and higher resolution.

UHPLC (ultra-high-performance liquid chromatography) and UPLC are both liquid chromatography (LC) techniques used to separate the different components found in mixtures. UHPLC and UPLC are the same techniques. In 2004, Waters launched the technique known as UPLC based upon sub 2-micron porous particles and trademarked the acronym UPLC. Other vendors entered the market with similar instruments named UHPLC as a way to refer to instruments similar to UPLC. The technology behind both UHPLC and UPLC instruments is sub 2-micron particles. Most HPLC methods are developed on particles between 2.5 and 5 microns in size. The smaller sub 2-micron particles require a higher pressure to work. UHPLC or UPLC systems need to be capable of operating above 6000 PSI.

Since sub 2-micron particles are so small the diffusion path between the sample analytes and the stationary phase is shorter and the efficiency is higher. HPLC, UHPLC, and UPLC rely on pumps, which pass a pressurized liquid containing a sample mixture through absorbent materials.

Here we consider both terms, UHPLC and UPLC, to be synonymous.

12.1.2 Anthocyanins

Analytical techniques for profiling anthocyanins have been widely reported in the last decade for in vitro and in vivo studies. A number of important technological advances in HPLC and the introduction of the UHPLC have achieved significant gains in both the speed and the separation of anthocyanins in the past decade. Advancements in mass spectrometry (MS), such as high-resolution and sequential collision MS (HRMS and MSn) allow fast structural elucidation of food anthocyanins that play a very important role in food anthocyanin research. A significant increase in the number of

DOI: 10.1201/9781003453260-14

studies has been reported for tandem MS-, HRMS-, and MSn-based approaches for food analysis. The article of Sun et al. (1) provides an updated review of the application of LC (HPLC and UHPLC) in combination with MS for the study of food anthocyanins in recent years.

Flavonoids can be separated on reverse phase (RP) columns due to their differential polarities resulting from their different functionalization of their central ring structures via hydroxylation, methylation, and glycosylation (2,3). The general elution order observed for reverse columns is glycosylated flavonoids first/earlier, followed by flavonoid aglycones and then methylated flavonoid aglycones. Flavonoid aglycones absorb UV light and thus can be detected using an UV absorbance or photodiode array (PDA) detector. Their UV spectra are typically characterized by two absorption maxima (i.e., Band II and Band I) (4). The absorption maximum at around 240–290 nm (Band II) is due to the A ring and its conjugation and functionalization. The other absorbance maximum (Band I) originates from the conjugation of B and C rings via the C2 and C3 double bond and appears around 300–550 nm for most flavonoids. For anthocyanins, the Band I absorption maximum is at approximately 460–560 nm, and for flavones and flavonols, it is around 310–370 nm (5). Glycosylation decreases the Band I absorption maximum.

The best separation of flavonoids can be obtained with a Waters Corp Acquity UPLC system equipped with a C18 reversed-phase column (Acquity UPLC BEH C18 Column, 130 Å, 1.7 μm particle, 2.1 mm diameter × 150 mm length). The gradient separations use a column temperature of 60°C, mobile phase A: 0.1% formic acid, B: acetonitrile, and flow rate: 0.56 mL/min. The linear gradient includes mobile B increased from 5 to 70% over 30 min, then to 95% over 3 min, hold at 95% for 3 min, then returned to 5% for re-equilibrium for 5 min. Injection volume is 2 μL.

Flavonoids are detected using PDA detectors with a wavelength range from 190 to 400 nm. Scanning up into the visible light range between 400 and 600 nm will also enable observation of colored flavonoids and aurones.

Flavonoids can be detected with MS and MS/MS (3). Most are preferentially ionized in negative electrospray ionization mode (ESI) due to their acidic phenolic hydroxyl groups. Permethylated flavonoids and anthocyanins can be analyzed in positive ESI mode. Fragmentation of flavonoids in MS/MS is structure-specific and can be used to identify and/or elucidate flavonoid structures. Aglycone fragmentation may proceed through C-ring opening or rearrangement, depending on the hydroxylation patterns (3). A protocol is optimized for a Waters Acquity UPLC system coupled to a Bruker Impact quadrupole time of flight mass spectrometer or a Bruker Impact II quadrupole time of flight mass spectrometer.

12.2 UPLC of Anthocyanins in Plasma and Urine

Numerous studies on anthocyanins confirm health-promoting effects in humans. Daily anthocyanin intake can be estimated via food databases, but the amount absorbed by the organism still remains uncertain because anthocyanin bioavailability is yet to be elucidated in its entirety. For this purpose, suitable and validated methods of sample preparation and analysis are needed. A sample preparation method for anthocyanin metabolite analysis in plasma was successfully established and validated by Kaiser et al. (6). The validation yielded acceptable results for the anthocyanins in terms of recovery (54–108%) and precision (coefficient of variation (CV) < 15%).

The UHPLC-MS method used in the consecutive reaction monitoring (CRM) mode was sufficiently sensitive, resulting in limits of detection <8.1 ng/mL.

The UHPLC consisted of an Acquity I-Class UPLC Binary Solvent Manager, an Acquity UPLC Sample Manager–FL, and an Acquity UPLC PDA eλ Detector from Waters each. The autosampler was cooled at 10°C, and the column oven was set at 40°C. For chromatographic separation, an HSS T3 C18 column (2.1 mm × 150 mm; 1.8 μm) equipped with a VanGuard HSS T3 C18 precolumn (5 mm × 2.1 mm, 1.7 μm) from Waters was used. Eluents were 3% formic acid in water (A) and acetonitrile (B); the flow rate was set at 0.4 mL/min, and the injection volume was 5 μL for all methods. The linear gradient to analyze the anthocyanins was as follows: 0 min 4% B, 7 min 8% B, 13 min 10% B, 19 min 17% B, 23 min 30% B, 23.3 min 100% B, 25.3 min 100% B, and 25.3–25.8 min 4% B. The linear gradient for the analysis of anthocyanidins was 0 min 15% B, 8.5 min 25% B, 9.5 min 100% B, 11 min 100% B, and 11–13 min 15% B. The MS system was an LTQ XL Iontrap MSn from Thermo Scientific equipped with an electrospray ionization interface operating in positive mode. The spray voltage was 4 kV, and the capillary temperature was set at 325°C; for the anthocyanidins, the capillary voltage was 9 V with the tube lens voltage at 40 V, and, for the anthocyanins, the capillary voltage was 15 V with the tube lens voltage being 60 V. CRM with collision-induced dissociation (CID) was used to record the abundance of the analytes. The instrument was tuned automatically with delphinidin-3-O-glucoside for the anthocyanins and pelargonidin for the aglycones. Xcalibur software version 2.2.0.48 from Thermo Fisher Scientific was used.

A single-dose pharmacokinetic trial was conducted in six adults to evaluate the bioavailability of anthocyanins and colonic polyphenol metabolites after consumption of 500 mg aronia berry extract (7). UHPLC-MS methods were developed to quantitate aronia berry polyphenols and their metabolites in plasma and urine. While anthocyanins were bioavailable, microbial phenolic catabolites increased 10-fold more than anthocyanins in plasma and urine. Among the anthocyanins, cyanidin-3-O-galactoside was rapidly metabolized to peonidin-3-O-galactoside. Aronia polyphenols were absorbed and extensively metabolized with t_{max} of anthocyanins and other polyphenol catabolites from 1.0 to 6.33 h in plasma and urine.

Dried residue from solid-phase extraction clean-up of urine or plasma was reconstituted with 100 μL methanol. UHPLC-MS analysis was conducted using a Shimadzu Nexera UHPLC equipped with DGU-20A5 Prominence degasser, SIL-30AC Nexera autosampler, CTD-30A Nexera column oven, two LC-30AD pumps, LCMS-2020 mass spectrometer, and a Kinetex Pentafluorophenyl (PFP) column (1.7 μm, 100 Å, 2.1 mm–50 mm). Chromatographic methods were developed to optimize resolution and detection of aronia polyphenols and metabolites in the reconstituted samples. For all of the chromatographic methods, the mass spectrometer was set to a nebulizing gas flow of 1.5 l/min, a drying gas flow of 15.0 l/min, the desolvation line at 250°C, the heat block at 500°C, and the interface temperature at 350°C. Other detector parameters were set to optimize the response of individual compounds of interest.

Cyanidin-3-O-glucoside, cyanidin-3-O-galactoside, cyanidin-3-O-arabinoside, and peonidin-3-O-galactoside were resolved by a gradient of water with trifluoroacetic acid (99.9:0.1, v/v) (A) and methanol (B) that started at 80% A and 20% B, then increased to 40% B over the next 9 min, then increased to 100% B at 12 min, and returned to 20% B at 18 min, and was held until the stop time of 20 min. The flow was maintained at 0.2 mL/min and the sample injection volume was set to 5 μL.

Red raspberries, containing ellagitannins and cyanidin-based anthocyanins, were fed to volunteers and metabolites appearing in plasma and urine were analyzed by UHPLC-MS (8).

The results showed an extensive degradation of raspberry anthocyanins and ellagitannins as they pass through the body with a great variety of metabolites, including methyl, glycine, glucuronide, and sulfate derivatives of phenolic acids, phenylpropanoids, ellagic acid, and urolithins.

Qualitative and quantitative analysis of anthocyanins in urine and plasma samples was performed by UHPLC-MS. Anthocyanins were analyzed using an Accela UHPLC 1250 with a LTQ XL linear ion trap-mass spectrometer fitted with a heated ESI probe (Thermo Scientific). Separation was performed with a Kinetex PFP (50 mm × 2.1 mm), 2.6 μm particle size column (Phenomenex). The volume injected was 5 mL and column oven was set to 30°C. Elution was carried out at a flow rate of 0.2 mL/min. The gradient started with 10% acetonitrile in 0.1% aqueous formic acid, and after 1 min a 9-min linear gradient of 10–60% acetonitrile was applied. From 10 to 11 min the acidified acetonitrile increased to 80%, and after 2 min was reduced to 10% acetonitrile to re-equilibrate the column. The MS was operated in positive ionization mode with a capillary temperature of 275°C, with the source temperature of 300°C. Analyses were carried out using full scan, data-dependent MS3 scanning from m/z 100 to 600, with CID of 35 (arbitrary units). Helium gas was used for CID. Data processing was performed using Xcalibur software (Thermo Scientific). Anthocyanins were quantified using a cyanidin-3-O-glucoside calibration curve.

12.3 UPLC of Anthocyanins in Juices

Individual anthocyanin analysis was carried out by ultra HPLC with PDA detection and electrospray ionization/MS (UHPLC-PDA-ESI/MSn) analysis (9). Five milliliter samples of centrifuged juice were loaded onto C18 Sep-Pak cartridges (Waters) that were previously conditioned with 5 mL of methanol and 5 mL of pure water. The anthocyanins and other polyphenols were adsorbed by these columns; sugars, acids, and other soluble compounds were removed by washing the cartridges with water. Anthocyanins were eluted with methanol containing 1% formic acid. The acidified methanol solutions were evaporated to dryness, and the dried fractions were redissolved in 7% aqueous formic acid. Similarly, young shoots/flowers/peel/stigma methanolic extracts were dried under vacuum distillation, resuspended in an aqueous acidic solution (1% formic acid) and then loaded onto C18 Sep-Pak cartridges that had been pre-washed with methanol and pre-equilibrated with water. The cartridges were eluted with acidified methanol (1% formic acid). The eluted methanolic solutions were evaporated to dryness, and the dried fractions were redissolved in 7% aqueous formic acid. Then, the samples were filtered through a 0.45 μm membrane filter and injected into the UHPLC-MSn chromatographic system to identify the individual anthocyanins.

Anthocyanin separation was conducted on a Chromolith Performance RP-18 end-capped column (100 mm × 3.0 mm inner diameter, monolithic particle size; Merck) using an Ultra-Fast HPLC system coupled to a PDA detector and a Finnigan LXQ ion trap equipped with an ESI interface in series configuration (Thermo Electron). A binary gradient composed of water containing 7% formic acid (solvent A) and methanol (solvent B) was used. The gradient was run as: (1) B from 5 to 40% for 20 min, (2) isocratic for 7 min, and (3) re-equilibration to initial conditions. The flow rate was 300 µL/min, the column temperature was 30°C and the injection volume was 20 µL. The range of wavelengths was set between 210 and 700 nm, and the chromatograms were recorded at 520 nm.

MS full-scan acquisition (scan range m/z 200–2000) was first performed in positive ion mode. Selected peaks were isolated in the ion trap and fragmented by MSn scan acquisition. Anthocyanins were identified using retention time (T_R), MS and MSn data in positive ion mode, and also by comparison of the MS data with those of pure standards and/or literature values. The relative compositions (%) of individual anthocyanins were calculated from peak areas at 520 nm using Xcalibur, version 2.0.7 (Thermo Electron).

12.4 UPLC of Anthocyanins in Wines

The metabolic pigment composition of Sangiovese wines produced from grapes harvested at 20 different vineyards in Montalcino over three consecutive years (2008–2010) on a semi-industrial scale and of 55 commercial Brunello di Montalcino wines (2004–2007) was studied, using a targeted method capable of analyzing 90 pigments in an 11 min UHPLCMS/MS chromatographic run (10). Interesting correlations were shown between various pigments formed during wine aging and those present in Sangiovese grapes. Vitisin B-like pigment and vitisin A-like pigment concentrations would seem to have a good correlation with ethyl-linked and direct-linked flavanol–anthocyanin concentrations, respectively. Moreover, the anthocyanic pattern recognition, genetically controlled by the plant variety, was shown to be inherited by the pigments formed during wine aging.

Analysis of the samples studied was performed with an Acquity UPLC system (Waters) coupled to a Xevo TQ MS system (Waters) operating under MassLynx XS software. All samples were analyzed on a RP Acquity UPLC BEH C18, 1.7 µm, 2.1 mm × 150 mm column (Waters), protected with an Acquity UPLC BEH C18, 1.7 µm, 2.1 mm × 5 mm precolumn (Waters), at 40°C and under a mobile phase flow rate of 0.4 mL/min. Water was used as weak eluting solvent (A) and methanol as strong elution solvent (B); formic acid 5% v/v was used as additive in both eluents. The multistep linear gradient used was as follows: from 95 to 60% of A for the first 4 min, from 60 to 45% A from 4 to 9 min, from 45 to 5% A from 9 to 11 min, and an isocratic hold for 3 min to clean the column. The equilibration time was 4 min, and the injection volume was 2 µL. All of the samples were analyzed in triplicate. The column eluent was directed to the mass spectrometer, and analyte detection was performed by multiple reaction monitoring (MRM). Electrospray positive ionization mode (ESI) was applied for all compounds.

Pigments have very similar structural characteristics to one another and are already charged, so their detection using a mass spectrometer occurs under very similar conditions. The chromatographic data reported in the literature were also used for identification. Quantification of individual compounds was performed using UHPLC-MS/MS. When the authentic standard was not available, the analytes were quantified relative to malvidin 3-glucoside, by the malvidin 3-glucoside calibration curve. Data processing was performed using the MassLynx TargetLynx Application Manager (Waters).

Hyphenation of micro-liquid and UPLC with a hybrid (QqTOF) tandem MS was studied for monitoring of anthocyanin dyes in various cultivars and different vintages of red wine (11). UHPLC system provided lower LOD and LOQ values as well as higher productivity and precision of retention parameters and peak areas with respect to micro-LC method.

On the other hand, micro-LC method offers significant savings of mobile phase, less contamination of ion source of mass spectrometer and makes the nebulization process in electrospray much easier.

In combination with principal component analysis (PCA), the method allows to elucidate long-term relations in a complete set of certified red wine cultivars grown in South Moravia (certified in Czech Republic until year 2005). Among the studied varieties Rubinet exhibit a specific anthocyanin profile. Especially interesting is a high content of 3-coumaroylglucoside-5-glucosides of malvidin and peonidin. Those dyes can be used as markers of artificial color enhancement. PCA applied to data processed with regard to enzymatic activity markedly facilitate classification among varieties.

Analyses were performed on two liquid chromatographs CapLC XE System (Waters) and PLATINblue (Knauer).

μLC separation of anthocyanins was achieved on a CapLC XE System using a C18 microcolumn (Gemini C18, 5 μm, 150 mm × 300 μm, Phenomenex) at room temperature.

A Waters CapLC 2996 PDA was used as the UV–vis detector prior to connection to the mass spectrometer.

Analyses of the last vintage (2009) were performed on PLATINblue UHPLC system. The separation was carried out using a C18 column (Acquity BEH, 1.7 μm, 50 mm × 2.1 mm, Waters), also at room temperature.

Mass spectrometric detection was performed on QqTOF mass spectrometer (Q-Tof Premier, Waters) in a positive electrospray (ESI) mode. For the μLC/MS analysis, the capillary voltage was kept at 3.2 kV, the temperature of the source was 120°C and the desolvation gas was set at 300°C with flow rates adjusted to 300 L/h. UHPLC/MS analysis required higher flow rates of the desolvation gas (550 L/h). Also, to facilitate the ionization process, a usage of a flow-splitting device was necessary. The chromatographic and mass spectrometric processing software used for data evaluation was MassLynx 4.0 and Masslynx 4.1.

Malvidin-3-O-glucoside (MvG), malvidin-3-O-(6-O-acetyl)-glucoside (MvAG), and malvidin-3,5-O-diglucoside (MvDG) are three representative malvidin-type anthocyanins in red wine. In the study of (12), the influence of structural differences on the formation efficiency of two types of derived pigments ((–)-epicatechin-ethyl-anthocyanins and pyranoanthocyanins) was investigated in model solutions using UHPLC-MS. The results showed that the yields of MvAG were higher than those of MvG to form both types of derived pigments, and the formation rate of pyranoanthocyanin was also relatively higher. In contrast, acetylation slowed the formation of (–)-epicatechin-ethyl-anthocyanins, indicating that the rate of covalent reactions may be linked to the affinity of (–)-epicatechin to copigment with anthocyanins. The condensation rate of MvDG with (–)-epicatechin, mediated by acetaldehyde, was much lower than that of the two monoglucosidic anthocyanins and also exhibited lower yields. In addition, pyranoanthocyanin was not generated from MvDG due to the absence of a free hydroxyl group at the C5 position.

The samples for the formation of (–)-epicatechin-ethyl-anthocyanins were taken at 0, 1, 2, 3, 4, 5, 6, and 7 days, and the samples for the formation of pyranoanthocyanins were taken at 0, 5, 10, 15, 20, 25, and 30 days. All the samples were transferred into 150-μL glass inserts in brown HPLC vials and stored at −40°C until further detection. Phenolic analysis was performed on an Agilent 1290 II series UHPLC system, combined with a 6545 series quadrupole time-of-flight MS (Agilent Technologies), using a Zorbax SB-C18 Rapid Resolution HD column (150 mm ×2.1 mm, 1.8 μm, Agilent Technologies) for separation. The mobile phase consisted of water/formic acid (100/1, v/v) (A) and methanol/acetonitrile/formic acid (50/50/1, v/v/v) (B). The other HPLC and MS conditions were based on reference (13). Qualitative analysis of anthocyanins and their derivatives was accomplished using UHPLC-MS according to the elution order, spectrum, molecular mass, and fragmentation pattern reported in refs 14 and 15 (14,15).

12.5 UPLC of Anthocyanins in Fruits

Euterpe oleracea fruits have gained much attention because of their phenolic constituents that have shown potential health benefits. The aim of the work of Dias et al. (16) is to develop a validated UHPLC–PDA method for major anthocyanins quantification in *E. oleracea* fruits. The separation was performed on HSS C18 column (1.8 μm) using a gradient elution with acetonitrile and 5% formic acid in a total run time of only 17 min. Total error and accuracy profiles were used as criteria for the validation process. Calibration in the matrix was found to be more accurate than calibration without matrix. Trueness (<6.76% relative bias), repeatability (<4.6% RSD), intermediate precision (<5.3% RSD), selectivity, response function, and linearity for major anthocyanins, cyanidin-3-glucoside and cyanidin-3-rutinoside, were evaluated. The concentration range validated was 1–48 g/mL for both compounds. In addition two cyanidin-di-O-glycosides were detected for the first time in this fruit. A first extraction of the fruits with ethyl acetate removes the lipophilic compounds and allows an easier extraction by methanol and quantification of anthocyanins in this extract.

The analysis were performed on a Accela UHPLC system acquired from Fisher Scientific (Thermo Fisher Scientific) consisting of a PDA detector, an autosampler equipped with a column oven and a tray compartment cooler and a quaternary pump with a built-in solvent degasser, all piloted by ChemoQuest software. The chromatographic separation was performed on an Acquity UPLC HSS C18 column 1.8 μm, 100 mm × 2.1 mm ID (Waters) equipped with a VanGuard UPLC HSS C18 precolumn 1.8 μm, 5 mm × 2.1 mm ID (Waters). 5 μL of samples was injected in a full loop injection mode. The separations were performed with a constant flow rate of 400 μL/min with the eluents being (A) H_2O/HCOOH (95/5) and (B) ACN with the following gradients: 0–10 min: 5–15% B; 10–10.1 min: 15–95% B; 10.1–12 min: 95% B; 12–12.1 min: 95–5% B; 12.1–17 min: 5% B. The column oven and tray cooler temperatures were set to 30 and 4°C, respectively. For quantitative analysis the PDA detector was used at 515 nm. The quantification was performed using the external calibration method. The Accela UHPLC system was hyphenated with a LTQ-Orbitrap XL MS system (Thermo Fisher Scientific). The system was equipped with an ESI interface that was used in positive ionization mode.

The compounds were identified by comparison with reference compounds from retention times, MS and MS/MS analysis. For the MS experiments two scan events were applied. The first was a full MS scan, for which the spectra were recorded in the range of m/z 100–1000 with a resolution of 30,000. The second was a data-dependent scan that selected the most intense ion or specified ions in another setting from the first scan event for the acquisition of MS/MS spectra. The CID activation was set to a normalized collision energy of 20 or 15%.

The major non-anthocyanin flavonoids (NAF) in the juice of *E. oleracea* were quantified by an accurate method coupling UHPLC with a linear ion trap-high-resolution Orbitrap MS system (UHPLC-LTQ-Orbitrap MS) (17). Fruits were processed to juice, and then the juice was lyophilized and defatted. The residue was then extracted in the presence of methanol by sonication. The extraction time was optimized and recovery rates of the extraction were >90%. The extracts were dried and solubilized again in 40% MeOH, which showed the best compromise for MS detection. For the UHPLC quantification, a HSS C18 column (1.8 µm) was used with a gradient elution of methanol and water both with 0.1% formic acid. Seven compounds and their isomers were successfully separated, including the major NAF. Calibration in the matrix was found to be more accurate than calibration without matrix.

The UHPLC analysis was performed on an Accela UHPLC system acquired from Fisher Scientific (Thermo Fisher Scientific) that consisted of an autosampler equipped with a column oven, a tray compartment cooler, and a quaternary pump with a built-in solvent degasser, all piloted by Xcalibur software. The chromatographic separation was performed on an Acquity UPLC HSS C18column, 1.8 µm, 100 mm × 2.1 mm i.d. (Waters), equipped with a VanGuard UPLC HSS C18 precolumn, 1.8 µm, 5 mm × 2.1 mm i.d., from the same supplier. Five µL of samples were injected in a full loop injection mode. The separations were performed with a constant flow rate of 400 µL/min. The eluents were (A) 0.1% formic acid in water and (B) 0.1% formic acid in methanol. They were eluted with the following gradient: 0–2 min: 15% B; 218 min: 15–27% B; 18–28 min: 27–95% B; 28–30 min: 95% B; 30–30.1 min: 95–15% B; 30.1–35 min: 15% B. The column oven and tray cooler temperatures were set to 30 and 4°C, respectively. The Accela UHPLC system was hyphenated with a LTQ Orbitrap XL mass spectrometer (Thermo Fisher Scientific). The system was equipped with an ESI source that was operated in negative ion. Spectra were recorded in the range of m/z 100–1000 with a resolution of 30,000. For identification purposes, two scan events were applied for the MS experiments. The first was in a full scan MS mode and the second was a data-dependent scan that selected the most intense ion or specified ions in another setting from the first scan event for the acquisition of MS/MS spectra. The CID activation was set to normalized collision energy of 25%, except for quercetin and luteolin (35%), vanillic, caffeic, and protocatechuic acids (20%). The compounds were identified by comparison with reference compounds when available and with literature data, from retention times, MS, MS/MS, and MSn analysis.

The phenolic composition of grape berries grown in Italy (Cabernet Cortis, Johanniter, Solaris, Phoenix, and Regent) and in Germany (Regent and Phoenix) was analyzed using two complementary LC-MS/MS (LC-Tandem Quadrupole MS) methods (18). These five cultivars belong to fungus-resistant grape varieties (PIWI) and little is known about their phenolic profile, although phenolics such as phytoalexins have been recognized, playing a key role in the resistance mechanism. A triple quadruple mass spectrometer detector was used for both identification and quantification, and analytical tools from untargeted metabolomics were applied to check and control the quality of the results. Specifically, biological, technical, and instrumental replications were included in order to study the variability at different levels. The results allowed tuning of the sampling protocol and provided a rich phenolic metabolite profile for the aforementioned PIWI varieties, as compared with *Vitis vinifera* cultivars, especially in the class of stilbenes.

Analysis was carried out on a Waters Acquity UPLC system, consisting of a binary pump, cooling autosampler, on-line vacuum degasser, and column oven, coupled to a Waters Xevo triple-quadrupole mass spectrometer detector. Flow injections of each individual metabolite were used to optimize the MRM conditions. For the majority of the metabolites, this was done automatically by the Waters Intellistart software, whereas for some compounds, the optimal cone voltages and collision energies were identified during CID experiments and manually set.

Analysis of phenolic metabolites (except anthocyanins) was performed according to Vrhovsek et al. (19), with a Waters Acquity HSS T3 column (1.8 µm, 100 mm × 2.1 mm). The injection volume was 10 µl. For the triple-quadrupole mass spectrometer, capillary voltage was 3.5 kV in positive mode and 2.5 kV in negative mode; the source was kept at 150°C; desolvation temperature was 500°C; cone gas flow, 50 l/h; and desolvation gas flow, 800 l/h. Unit resolution was applied to each quadrupole. The injection of samples was performed following the same randomized sequence used for extraction, and the QC sample was injected twice at the beginning of the sequence and after every six real sample injections. Analysis of anthocyanins was carried out according to Arapitsas et al., (20) with a few minor modifications. A reversed-phase Acquity UPLC BEH C18 (1.7 µm, 2.1 mm × 150 mm) column, protected with an Acquity UPLC BEH C18 precolumn (1.7µm, 2.1 mm × 5 mm) was used and kept at 40°C. Mobile phase flow rate was 0.4 mL/min, using water containing 5% v/v formic acid (A) and methanol containing 5% v/v formic acid as eluting solvents. The gradient was set as follows: 0 min 5% B, from 0 to 4 min linear to 40% B, from 4 to 9 min linear to 55% B, from 9 to 11 min linear to 95% B, and from 11 to 14 min isocratic 95% B for washing and back to the initial conditions. Injection volume was 2 µl, and samples were analyzed in duplicate according to two different randomized sequences. Electrospray positive ionization mode was applied for all compounds. Once again, a QC sample was injected after every six real sample injections. The same injection procedure was also carried out by diluting the samples 10 times due to the signal above the linearity range in a few cases.

Identification was carried out on the basis of authentic standards by checking the related retention time and considering both qualifier and quantifier ions. For quantification, calibration linear curves were built according to Vrhovsek et al. (19) and Arapitsas et al. (20) for metabolites with a signal-to-noise ratio higher than 10 (limit of quantification–LOQ s/n=10). The limit of detection was given by s/n=3. LOD and LOQ were determined according to the results of real injections and not calculated with mathematic formulas, so for some compounds, LOD and LOQ were identical. Data processing was carried out using Waters MassLynx version 4.1 and TargetLynx software. PCA was performed using IBM SPSS Statistics for Windows, Version 19.0.

Avula et al. (21) developed a simple, quick, and reliable UHPLC-PDA-MS method to identify chemical markers for authentication and compare fruit samples of four elder species, *Sambucus nigra*, *Sambucus ebulus*, *Sambucus canadensis*, and *Sambucus cerulea*. The UHPLC method showed an adequate separation of 11 compounds including anthocyanins, phenolic acids, and flavonoids in berries of the four Sambucus species. Quantitation, detection limits, precision, accuracy as well as robustness were determined. The analysis of 55 different samples of four Sambucus species showed distinct phytochemical fingerprints. Thirty-one dietary supplements claiming to contain *S. nigra* berries extract were analyzed to determine authenticity. Ten (32%) of the 31 products complied with the label and contained only elderberry. The application of UHPLC-QToF provided useful information to characterize 73 compounds that can be used to distinguish among the four Sambucus species and also help to determine the presence of adulterants. Extracts of black rice (*Oryza sativa*) were identified as the main adulterants in the dietary supplements, although other undeclared ingredients such as extracts of purple carrot (*Daucus carota* ssp. *sativus* var. *atrorubens*) and flowers of *S. nigra* were also found instead of elderberry in two products.

All analyses were performed on a Waters Acquity UPLC system (Waters) including quaternary solvent manager, sample manager, column compartment, PDA (Waters Acquity model) connected to Waters Empower 3 data station. An Acquity UPLC HSS C18 column (100 mm × 2.1 mm I.D., 1.8 µm), also from Waters, was used for the separation. The column and sample temperature were maintained at 45°C and 15°C, respectively. The column was equipped with a LC-18 guard column (Vanguard 2.1 mm ×5 mm, Waters). The mobile phase consisted of water (A), and acetonitrile (B) both containing 1% formic acid at a flow rate of 0.135 mL/min with a gradient of % B: initial, 11%; 8 min, 23%; 13 min, 35%; 18 min, 100% (run time 20 min). Each run was followed by a 3 min wash with 100% B and an equilibration period of 5 min with 89% A/11% B. Strong (95/5; acetonitrile/water) and weak (10/90; acetonitrile/water) needle wash solutions were used to reduce the risk of carryover. Two µL of the sample was injected and peak identities were assigned based on the mass of the compounds, comparison of the retention times with standard compounds, and by comparison of the UV/Vis spectra to those of reference compounds. The effluent from the LC column was directed from the PDA into the ESI probe. Data acquisition with a QDa single quadrupole mass detector was performed in both positive and negative modes. Compounds were monitored in selected ion recording (SIR) mode.

For the UPLC-MS/MS method, the liquid chromatographic system was an Agilent Series 1290 and the separation was achieved on an Acquity UPLC HSS C18 column (100 mm × 2.1 mm I.D., 1.8 µm). The mobile phase consisted of water with 0.1% formic acid (A) and acetonitrile with 0.1% formic acid (B) at a flow rate of 0.15 mL/min. Analysis was performed using the following gradient elution: 11–23% B in 8 min, then to 35% B in the next 5 min, and finally to 100% B at 18 min. Each run was followed by a 3 min wash with 100% B and an equilibration period of 5 min with 89% A/11% B. Two microliters of the sample were injected. The column temperature was 45°C. The mass spectrometric analysis was performed with a QToF-MS-MS (Model #G6545B, Agilent Technologies) equipped with an ESI source with Jet Stream. All the operations, acquisition and analysis of data were controlled by Agilent MassHunter Acquisition software Ver. A.10.1 and processed with MassHunter Qualitative Analysis software Ver. B.07.00.

Pistacia lentiscus L. is a shrub of the Anacardiaceae family whose fruits are used in Tunisian and Algerian diets. The phenolic composition at 5 different physiological stages of the fruit was investigated using two different targeted methodologies: UHPLC–UV/visible detection (UHPLC-UV/vis) for anthocyanins and UHPLC coupled with tandem MS (UHPLC-MS/MS) for the other polyphenols (22). For the specific analysis of anthocyanins, compound identification was confirmed by UHPLC-MS/MS and LC-NMR analysis. This study revealed the identification of 30 phenolic compounds including 9 anthocyanins, 7 flavanols, 7 flavonols, 2 phenolic acids, 1 stilbene, 2 flavanones, 1 flavanonol, and 1 dihydrochalcone. Quantification showed significant qualitative and quantitative variation in phenolic content during the ripening of *P. lentiscus* fruits, flavonols being the main compounds for the unripe berries and anthocyanins for ripe berries. This study (22) reports the presence of piceid and protocatechuic acid in *P. lentiscus* L. fruits, as well as several anthocyanins in Pistacia, for the first time.

Individual anthocyanin identification was performed by a combination of HPLC-MS and HPLC-NMR analysis following the method developed by Acevedo De la Cruz et al. (23). Individual anthocyanin quantification was performed on a UHPLC-MWL-MS/MS system (Thermo Scientific) composed of an Accela 1250 system coupled to a TSQ Quantum Access Max triple quadrupole equipped with an H-ESI ion source. Anthocyanins were separated on a Zorbax SB-C18 column (2.1 mm × 100 mm, 1.8 µm, Agilent) with 1 µL of volume injection. Mobile phase A was deionized water containing 2.5% formic acid (v/v) and mobile phase B was methanol (UHPLC grade) containing 2.5% formic acid (v/v). The elution gradient was for solvent B: 2.5% (0 2 min); 15% (7 min); 23% (9.3 min); 55% (10 min); 95% (11–12.5 min); 2.5% (13–15 min). The temperature and the flow were fixed to 35°C and 0.4 mL/min, respectively. The mass detector operated in the positive mode. All compounds were quantified using the MWL detector at 520 nm using a calibration curve built with Dp3glu (range 0.97 1000 mg/L). Results were expressed as mg delphinidin 3-O-glucoside (Dp3glu) equivalents per 100 g of dry-weight fruits (mg DGE/100 g DW). The limits of detection (LOD =0.08 mg DGE/100 g DW) and quantification (LOQ = 0.23 mg DGE/100 g DW) were calculated as LOD =3.3 × s/b and LOQ = 10 × s/b, where "b" is the slope of the curve and "s" the standard deviation of the signal determined as the residual standard deviation of the calibration line in the LOD region (24).

An UHPLC-UV/vis-MS/MS methodology was used to obtain more details of the individual anthocyanin composition of the fruits.

Gaultheria pumila (Ericaceae) (known as Chaura or Mutilla) is a Chilean native small shrub that produces berry fruits consumed by local Mapuche people. In the study of Fernández-Galleguillos et al. (25), the chemical fingerprinting and antioxidant, enzyme inhibition, and antiproliferative activities of the berries were investigated for the first time. Thirty-six metabolites were identified in the fruits by UHPLC- PDA detection, hyphenated with Orbitrap MS analysis (UHPLC-DAD-Orbitrap-MS). Metabolites included anthocyanins, phenolic acids, flavonoids, iridoids, diterpenes, and fatty acids.

LC was performed using a UHPLC C18 column (Acclaim, 150 mm × 4.6 mm ID, 2.5μm; Thermo Fisher Scientific) operating at 25°C. The detection wavelengths were 280, 254, 330, and 354 nm, and PDA detectors were set from 200 to 800 nm. Mobile phases were 1% formic aqueous solution (A) and acetonitrile 1% formic acid (B). The gradient program started at 5% B at time zero; was maintained at 5% B for 5 min; went to 30% B for 10 min; was maintained at 30% B for 15 min; went to 70% B for 5 min; was maintained at 70% B for 10 min; and was finally returned to the initial conditions within 10 min and 12 min for column equilibration prior to each injection. The flow rate was 1.00 mL/min and the injection volume was 10 μL. The standards and lyophilized decoction dissolved in methanol were kept at 10°C during its storage in the autosampler.

The objectives of the work of Feuereisen et al. (26) were to determine the phenolic profile of *Schinus terebinthifolius* and *Schinus molle* fruits and to develop a reliable method for the differentiation of these two similar spices both known as pink pepper. Anthocyanins, bioflavonoids, and gallotannins, some of which are reported for the first time in these species, were identified by UHPLC–UV/vis–MS/MS. Consideration of the relative and absolute amounts of phenolics as well as indicator compounds from 18 samples revealed that the relative amounts of anthocyanins and bioflavonoids are the most trustworthy parameters. PCA and cluster analysis (CA) allowed a grouping of the samples according to their species, showing that the anthocyanins are most important for the identification of species.

Two different LC–MS systems were used for the analysis of phenolic compounds. The MS of the first system is more sensitive and therefore particularly suitable for the detection of minor compounds, whereas the UV detector of the second, more acid-resistant system has a higher sensitivity and is therefore preferable for the purposes of UV/vis quantitation. For the identification of phenolic compounds, an Acquity UPLC system (Waters) was used as described previously. Also, the methods were the same with the exception of the eluents and the gradient used for anthocyanin analysis. For enhanced separation, 5 and 3% formic acid in water and acetonitrile, respectively, were used. The gradient for anthocyanin analysis was slightly modified as follows: 0.0 min, 0% B; 0.2 min, 5% B; 12 min, 16.9% B; 1 min, 25% B; 17 min, 98% B; 19 min, 98% B; 20 min, 1% B; 22 min, 0% B. Furthermore, the gradient used for bilavonoid analysis was extended for more efficient detection of hinokiflavone: 0 min, 2% B; 24 min, 71.6% B; 24.3 min, 100% B; 25.3 min, 100% B; 26 min, 2% B; 27.5 min, 2% B. Quantitation and determination of relative amounts of anthocyanins and bioflavonoids were performed with an Acquity I-Class UHPLC system (Waters) equipped with a binary pump (BSM), an autosampler (SM-FL) cooled at 10°C, a column oven (CM) set at 40°C, a diode array detector (PDA eλ) scanning from 200 to 700 nm for the anthocyanins and from 200 to 480 nm for the biflavonoids, and an LTQXL linear ion trap mass spectrometer (Thermo Scientific) with an electrospray interface. The mass spectrometer was tuned using a solution of cyanidin-3-O-glucoside for the anthocyanins and I3,II8-biapigenin for the bioflavonoids. The MS scan was performed in full scan mode, and MS2 and MS3 spectra were acquired. Xcalibur version 2.2 (Thermo Electron) was used to control the UHPLC–MS system and to evaluate the results. The bioflavonoids and anthocyanidins were quantified by a calibration curve using amentoflavone and malvidin-3-O-glucoside as external standards.

Vortex-assisted matrix solid-phase dispersion (VA-MSPD) is used to determine phenolic compounds in the edible part of *Myrciaria cauliflora* combined with UHPLC coupled to tandem MS (UHPLC-MS/MS) analysis (27).

For the chromatographic analysis of the 10 selected phenolic compounds, which were chosen since they have been identified in *M. cauliflora* fruit (28–30), a UHPLC-MS/MS was developed. For this, a UPLC Acquity system coupled to a triple quadrupole mass spectrometer equipped with an electrospray ionization source (ESI, Waters Z spray™, Waters) was used. The mobile phase consisted of ultrapure water acidified with 0.1% formic acid (A), and methanol (B), using a gradient program elution. The percentage of the organic solvent was changed linearly as follows: 10% B (0–4 min), 70% B (4–7 min), 100%B (7–8 min). The system was re-equilibrated with the initial conditions of the elution program for 6 min prior to the next injection. A column Acquity UPLC BEH C18 (2.1 mm × 50 mm i.d., 1.7 µm particle size) from Waters was used for the separation at a flow rate of 0.150 mL/min. The column temperature was held at 30°C and the injection volume was 1.5 µL. The mass spectrometer was operated using an electrospray (ESI) source in negative ion mode. The mass spectrometer was operated in MS/MS mode using selected reaction monitoring (SRM). Two characteristic ions were selected for identification and quantification for each analyte. The corresponding cone voltage and collision energy were optimized with maximum intensity. MassLynx™ v.4.1 software (Waters) was used for instrument control, data acquisition, and processing.

Myrtus communis L.(Myrtaceae) is a self-seeded shrub, widespread in Sardinia, with anti-inflammatory, antiseptic, antimicrobial, hypoglycemic, and balsamic properties. Its berries, employed for the production of sweet myrtle liqueur, are characterized by a high content of bioactive polyphenols, mainly anthocyanins. Anthocyanin composition is quite specific for vegetables/fruits and can be used as a fingerprint to determine the authenticity, geographical origin, and quality of raw materials, products, and extracts. To rapidly analyze and determine anthocyanins in 17 samples of *M. communis* berries, a platform based on the integration of UHPLC–MS/MS quantitative data and multivariate analysis with the aim of extracting the most information possible from the data was developed (31).

UHPLC-ESI-MS/MS methods, working in positive ion mode, were performed for the detection and determination of target compounds in MRM mode. Optimal chromatographic conditions were achieved using an XSelect HSS T3 column and a gradient elution with 0.1% formic acid in water and 0.1% formic acid in acetonitrile. PCA was applied to the quantitative data to correlate and discriminate 17 geographical collections of *M. communis*.

The developed quantitative method was reliable, sensitive and specific and was successfully applied to the quantification of 17 anthocyanins. Peonidin-3-O-glucoside was the most abundant compound in all the extracts investigated.

Full scan electrospray ionization MS (ESI-MS), MS/MS, and MS^3 analyses of standards and samples were performed on an ABSciex API 4000 Q-Trap spectrometer. The analytical parameters were optimized by infusing each standard solution (1 µg/mL in methanol 50%) in to the source at a flow rate of 10 µL/min. Data were acquired in the positive ion MS, MS/MS and MS^3 modes. Direct infusions of each individual standard compound were used to optimize the MRM conditions, which was done automatically by the Analyst 1.6.2 software, and then manually checked for selection of the quantifier and qualifier ions. Qualitative on-line HPLC-ESI-MS/MS analyses of extracts were performed using a UHPLC system interfaced to an ABSciex API 4000 Q-Trap instrument in ion trap mode. LC analyses were conducted using a system equipped with a Flexar UHPLC AS system (Perkin–Elmer) consisting of degasser, Flexar FX-10 pump, autosampler, and PE200 column oven. Samples were injected (5 µL) in to an XSelect HSS T3 column (Waters) (100 mm × 2.1 mm i.d., 2.5 µm). Mobile phase A was water containing 0.1% formic acid while mobile phase B was acetonitrile containing 0.1% formic acid. Elution was carried out at 40°C according to the following gradient: 0–0.5 min, isocratic 5% B; 0.519 min, linear gradient 5–95% B. The flow rate was 300 µL/min and was injected from the chromatograph directly into the ESI source. Qualitative analysis of the compounds was performed using IDA (information-dependent acquisition).The IDA method created included an IDA criteria (specify the charge state, mass range), enhanced MS scan, enhanced resolution, enhanced product ion scan, or MS/MS scan. Enhanced MS (EMS) was conducted with mass range from m/z 200.0 to 800.0 with a scan rate of 10000 amu/s.

MS data were acquired using the software provided by the manufacturer (Analyst software 1.6.2), and extracted ion fragmentograms (XIC) were elaborated in order to identify anthocyanins from their protonated molecular ions and retention time. In the product ion spectra obtained, the predominant fragments relative to anthocyanin compounds were chosen to develop the MRM method for quantification.

Quantitative on-line HPLC-ESI-MS/MS analyses were performed using the same LC-ESI-MS/MS equipment but with the mass spectrometer having the triple quadrupole analyzer in MRM mode. Elution was carried out at 41°C according to the following flow and solvent gradient: 0–4 min, isocratic 0% solvent B and the flow changes from 300 µL to 350 µL; 4–6 min, linear gradient 0–12% B and the flow achieves 400 µL/min; 6–12 min, linear gradient 12–20% B and flow constant at 400 µL/min; 16–17min, linear gradient 20–100% B and flow retrieves to 300 µL/min. The API 4000 ES source was operated in positive ion mode and was tuned by infusing solutions of standards (1 µg/µL in methanol 50%) in to the source at a flow rate of 10 µL/min. Data acquisition and processing were performed using Analyst software 1.6.2.

Anthocyanins are the major polyphenolic compounds in strawberry fruit responsible for its color. Due to their sensitivity, they are affected by food processing techniques such as fermentation that alters both their chemical composition and organoleptic properties. The work of Hornedo-Ortega et al. (32) aims to evaluate the impact of different fermentation processes on individual anthocyanins compounds in strawberry wine and vinegar by UHPLC-MS/MS Q Exactive analysis.

Nineteen, 18, and 14 anthocyanin compounds were identified in the strawberry initial substrate, strawberry wine, and strawberry vinegar, respectively. Four and 8 anthocyanin compounds were tentatively identified with high accuracy y for the 1st time to be present in the beverages obtained by alcoholic fermentation and acetic fermentation of strawberry, respectively. Both, the total and the individual anthocyanin concentrations were decreased by both fermentation processes, affecting the alcoholic fermentation to a lesser extent (19%) than the acetic fermentation (91%). Indeed, several changes in color parameters have been assessed. The color of the wine and the vinegar made from strawberry changed during the fermentation process, varying from red to orange color, this fact is directly correlated with the decrease of anthocyanins compounds.

An UHPLC Dionex Ultimate 3000 system (Thermo Scientific), coupled to a Thermo Scientific Q-Exactive TM hybrid quadrupole-orbitrap mass spectrometer was used. The UHPLC system consisted of consisting of a quaternary Rs Pump Dionex Ultimate 3000 and Rs autosampler Dionex Ultimate 3000, connected to a quadrupole-orbitrap (Q Exactive) hybrid mass spectrometer with heated-electrospray ionization probe (HESI-II, Thermo Fisher Scientific). The analytical method was previously published (Nati' c and others 2015). Separation was performed on a column SB-C18 (2.1 mm × 100 mm, 1.8 µm) set at 40°C (Agilent). Injection volume was 1 µL and flow rate was 0.4 mL/min. The solvents used as a mobile phase were solvent A (water/formic acid 95:5 v/v) and solvent B (acetonitrile/formic acid 95:5 v/v), scheduled in the following gradient: 0.0–2.0 min 5% B, 2.0–12.0 min from 5 to 100% B, 12.0–13.0 min from 100 to 5% B, then 5% B up to 15.0 min.

Anthocyanin identification and quantitation were acquired in positive mode by full-range acquisition covering m/z 100–1500 at 35,000 resolution and by targeted MS2 normalized Higher Energy Collision Dissociation (HCD).

Xcalibur software (version3.0.63) was used for instrument control, data acquisition, and data analysis. Compounds were identified according to their calculated mass, accurate mass, mass spectra, characteristic fragmentation data, and retention time. Pelargonidin 3-glucoside, cyanidin 3-glucoside, delphinidin 3-glucosise, and peonidin 3-glucoside available standards were used both for positive identification and quantification purposes. Anthocyanin compounds were quantified using the areas of the aglycone counterparts.

Sweet cherries are highly appreciated by consumers worldwide and are usually cold-stored during postharvest to prevent over-ripening before distribution to the market. Sweet cherry is a non-climacteric fruit, for which ripening is known to be regulated by abscisic acid. The hormone profiles, including measurements of abscisic acid, auxins, cytokinins, and gibberellins by UHPLC coupled to tandem MS (UHPLC–MS/MS) were examined (33), in relation to variations in sugar and anthocyanin contents, during growth and ripening of this fruit. Hormonal profiling revealed that indole-3-acetic acid, GA_1, and trans-zeatin levels decreased at early stages of fruit development, while GA_3 levels decreased at early stages but also later, once anthocyanin accumulation started. Conversely, abscisic acid levels rose significantly once the fruit started to synthetize anthocyanins, and isopentenyl-adenosine levels also increased during the ripening of sweet cherries. A strong negative correlation was found between GA_4 levels and both fruit biomass and anthocyanin levels, and between the levels of trans-zeatin and both fruit biomass and total sugar contents. In contrast, abscisic acid and isopentenyladenosine levels correlated positively with fruit biomass, anthocyanin, and total soluble sugar content. Results suggest that auxins, cytokinins, and gibberellins may act coordinately with abscisic acid in the regulation of sweet cherry development and ripening. Furthermore, it is shown that hormonal profile measurements by UHPLC–MS/MS may be a helpful tool to elucidate the timing of action of each specific hormonal compound during ripening, which has important applications in the agri-food biotechnological sector.

The system consisted of an Acquity UPLC System (Waters) quaternary pump equipped with an autosampler. An HALO C18 (Advanced Materials Technology) column (2.1 mm × 75 mm, 2.7 µm) was used. Solvent A was water with 0.05% glacial acetic acid and solvent B was acetonitrile with 0.05% glacial acetic acid. Flow rate was set at 0.6 mL/min. Quantification was made considering recovery rates for each sample by using the deuterium-labeled internal standards.

Response surface methodology was employed to investigate the effects of pressurized liquid extraction (PLE) parameters on the recovery of phenolic compounds (anthocyanins, biflavonoids) from Brazilian pepper (*S. terebinthifolius* Raddi) fruits (34). The effects of temperature, static time, and ethanol as well as acid concentration on the polyphenol yield were described well by quadratic models (p < 0.0001). A significant influence of the ethanol concentration (p < 0.0001) and several interactions (p < 0.05) were identified. Identification of the biflavonoid I30,II8-binaringenin in drupes of *S. terebinthifolius* was achieved by UHPLC-MS^2. Interestingly, at high extraction temperatures (>75°C), an artifact occurred and was tentatively identified as a diastereomer of I30,II8-binaringenin. Multivariate optimization led to high yields of phenolic compounds from the exocarp/drupes at 100/75°C, 10/10 min, 54.5/54.2% ethanol, and 5/0.03% acetic acid. This study demonstrates that PLE is well-suited for the extraction of phenolic compounds from *S. terebinthifolius* and can efficiently be optimized by response surface methodology.

The HPLC system was equipped with a quaternary pump, a degasser, an autosampler, and a column oven (e2965 Separation module, Waters). UV/Vis spectra were recorded from 210 to 800 nm using a Waters 996 PDA detector. A NUCLEODUR 100-5 C18 ec analytical column (150 mm, 2 mm i. d.; 3 µm, Macherey-Nagel) with a SecurityGuard Guard Cartridge (AQ C18, 4 mm, 2 mm, Phenomenex) was used for the separation of the phenolic compounds. Elution was conducted at 35°C with 1% (v/v) acetic acid in water (solvent A) and 1% (v/v) acetic acid in acetonitrile (solvent B) at a constant flow rate of 0.3 mL/min. The gradient program was as follows: 0–40% B in 80 min, 40–100% B in 10 min, 100–0% B in 10 min, and 0% B for 10 min for equilibration. The injection volume was 20 µL for the exocarp samples and 10 µL for the drupe samples. The extracts were filtered through 0.2 µm Chromafil RC-20/15 MS filters (Macherey-Nagel) prior to injection. According to their respective absorption maxima, the peak integration was done at a wavelength of 500 nm for peaks 1–3, 330 nm for peaks 4–6, and 290 nm for peak 7. For quantification, a calibration was compiled with malvidin for the anthocyanins and with amentoflavone and naringenin for the biflavonoids. The characterization of the phenolic compounds was performed using an Acquity UHPLC-DAD-MS/MS system (Waters) as described in our previous work (35).

The majority of components in fruits are sensitive to heat-processing. Nevertheless, fruits are becoming popular ingredients in processed foods, like bakery foods. Therefore, the fate of the components in the fruit-involved food during thermal processing is important for the assessment of their nutritional values and sensory properties. Unfortunately, comprehensive knowledge of the compositional alteration in real food products during processing is limited. In the study of Zheng et al. (36), a popular bakery food, blueberry-filled pastry, was taken as the object, and a widely targeted metabolomic approach was applied to investigate the holistic compositional variation of blueberry filling during pastry preparation. Amongst the total of 630 chemicals identified, 288 chemicals were screened as differential compounds between samples collected at different processing stages. The most variation of the chemicals was observed during the process of stir-frying. A total of 197 chemicals varied significantly in concentrations during stir-frying, while only 75 chemicals altered significantly in contents during baking. Amongst 288 differential compounds, 117 belonged to the group of phenolic compounds, with the others found to be sugars and organic acids, amino acids, lipids, nucleotides, etc. The possible mechanisms of the chemical alterations during thermal processing were also discussed. The data provide comprehensive information on the compositional changes in berry-containing fillings during thermal processing, and hints, and emphasis for further investigation of the underlying mechanisms.

Exactly 2 μL of sample were injected and analyzed using a Shim-pack UFLC Shimadzu CBM 30A system (Shimadzu Corporation) interfaced to an Applied Biosystems 6500 QTRAP mass spectrometer (ABSciex). The sample was separated on a Waters Acquity UPLC HSS T3 C18 column (100 mm × 2.1 mm, particle size 1.8 μm) (Waters Corporation, Milford, MA, USA) by using water (0.04% acetic acid) as solvent A and acetonitrile (0.04% acetic acid) as solvent B. The eluting gradient program was as follows: 0.0–11.0 min, 5–95% B; 11.0–12.0 min, 95% B; 12.0–12.1 min, 95–5% B; and 12.1–15.0 min, 5% B. The flow rate was 0.4 mL/min. The temperature of column oven was kept at 40°C during the analysis. The effluent was connected to an ESI-triple quadrupole linear ion trap (QQQ-LIT) mass spectrometer equipped with an ESI Turbo Ion-Spray interface operating in both positive and negative ion modes and controlled by Analyst 1.6.3 software. Instrument tuning and mass calibration were performed with 10 and 100 μmol/L polypropylene glycol solutions in QQQ and LIT modes, respectively. QQQ scans were acquired as MRM.

References

1. Sun, J., Lin, L., & Chen, P. (2013). Recent applications for HPLC-MS analysis of anthocyanins in food materials. *Current Analytical Chemistry, 9*(3), 397–416.
2. Lei, Z., Sumner, B. W., Bhatia, A., Sarma, S. J., & Sumner, L. W. (2019). UHPLC-MS analyses of plant flavonoids. *Current Protocols in Plant Biology, 4*(1), e20085.
3. Lei, Z., Jing, L., Qiu, F., Zhang, H., Huhman, D., Zhou, Z., & Sumner, L. W. (2015). Construction of an ultrahigh pressure liquid chromatography-tandem mass spectral library of plant natural products and comparative spectral analyses. *Analytical Chemistry, 87*(14), 7373–7381.
4. de Rijke, E., Out, P., Niessen, W. M. A., Ariese, F., Gooijer, C., & Brinkman, U. A. T. (2006). Analytical separation and detection methods for flavonoids. *Journal of Chromatography A, 1112*(1), 3163.
5. Santos-Buelga, C., Garcia-Viguera, C., & Tomás-Barberán, F. A. (2003). On-line identification of flavonoids by HPLC coupled to diode array detection. (C. Santos-Buelga & G. Williamson Eds.). Royal Society of Chemistry (Great Britain): United Kingdom at the University Press, Cambridge.
6. Kaiser, M., Müller-Ehl, L., Passon, M., & Schieber, A. (2020). Development and validation of methods for the determination of anthocyanins in physiological fluids via UHPLC-MSn. *Molecules, 25*(3), 518.
7. Xie, L., Lee, S. G., Vance, T. M., Wang, Y., Kim, B., Lee, J. Y., …, & Bolling, B. W. (2016). Bioavailability of anthocyanins and colonic polyphenol metabolites following consumption of aronia berry extract. *Food Chemistry, 211*, 860–868.
8. Ludwig, I. A., Mena, P., Calani, L., Borges, G., Pereira-Caro, G., Bresciani, L., …, & Crozier, A. (2015). New insights into the bioavailability of red raspberry anthocyanins and ellagitannins. *Free Radical Biology and Medicine, 89*, 758–769.
9. Fabroni, S., Ballistreri, G., Amenta, M., & Rapisarda, P. (2016). Anthocyanins in different Citrus species: An UHPLC-PDA-ESI/MS n-assisted qualitative and quantitative investigation. *Journal of the Science of Food and Agriculture, 96*(14), 4797–4808.
10. Arapitsas, P., Perenzoni, D., Nicolini, G., & Mattivi, F. (2012). Study of Sangiovese wines pigment profile by UHPLC-MS/MS. *Journal of Agricultural and Food Chemistry, 60*(42), 10461–10471.
11. Papoušková, B., Bednář, P., Hron, K., Stávek, J., Balík, J., Myjavcová, R., …, & Lemr, K. (2011). Advanced liquid chromatography/mass spectrometry profiling of anthocyanins in relation to set of red wine varieties certified in Czech Republic. *Journal of Chromatography A, 1218*(42), 7581–7591.

12. Zhao, X., Zhang, N., He, F., & Duan, C. (2022). Reactivity comparison of three malvidin-type anthocyanins forming derived pigments in model wine solutions. *Food Chemistry, 384*, 132534.

13. Zhao, X., He, F., Zhang, X.-K., Shi, Y., & Duan, C.-Q. (2021). Impact of three phenolic copigments on the stability and color evolution of five basic anthocyanins in model wine systems. *Food Chemistry*, 131670.

14. Li, S. Y., He, F., Zhu, B. Q., Wang, J., & Duan, C. Q. (2017). Comparison of phenolic and chromatic characteristics of dry red wines made from native Chinese grape species and *Vitis vinifera*. *International Journal of Food Properties, 20*(9), 2134–2146.

15. Zhang, X. K., Li, S. Y., Zhao, X., Pan, Q. H., Shi, Y., & Duan, C. Q. (2020). HPLC-MS/MS based targeted metabolomic method for profiling of malvidin derivatives in dry red wines. *Food Research International, 134*, Article 109226.

16. Dias, A. L. S., Rozet, E., Chataigné, G., Oliveira, A. C., Rabelo, C. A. S., Hubert, P., …, & Quetin-Leclercq, J. (2012). A rapid validated UHPLC–PDA method for anthocyanins quantification from *Euterpe oleracea* fruits. *Journal of Chromatography B, 907*, 108–116.

17. Dias, A. L., Rozet, E., Larondelle, Y., Hubert, P., Rogez, H., & Quetin-Leclercq, J. (2013). Development and validation of an UHPLC-LTQ-Orbitrap MS method for non-anthocyanin flavonoids quantification in *Euterpe oleracea* juice. *Analytical and Bioanalytical Chemistry, 405*, 9235–9249.

18. Ehrhardt, C., Arapitsas, P., Stefanini, M., Flick, G., & Mattivi, F. (2014). Analysis of the phenolic composition of fungus-resistant grape varieties cultivated in Italy and Germany using UHPLC-MS/MS. *Journal of Mass Spectrometry, 49*(9), 860–869.

19. Vrhovsek, U., Masuero, D., Gasperotti, M., Franceschi, P., Caputi, L., Viola, R., & Mattivi, F. (2012). A versatile targeted metabolomics method for rapid quantification of multiple classes of phenolics in fruit and beverages. *Journal of Agricultural and Food Chemistry, 60*, 8831.

20. Arapitsas, P., Perenzoni, D., Nicolini, G., & Mattivi, F. (2012). Study of Sangiovese wines pigment profile by UHPLC-MS/MS. *Journal of Agricultural and Food Chemistry, 60*, 10461.

21. Avula, B., Katragunta, K., Wang, Y. H., Ali, Z., Srivedavyasasri, R., Gafner, S., …, & Khan, I. A. (2022). Chemical profiling and UHPLC-QToF analysis for the simultaneous determination of anthocyanins and flavonoids in *Sambucus* berries and authentication and detection of adulteration in elderberry dietary supplements using UHPLC-PDA-MS. *Journal of Food Composition and Analysis, 110*, 104584.

22. Aissat, A. K., Chaher-Bazizi, N., Richard, T., Kilani-Atmani, D., Pedrot, E., Renouf, E., …, & Fonayet, J. V. (2022). Analysis of individual anthocyanins, flavanols, flavonols and other polyphenols in *Pistacia lentiscus* L. fruits during ripening. *Journal of Food Composition and Analysis, 106*, 104286.

23. De la Cruz, A. A., Hilbert, G., Rivière, C., Mengin, V., Ollat, N., Bordenave, L., … & Richard, T. (2012). Anthocyanin identification and composition of wild Vitis spp. accessions by using LC–MS and LC–NMR. *Analytica Chimica Acta, 732*, 145–152.

24. Kruve, A., Rebane, R., Kipper, K., Oldekop, M.-L., Evard, H., Herodes, K., Ravio, P., & Leito, I. (2015). Tutorial review on validation of liquid chromatography-mass spectrometry methods: Part I. *Analytica Chimica Acta, 870*, 29–44.

25. Fernández-Galleguillos, C., Quesada-Romero, L., Puerta, A., Padrón, J. M., Souza, E., Romero-Parra, J., & Simirgiotis, M. J. (2021). Uhplc-ms chemical fingerprinting and antioxidant, antiproliferative, and enzyme inhibition potential of gaultheria pumila berries. *Metabolites, 11*(8), 523.

26. Feuereisen, M. M., Zimmermann, B. F., Schulze-Kaysers, N., & Schieber, A. (2017). Differentiation of Brazilian peppertree (*Schinus terebinthifolius* Raddi) and Peruvian peppertree (*Schinus molle* L.) fruits by UHPLC–UV–MS analysis of their anthocyanin and biflavonoid profiles. *Journal of Agricultural and Food Chemistry, 65*(26), 5330–5338.

27. Senes, C. E. R., Rodrigues, C. A., Nicácio, A. E., Boeing, J. S., Maldaner, L., & Visentainer, J. V. (2021). Determination of phenolic acids and flavonoids from *Myrciaria cauliflora* edible part employing vortex-assisted matrix solid-phase dispersion (VA-MSPD) and UHPLC-MS/MS. *Journal of Food Composition and Analysis, 95*, 103667.

28. Borges, L. L., Conceição, E. C., & Silveira, D. (2014). Active compounds and medicinal properties of *Myrciaria* genus. *Food Chemistry, 153*, 224–233.

29. Seraglio, S. K. T., Schulz, M., Nehring, P., Betta, F. D., Valese, A. C., Daguer, H., Gonzaga, L. V., Fett, R., & Costa, A. C. O. (2018). Nutritional and bioactive potential of *Myrtaceae* fruits during ripening. *Food Chemistry, 239*, 649–656.

30. Wu, C. C., Hung, C. N., Shin, Y. C., Wang, C. J., & Huang, H. P. (2016). *Myrciaria cauliflora* extracts attenuate diabetic nephropathy involving the Ras signaling pathway in streptozotocin/nicotinamide mice on a high fat diet. *Journal of Food and Drug Analysis, 24*, 136–146.

31. Maldini, M., Chessa, M., Petretto, G. L., Montoro, P., Rourke, J. P., Foddai, M., …, & Pintore, G. (2016). Profiling and simultaneous quantitative determination of anthocyanins in wild *Myrtus communis* L. berries from different geographical areas in Sardinia and their comparative evaluation. *Phytochemical Analysis, 27*(5), 249–256.

32. Hornedo-Ortega, R., Álvarez-Fernández, M. A., Cerezo, A. B., Garcia-Garcia, I., Troncoso, A. M., & Garcia-Parrilla, M. C. (2017). Influence of fermentation process on the anthocyanin composition of wine and vinegar elaborated from strawberry. *Journal of Food Science, 82*(2), 364–372.

33. Teribia, N., Tijero, V., & Munné-Bosch, S. (2016). Linking hormonal profiles with variations in sugar and anthocyanin contents during the natural development and ripening of sweet cherries. *New Biotechnology, 33*(6), 824–833.

34. Feuereisen, M. M., Barraza, M. G., Zimmermann, B. F., Schieber, A., & Schulze-Kaysers, N. (2017). Pressurized liquid extraction of anthocyanins and biflavonoids from *Schinus terebinthifolius* Raddi: A multivariate optimization. *Food Chemistry, 214*, 564–571.

35. Feuereisen, M. M., Hoppe, J., Zimmermann, B. F., Weber, F., Schulze-Kaysers, N., & Schieber, A. (2014). Characterization of phenolic compounds in Brazilian pepper (*Schinus terebinthifolius* Raddi) exocarp. *Journal of Agricultural and Food Chemistry, 62*, 6219–6226.

36. Zheng, J., Wu, Z., Yang, N., Zhou, K., Hu, W., Ou, S., & Liu, P. (2020). Widely targeted UHPLC-MS/MS metabolomic analysis on the chemical variation in blueberry-filled pastries during processing. *Frontiers in Nutrition, 7*, 569172.

13

NMR Analysis of Anthocyanins

Shigenori Kumazawa

Nuclear magnetic resonance (NMR) is an essential analytical tool for determining the structures of anthocyanins. Here, the preparation of anthocyanin samples for NMR analysis and the procedures for interpreting the NMR data using an actual example are described.

13.1 Sample Preparation for NMR Analysis

13.1.1 Amount for NMR Measurement

^{1}H and ^{13}C nuclei are generally used for NMR analysis of the structures of anthocyanins. ^{1}H-NMR analysis can be performed with high sensitivity because the natural abundance of ^{1}H is close to 100%, whereas the sensitivity of ^{13}C-NMR is low due to the natural abundance of only 1.1%. The sample used for the NMR measurements must be of high purity. The samples must also be thoroughly dried in a desiccator. Depending on the molecular weight of the anthocyanin, approximately 1 mg of the compound is sufficient for obtaining a ^{1}H-NMR spectrum if the NMR instrument is operating at 400 MHz or higher. However, several milligrams or more are required to obtain ^{13}C-NMR and two-dimensional (2D) NMR spectra.

13.1.2 Solvent for NMR Measurement

NMR analysis of anthocyanins requires a solvent in which the sample dissolves well. However, the solvent must be the same as that used in literature studies to enable comparison of the data. A solvent containing 10% trifluoroacetic acid (TFA) is often used for NMR measurements because the aglycon structure of anthocyanins changes with pH. CD$_3$OD:TFA-d_1 (9:1) and DMSO-d_6:TFA-d_1 (9:1) are generally used as solvents for anthocyanin analysis. After dissolution in the deuterated solvent, NMR analysis of anthocyanins is performed immediately. Although some anthocyanins can be stably dissolved in the solvent for the NMR measurement, the sample should be stored in a cool and dark place away from light if measurement is not performed immediately.

13.2 Example for the Elucidation of NMR Data

In this section, the actual NMR spectral data of anthocyanins are presented, and the information obtained from each spectrum is described. The anthocyanin used as an example is alatanin E (3-O-(6-O-(3-O-(β-D-glucopyranosyl)-6-O-(E)-sinapoyl-D-glucopyranosyl)-

DOI: 10.1201/9781003453260-15

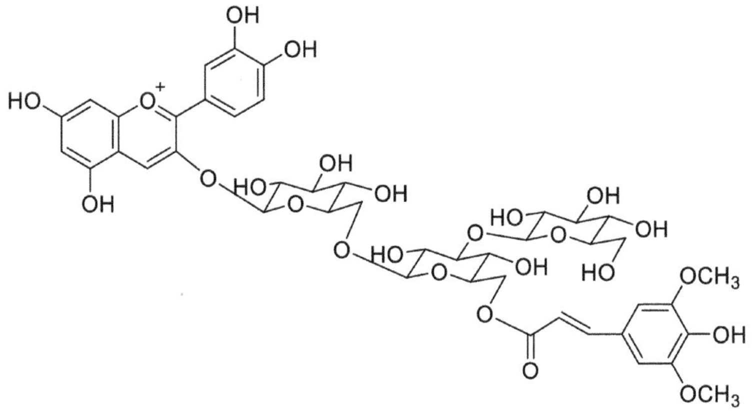

FIGURE 13.1
Structure of alatanin E.

β-D-glucopyranosyl)cyanidin) (Figure 13.1) [1]. This anthocyanin was isolated as a pigment from purple yam (*Dioscorea alata* L.) and its molecular formula was determined to be $C_{44}H_{51}O_{25}$ using high-resolution electrospray ionization mass spectroscopy (observed: m/z 979.2691, calculated: 979.2714 M+). The MS/MS analysis revealed the presence of cyanidin, one (*E*) sinapic acid unit, and three glucose units.

13.2.1 ^{1}H-NMR

Figure 13.2 shows the 400 MHz ^{1}H-NMR spectrum of alatanin E (5 mg) in CD_3OD/CF_3COOD (9:1).

The characteristic H-4 signal of anthocyanidin is observed at δ_H 8.62. The signals at 3.3–5.5 ppm are derived from the sugar moiety. The large coupling constant (J = 16 Hz) for the α/β positions of the sinapic group (δ_H 6.11/7.30) indicates that the sinapic acid moiety is in the (*E*) configuration. However, it is generally difficult to determine the full structure of anthocyanins using only ^{1}H-NMR spectra. Thus, ^{13}C-NMR and 2D NMR data were required.

13.2.2 ^{13}C-NMR

Figure 13.3 shows the 100 MHz ^{13}C-NMR spectrum of alatanin E (5 mg) in CD_3OD/CF_3COOD (9:1). The ^{1}H-NMR spectrum in Figure 13.2 was obtained within 20 minutes, whereas acquisition of the ^{13}C-NMR spectrum took 5 hours. It was impossible to quantify each type of carbon in the molecule using the ordinally proton-complete decoupling ^{13}C-NMR spectrum. However, the ^{13}C-NMR spectrum provides information about the skeleton of the molecule that is not obtained from ^{1}H-NMR spectrum.

Distortionless enhancement by polarization transfer (DEPT) measurements are often used to obtain information on which carbon atoms are attached to hydrogen atoms. In the DEPT-135 spectrum, the CH_3 and CH resonances were positive, the CH_2 resonances were negative, and quaternary carbons did not appear because these carbons are not directly attached to hydrogen. Another DEPT experiment was conducted using DEPT-90. DEPT-135 shows all the resonances of the protonated carbons, whereas DEPT-90 shows only the

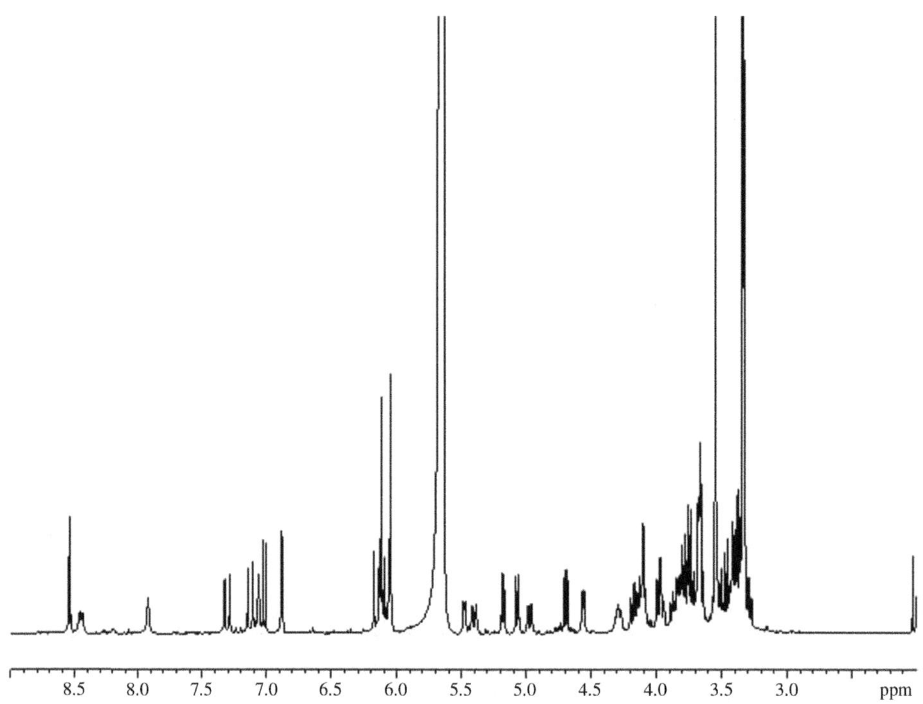

FIGURE 13.2
^{1}H-NMR spectrum of alatanin E (400 MHz, CD$_3$OD/CF$_3$COOD (9:1)).

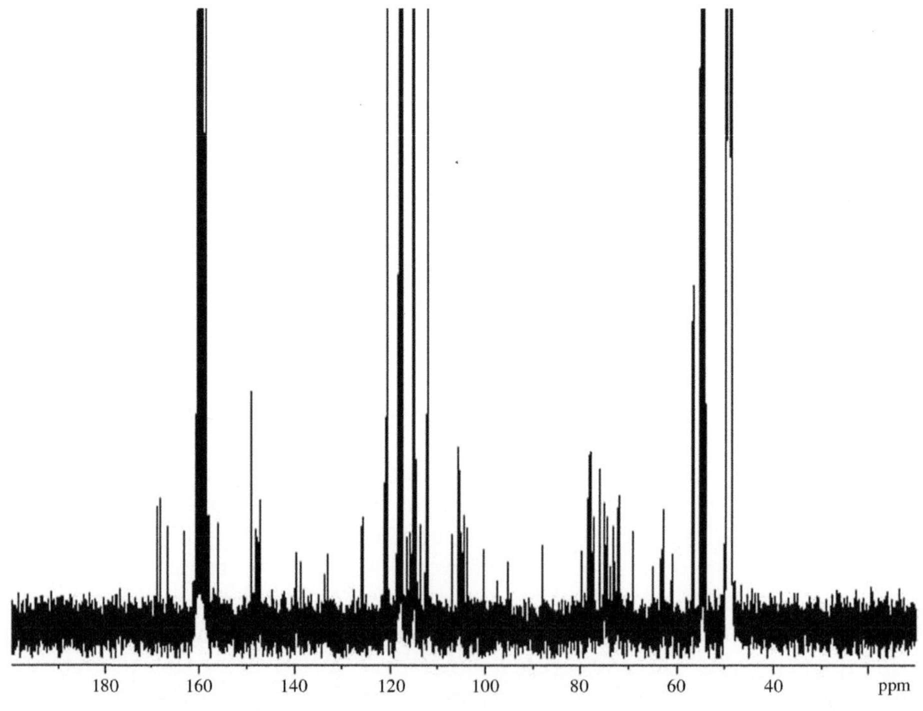

FIGURE 13.3
^{13}C-NMR spectrum of alatanin E (100 MHz, CD$_3$OD/CF$_3$COOD (9:1)).

CH peaks (positive). Recently, however, edited-heteronuclear single quantum correlation (edited-HSQC), 2D correlation spectroscopy of ^{1}H and ^{13}C combined with DEPT has often been used for anthocyanin analyses [2]. This method saves measurement time because there is no need to run a ^{13}C DEPT spectrum.

The seven, large splitting signals observed at δ_C 49 are derived from the carbon of CD_3 (CD_3OD). On the other hand, the four splitting signals at δ_C 112, 114, 116, and 118 are derived from the carbon of CF_3 (CF_3COOD), which is coupled with three F. The signal at δ_C 158 is derived from the carbonyl carbon of CF_3COOD, and this signal is also split into to four due to coupling with the three F. The ^{13}C-NMR spectrum was elucidated, except for the solvent signals.

The signals observed at 60–80 ppm and ~105 ppm are derived from 2 to 6-positions and 1-position of the sugars, respectively. The signals of the carbons at the 6- and 8-positions of the anthocyanidin were observed at δ_C 95 and 104, respectively. Although the ^{13}C signals derived from the benzene ring are usually observed at approximately 110–160 ppm, they are shifted according to the type of substituent. The signals at 165–170 ppm were derived from the carbonyl carbons. The ^{13}C signals cannot be definitively assigned for structural determination; however, it is insufficient to analyze 1D NMR spectra alone. Therefore, spectral elucidation using various 2D NMR techniques is required.

13.2.3 ^{1}H-^{1}H COSY

^{1}H-^{1}H correlation spectroscopy (COSY) is a homonuclear correlation NMR spectroscopy technique that enables determination of the proton-proton connectivity. Using this technique, we can identify protons that are directly coupled to each other within a molecule. Figure 13.4 shows the ^{1}H-^{1}H COSY spectrum of alatanin E.

In the ^{1}H-^{1}H COSY analysis, a 1D spectrum is displayed along each axis, with a contour projection of this spectrum along the diagonal axis. The off-diagonal peaks represent proton shift correlations (or proton couplings). The ^{1}H-^{1}H COSY correlations enabled assignment of the structural moieties. For example, a cross-peak was observed between the signals at δ_H 6.11 and δ_H 7.30, derived from the correlation between the α and β protons of sinapic acid. In some cases, it is difficult to elucidate cross-peaks in the ^{1}H-^{1}H COSY spectrum of compounds with sugar moieties, such as anthocyanins, due to the overlap of many peaks in the range of 3–4.5 ppm. Thus, double quantum filtered (DQF) ^{1}H-^{1}H COSY in phase-sensitive mode, not in magnitude mode, is useful for structure elucidation.

13.2.4 TOCSY

TOCSY (total correlation spectroscopy), previously called HOHAHA (homonuclear Hartmann-Hahn spectroscopy), provides correlations between all protons within a given spin system, and not just between geminal or vicinal protons. Correlations are usually observed between distant protons as long as there are couplings between protons. This information is extremely useful for identifying the protons on sugar rings. Figure 13.5 shows the 2D TOCSY spectrum of alatanin E.

In the TOCSY spectrum of anthocyanins, it is easy to distinguish the sugar signals because the cross-peaks from the 1- to 6-positions of the sugar appear in a straight line. 1D TOCSY, similar to 2D TOCSY, provides correlations between all the protons within a given spin system. For the 1D TOCSY analysis, one ^{1}H-NMR signal is selected (irradiated), and the signal is transferred to all coupled protons in a stepwise process. Instead of

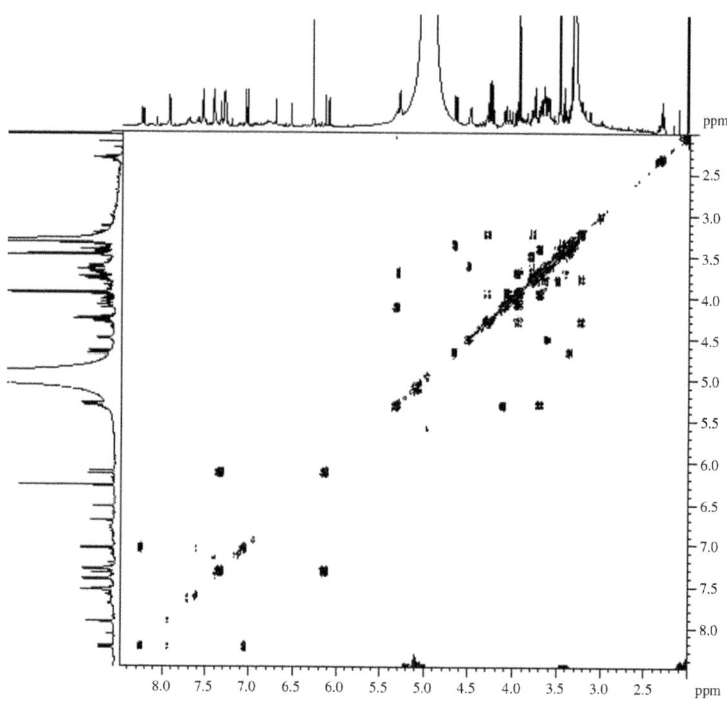

FIGURE 13.4
^{1}H-^{1}H COSY spectrum of alatanin E (400 MHz, CD_3OD/CF_3COOD (9:1)).

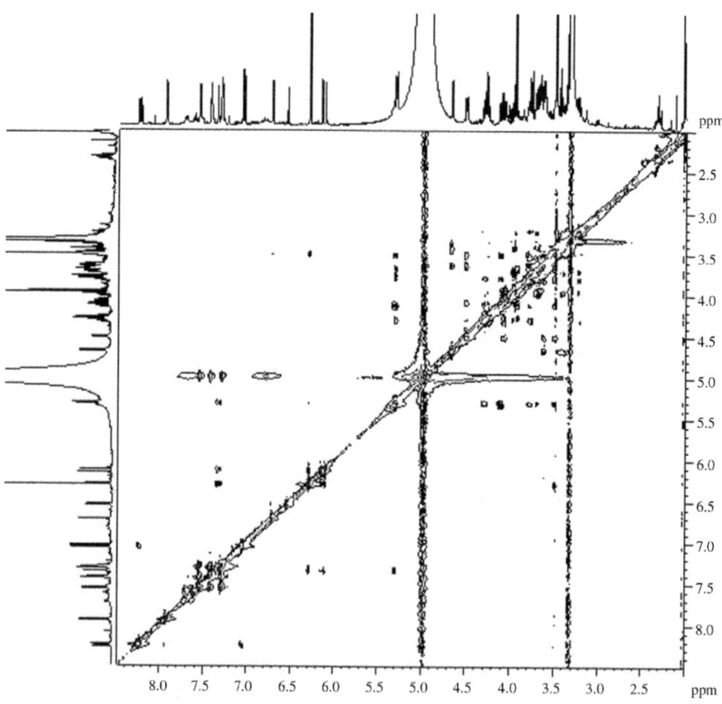

FIGURE 13.5
2D-TOCSY spectrum of alatanin E (400 MHz, CD_3OD/CF_3COOD (9:1)).

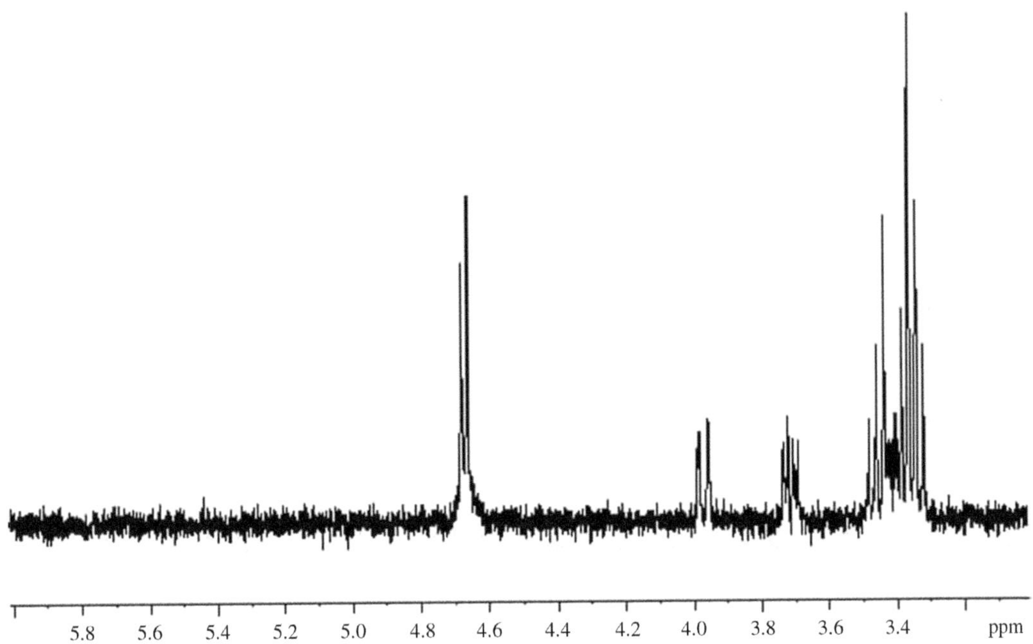

FIGURE 13.6
1D-TOCSY spectrum of alatanin E (400 MHz, CD$_3$OD/CF$_3$COOD (9:1)).

cross-peaks, the magnetization transfer is observed as an increase in the signal intensity. Figure 13.6 shows the 1D TOCSY spectrum irradiated δ_H 4.67, i.e., the 1-position of the sugar. One-dimensional (1D) TOCSY is useful as it can be used to give sub-spectra of only protons coupled to the chosen irradiated atom.

13.2.5 HSQC and HMQC

Heteronuclear single quantum correlation spectroscopy (HSQC) or heteronuclear multiple quantum correlation (HMQC) is generally used to determine which hydrogen in a molecule is directly bonded to which carbon. Figure 13.7 shows the HSQC spectrum of alatanin E.

To analyze the spectrum, each carbon signal was numbered and correlated with hydrogen using cross-peaks. However, because quaternary carbons do not have attached hydrogen, it is difficult to elucidate their detailed bonding information. Therefore, ^{1}H-^{13}C long-range correlation spectroscopy measurements are usually performed. By measuring the edited-HSQC with the functionality of DEPT-135, it is possible to simultaneously obtain information on the types of carbon and the direct bond between the hydrogen and carbon atoms.

13.2.6 HMBC

Heteronuclear multiple bond correlation (HMBC) is typically used to obtain information on ^{1}H-^{13}C long-range correlations. $^2J_{HC}$ and $^3J_{HC}$ can be determined from the HMBC spectra. Figure 13.8 shows the HMBC spectra of alatanin E.

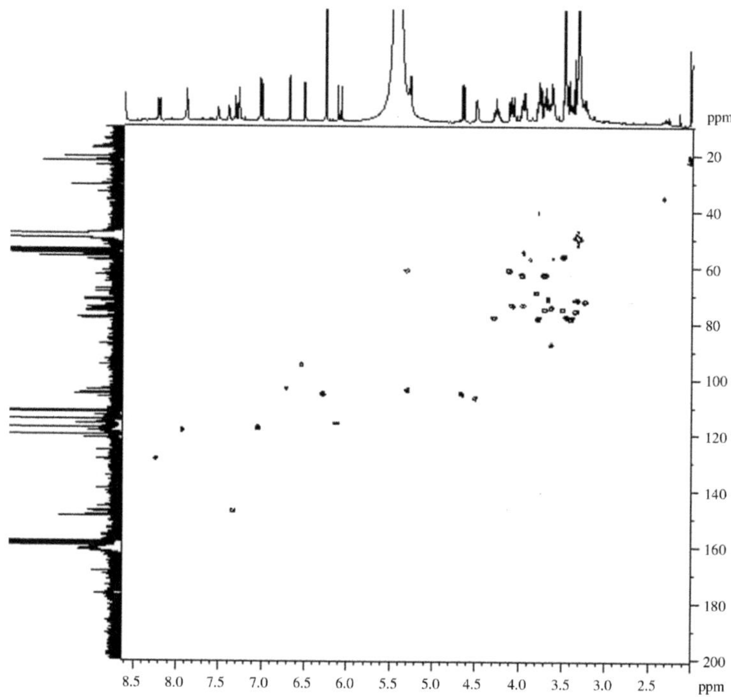

FIGURE 13.7
HSQC spectrum of alatanin E (400 MHz, CD_3OD/CF_3COOD (9:1)).

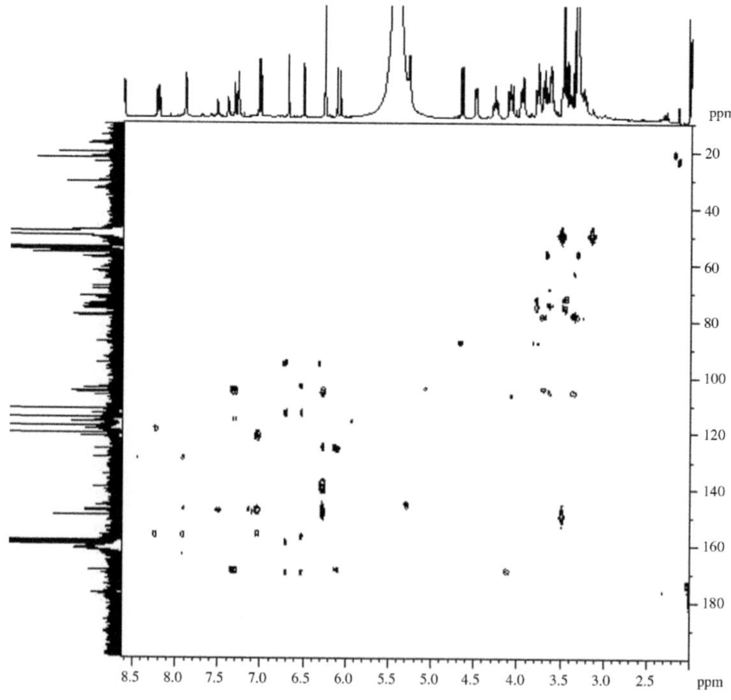

FIGURE 13.8
HMBC spectrum of alatanin E (400 MHz, CD_3OD/CF_3COOD (9:1)).

To analyze the HMBC spectrum, the relationship between the hydrogen and carbon atoms, which are two or three bonds apart, was clarified with reference to the $^1J_{HC}$ correlation information obtained from the HSQC or HMQC spectra. However, in the presence of a high-intensity ^1H signal, it is necessary to distinguish the peaks from the long-range couplings of ^1H-^{13}C, as the satellite peaks of $^1J_{HC}$ may be observed.

13.2.7 NOESY and ROESY

The nuclear Overhauser effect (NOE) is a phenomenon involving energy transfer between spatially close protons (within about 6 Å), which causes an increase or decrease in the ^1H-NMR signals. The 2D NOE spectrum was obtained using NOESY (NOE spectroscopy). The NOE intensity is related to the molecular weight of the compound and the magnetic field intensity of the NMR instrument. Therefore, rotating-frame nuclear Overhauser effect spectroscopy (ROESY) is used when NOE is difficult to observe. Figure 13.9 shows the ROESY spectrum of alatanin E.

NOESY and ROESY are often used to determine the steric structures of compounds. These techniques are used to determine the type of sugar or position of the bond in anthocyanins. In some cases, the structures of anthocyanins can be determined without NOE information.

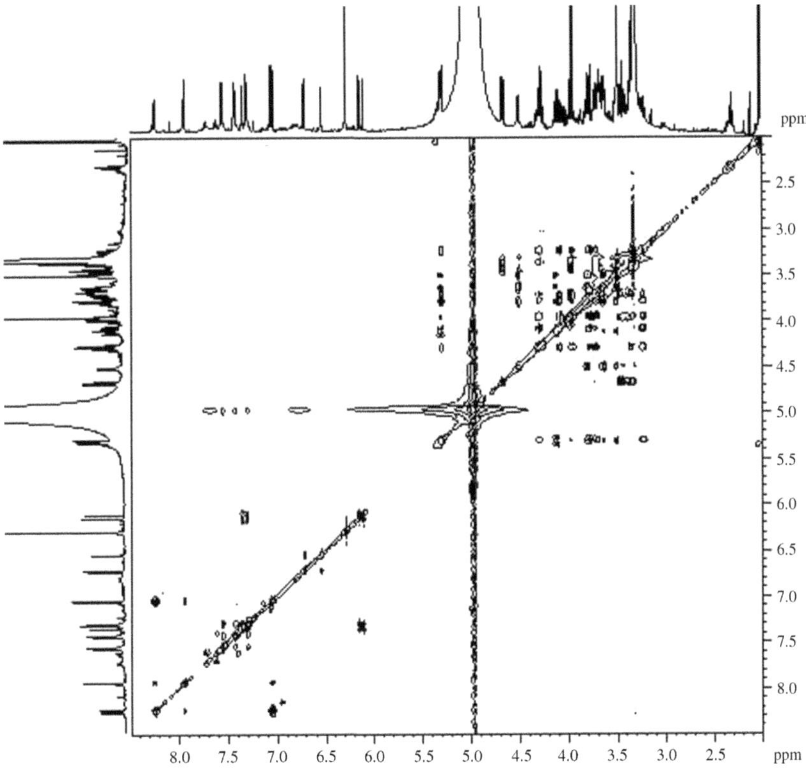

FIGURE 13.9
ROESY spectrum of alatanin E (400 MHz, CD_3OD/CF_3COOD (9:1)).

13.3 Summary for the NMR Analysis of Anthocyanins

Here various NMR methods for the structural determination of anthocyanins are described. Details of the NMR methodologies can be found in various reviews on the structural analysis of natural products [2–4]. The purity and quantity of the samples are important for obtaining good NMR data, not only for anthocyanins. Therefore, the adequate chromatographic techniques for the isolation of anthocyanins need to be developed. The acquisition of good NMR data requires understanding the meaning of pulse sequencing and parameter settings for each NMR method and the execution of NMR measurements under appropriate conditions.

Recently, quantitative analysis of anthocyanins using NMR has become commonplace [5, 6]. Metabolomic studies on anthocyanins using NMR have also been conducted [7, 8]. With advances in hardware and software, NMR continues to be an indispensable instrument for anthocyanin analysis.

References

1. Moriya, C., Hosoya, T., Agawa, S., Sugiyama, Y., Kozone, I., Shin-ya, K., Terahara, N., and Kumazawa, S. (2015). New acylated anthocyanins from purple yam and their antioxidant activity. *Biosci. Biotechnol. Biochem.*, **79**, 1484–1492.
2. Breton, R.C., and Reynolds, W.F. (2013). Using NMR to identify and characterize natural products. *Nat. Prod. Rep.*, **30**, 501–524.
3. Halabalaki, M., Vougogiannopoulou, K., Mikros, E., and Skaltsounis, A.L. (2014). Recent advances and new strategies in the NMR-based identification of natural products. *Curr. Opin. Biotechnol.*, **25**, 1–7.
4. Elyshberg, M. (2015). Identification and structure elucidation by NMR spectroscopy. *Trends Analyt. Chem.*, **69**, 88–97.
5. Teng, Z., Jiang, X., He, F., and Bai, W. (2020). Qualitative and quantitative methods to evaluate anthocyanins. *eFood*, **1**, 339–346.
6. Constantin, O.E., and Istrati, D.I. (2022). Extraction, quantification and characterization techniques for anthocyanin compounds in various food matrices – A review. *Horticulturae*, **8**, 1084.
7. Consonni, R. (2018). The potentiality of NMR-based metabolomics in food science and food authentication assessment. *Magn. Reson. Chem.*, **57**, 558–578.
8. Avula, B., Katragunta, K., Osman, A.G., Ali, Z., John Adams, S., Chittiboyina, A.G., and Khan, I.A. (2023). Advances in the chemistry, analysis and adulteration of anthocyanin rich-berries and fruits: 2000–2022. *Molecules*, **28**, 560.

14

Spectroscopic Methods in Analysis of Anthocyanins

Özlem ÇAĞINDI and Emine ERDAĞ AKCA

14.1 Introduction

Anthocyanins are a subclass of flavonoids that belong to phenolic compounds. They are the most widely existing natural food pigment with a huge scale of shiny orange, pink, red, violet and blue colors besides chlorophyll. The growing interest in anthocyanins is caused by not only their unique red-blue color but also their health benefit. There are more than 700 species of anthocyanin class, but approximately 90% of naturally found anthocyanins are represented by six members (cyanidin, pelargonidin, delphinidin, peonidin, petunidin, and malvidin) of this group. Several studies focused on qualitative and quantitative identification of anthocyanins, and it could be characterized by various chromatographic and spectroscopic analyses techniques. This analysis generally included pigment extraction followed by purification and application of spectroscopic methods. Traditional extraction methods, that have some disadvantages such as requiring more organic solvents and extraction times, lower extraction yields and selectivity, and thermal degradation, could be used for extracting pigments. However, new techniques getting more attractive to overcome the limitations of conventional extraction techniques and the most applied non-conventional methods such as ultrasound-assisted extraction (UAE), microwave-assisted extraction (MAE), supercritical fluid extraction (SFE), pressurized liquid extraction (PLE), pulsed electric fields (PEFE), enzyme-assisted extraction (EAE) have been explained in this research.

The science of spectroscopy has been continuously developed by scientists since the demonstration of the light spectrum by Isaac Newton. It could explain that the science of spectroscopy is analyze matter with the help of electromagnetic radiation. For food analysis, spectrometric methods have been widely used detection of geographic origin, identification of composition, authentication and quality of elements, and effect of processing and storage. Ultraviolet-visible (UV-Vis) spectroscopy is one of the most frequently used techniques for the analysis of compounds and is based on the measurement of the wavelength and intensity of ultraviolet and visible light absorbed by the matter. However, non-destructive spectroscopic techniques like Near-Infrared (NIR) Spectroscopy could be a solution for rapid and *in situ* measurements. Raman spectroscopy (RS) also could be a good alternative especially for reliable, fast, specific and sensitive measurements, that enable the result interaction of light with chemical bonds within a material for only Raman active substances. Nuclear Magnetic Resonance (NMR) Spectroscopy is another non-destructive method for anthocyanin analyses, although further development is needed to improve detection limits. Fluorescence Spectroscopy has been used for chemical fluorescence chemical fingerprinting that specifically identifies each compound. In short, there

DOI: 10.1201/9781003453260-16

is a wide variety of spectroscopic methods with advantages and disadvantages for the analysis of anthocyanins. This chapter presents current knowledge about anthocyanins and extraction methods and applications of spectroscopic analysis for the identification and characterization of anthocyanins.

14.2 Anthocyanins

Anthocyanins play an important role among water-soluble natural color pigments as they are widely distributed in nature, especially in roots, stems, tubers, leaves, fruits, and seeds. They are responsible for red-blue colors and their derivatives in the plants and these properties gain the attractiveness for fruit or vegetables (Mattioli et al., 2020; Enaru et al., 2021). It was reported that berries are the main source of anthocyanins, followed by red wine, vegetables, and other fruits. Especially elderberries, chokeberries, blueberries, pomegranates, and acai are rich in anthocyanins with their unique color. There are huge species of over 700 anthocyanins, but 6 of them represent about 90% of naturally found like cyanidin, pelargonidin, delphinidin, peonidin, petunidin, and malvidin (Wallace and Giusti, 2019; Gonçalves et al., 2021).

Academic interest in the research and/or use of anthocyanins is not only related to their coloring properties but also to their potential health benefits. Anthocyanins are considered health promoting due to their special chemical structure and fragrance. Moreover, their consumption has been associated with several health benefits, including the mitigation of cardiovascular disease, type 2 diabetes, non-alcoholic fatty liver disease, neurological disorder, obesity, and cancer (Cappellini et al., 2021; Oteiza et al., 2023). It has been highlighted by *in vitro* and *in vivo* studies that they have low toxicity and high safety (Liu et al., 2021; Tan et al., 2022).

Anthocyanins are a subclass of the flavonoids group of phenolic. The phenylpropanoid pathway produces flavonoid derivatives of anthocyanidins, which are known as anthocyanins. Although more than 700 subspecies of anthocyanins with 30 different core structures, the major 6 anthocyanidin class represent approximately 90% of all anthocyanins. The major anthocyanidin subclasses (anthocyanidin metabolites) are: *pelargonidin* (pelargonidin-3-glucoside, pelargonidin-3-rutinoside, pelargonidin-3-galactoside, pelargonidin-3-arabinoside), *cyanidin* (cyanidin-3-glucoside, cyanidin-3-rutinoside, cyanidin-3-galactoside, cyanidin-3-arabinoside), *delphinidin* (delphinidin-3-glucoside, delphinidin-3-rutinoside, delphinidin-3-galactoside, delphinidin-3-arabinoside), *peonidin* (peonidin-3-glucoside, peonidin-3-rutinoside, peonidin-3-galactoside, peonidin-3-arabinoside), *petunidin* (petunidin-3-glucoside, petunidin-3-rutinoside, petunidin-3-galactoside, petunidin-3-arabinoside), and *malvidin* (malvidin-3-glucoside, malvidin-3-rutinoside, malvidin-3-galactoside, malvidin-3-arabinoside). The chemical structure anthocyanidin is shown in Figure 14.1. The basic structure of the primary nuclei is based on a conjugated 2-phenylbenzopiran cation. The C6-C3-C6 skeleton is formed by bonding two benzene rings with three carbon atoms. According to the side chain differences in the carbon positions of the two benzene rings, subgroups of anthocyanidin are formed. Cyanidin R1=H, R2=OH, pelargonidin R1=H, R2=H, deltanidin R1=OH, R2=OH, peonidin R1=OCH$_3$, R2=H, petronidine R1=OH, R2=OCH$_3$, and malvidin R1=OCH$_3$, R2=OCH$_3$ are among the side chains mentioned (Houghton et al., 2021). The conversion of B rings to hydroxyl or methoxyl groups affects the maximum absorbance of anthocyanidins, so their color changes from orange (pelargonidin) to purple (malvidin) (Cai et al., 2022; Houghton et al., 2021).

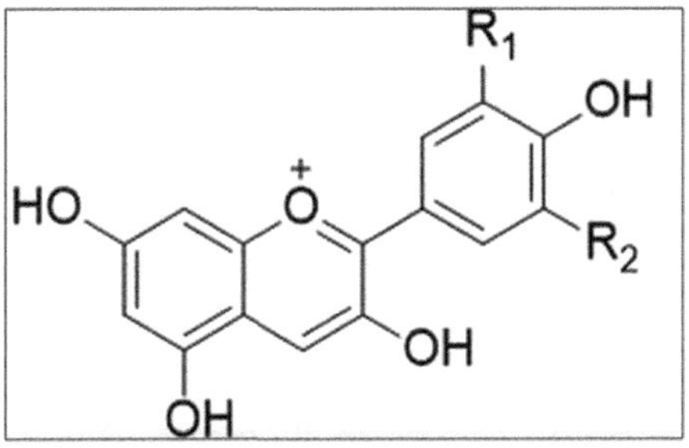

FIGURE 14.1
The chemical structure anthocyanidin.

Because of their susceptibility to degradation due to various factors of food processing and storage conditions, ensuring the chemical stability of anthocyanins has been a major challenge. Besides being unstable in their aqueous solution, they are quite vulnerable to environmental stresses such as relative humidity, oxygen, light, pH, temperature, sugars, ascorbic acid, sulfur dioxide or sulfites, enzymes, copigments, and metal ions. Table 14.1 represents the effects of major environment conditions on anthocyanin stability. Furthermore, the researchers have concluded that anthocyanins have less than 1% bioavailability due to their high sensitivity, and it is possible to observe an enhancement of bioavailability with the help of some ways like polymerization, cleavage, and derivatization (Yousuf et al., 2016; Enaru et al., 2021; Zang et al., 2022).

In the extraction of flavonoids, various techniques have been investigated to increase the extraction yields of major bioactive compounds. For this purpose, maceration, percolation, hydro-distillation, boiling, and Soxhlet are widely used. However, these traditional techniques show some disadvantages such as requiring large amounts of organic solvent,

TABLE 14.1

Effect of Some Environmental Conditions on Anthocyanin Stability

Environmental Condition	Effect on Anthocyanins
pH	Acidic conditions improve the anthocyanin stability pH < 2 Alkaline conditions adverse effects for anthocyanin stability pH > 8
Oxygen	Leading the opening-ring reaction for further anthocyanin degradation
Light	While plays an indispensable role in anthocyanin biosynthesis, it can also accelerate its degradation.
Temperature	High temperatures increase the rate of anthocyanins' degradation.
Ascorbic acid	Has an improved effect on anthocyanins stability and protects from enzymatic degradation.
Sugars	The high sugar content could help maintain stability due to limited water activity.
Copigmentation	It is known as interactions with other colorless organic compounds or metal ions and is often used to increase stability.
Enzymes	Active enzymes in the raw material could play a role in the degradation of anthocyanins. The most important of these are glycosidases, peroxidases (phenol oxidases) and phenolases (polyphenol oxidases), and these enzymes are commonly called anthocyanins.
Metal ions	Copigmentation can be created in metals such as copper (Cu), iron (Fe), magnesium (Mg), tin (Sn), and potassium (K), if there is no risk of metal contamination of the products and nontoxic for consumers. Such integration could be an effective tool for stability.

Source: Enaru et al. (2021); Cai et al. (2022).

low extraction efficiency, low selectivity, long extraction times, thermal degradation. Therefore, new extraction techniques have been developed to overcome such limitations of traditional methods (Chaves et al., 2020; Chemat et al., 2020). The most widely applied non-conventional techniques to extract anthocyanins are shown in Table 14.2.

There have been two types of anthocyanins including acylated and nonacylated forms. The acylated anthocyanins contain sugar that are bonded with an aliphatic/aromatic acid in their anthocyanidin structure. Additionally, it is known that more than 60% of naturally anthocyanins are the acylated form and these are more stable compared to nonacylated form. Many fruits and vegetables such as red cabbage, black carrots, red radishes, purple

TABLE 14.2

Some Non-Conventional Methods for Anthocyanin Extraction

Extraction Methods	Basic Principle	Advantage	Disadvantage	References
Ultrasound-assisted extraction (UAE)	The structure of the cell wall has been disrupted by the cavitation and strong shear forces created by ultrasound with 20–50 MHz frequency	Low cost Need for lower solvent Short time	The structure of anthocyanins could be disrupted by cavitation and mechanical effects.	Esclapez et al. (2011), Tan et al. (2022)
Microwave-assisted extraction (MAE)	Electromagnetic radiation energy in the microwave region is absorbed by polar molecules. This absorption causes the movement of molecules and ion migration. This process evaporates the matrix water and creates high pressure in the cell wall. Thus, mass transfer of the target component occurs	Employment of a wider range of solvents	Local excessive vibration and high temperature could lead to structure degradation	Farooq et al. (2020), Tena and Asuero (2022), Tan et al. (2022)
Supercritical fluid extraction (SFE)	The supercritical fluid was applied above its critical pressure and temperature, thus gaining gas-like diffusivity and liquid-like viscosity. This provides more easier penetration into solid matrices easier to penetration. Supercritical CO_2 has been commonly used due to its nontoxicity, cheapness, and protection of the extracts from atmospheric oxidation.	Sustainable green technologies High efficiency Absence of light and oxygen Low temperature	High cost of equipment Requiring more technical investment	Cheah et al. (2010), Geow et al. (2021), Tena and Asuero (2022)

(Continued)

TABLE 14.2 *(Continued)*

Some Non-Conventional Methods for Anthocyanin Extraction

Extraction Methods	Basic Principle	Advantage	Disadvantage	References
Pressurized liquid extraction (PLE)	It is based on the use of liquid solvent at temperatures 50–20°C and 3.5–20 MPa. The solvent maintains the liquid form even at temperatures above the boiling point with pressure and diffuses into the tissues through pores.	Sustainable green technologies Reduce cost and time (5–20 min)	High costs for instrumentation and energy requirements	Andrade et al. (2021), Perez-Vazquez et al. (2023), Višnjevec et al. (2024)
Pulsed electric fields (PEFE)	The electric fields have been applied to break down the cell wall by electroporate.	Classified as a non-thermal extraction Less solvent and time requirement	High cost of the equipment	Lončarić et al. (2020), Tena and Asuero (2022), Choton et al. (2023)
Enzyme-assisted extraction (EAE)	It relies on enzyme power that breaks down the cell wall and releases targeted compounds. cellulase, α-amylase, β-glucosidase, protease, xylanase, or pectinase have often been used for the plant cell.	Usually not necessary for chemical solvent Lower time and temperature	Expensive cost of enzymes Required special condition for every enzyme activity	González et al. (2022), Tena and Asuero (2022), Amulya and ul Islam (2023)

potatoes, or purple corn contain acylated anthocyanins. It is called nonacylated anthocyanins when they do not contain bonded sugar (Saha et al., 2020; Dini et al., 2020).

14.3 Analysis of Anthocyanins by Spectroscopic Methods

Anthocyanins can be characterized using different chromatographic and spectroscopic analyses techniques. It is noteworthy that the most widely used method for the analysis of anthocyanins (especially color of blue and purple samples) was conducted with beginning of the pigment extraction, usually with weakly acidified alcoholic solvents, followed by purification and application of spectroscopic techniques (Bruni et al., 2023).

The science of spectroscopy has been continuously developed by scientists since the demonstration of the spectrum of light by Isaac Newton. This was followed by the discovery by William Wollaston (in 1802) and Joseph Fraunhofer (in 1814) that some dark lines in the solar spectrum were recognized as the basis for spectroscopic analysis. Another milestone was the invention of a basic spectroscope for the systematic study of spectra by Robert Bunsen and Gustav Kirchhoff in 1859–1860. This device works on the principle that light from a source is projected onto a screen by passing it through a sample and a prism, respectively. Bunsen and Kirchhoff studied various samples in the spectroscope

TABLE 14.3

The Main Features of Spectroscopic Methods

Methods	Wavelength Limits	Light Source	Signal Source
UV-Vis	190–750 nm	Deuterium/tungsten halogen	Absorption
NIR	750–2500 nm	Tungsten	Absorption
Raman	750–1064 nm	Laser	Scattering
NMR	1–1000 m		
Fluorescence	250–750 nm	Laser	Emission

and observed that each element exhibited a characteristic series of wavelengths absorbed by light. Thus, these scientists discovered spectroscopy as a method to find elements in samples (Ball, 2001; Nelson, 2017).

Spectroscopy is basically the analysis of matter using electromagnetic radiation (Ball, 2001). Historically, spectroscopic techniques have been seen as attractive and promising for the fields of chemistry, pharmaceuticals, agri-food, and environmental analysis because they require minimal sample preparation steps, rapid and online analysis, and potentially multiple tests on a single sample (Baeten and Dardenne, 2002; Nawrocka and Lamorska, 2013). Spectrometric methods for food analysis are widely used for geographical origin determination, composition identification, elemental authentication and quality, and determination of the effect of processing and storage (Kharbach et al., 2023).

Spectroscopic methods are considered an effective way for both quantitative and qualitative identification. Multiple spectroscopic methods are available for solving a variety of analytical problems and categorized as the species to be analyzed (molecular or atomic spectroscopy), the type of radiation-matter interaction to be monitored (absorption, emission, or diffraction), and the region of the electromagnetic spectrum (Penner, 2017). The main features of spectroscopic methods are shown in Table 14.3 (Hassoun et al., 2020; Tombuloğlu et al., 2023).

14.3.1 Ultraviolet-visible (UV-Vis) Spectroscopy

UV-Vis spectroscopy is one of the most frequently preferred techniques for the analysis of compounds and based on measurement the wavelength and intensity of ultraviolet and visible light absorbed by the matter. For this purpose, it is applied that the UV radiation (ultraviolet) at a wavelength of 180–380 nm or visible (visible light) at a wavelength of 380–780 nm in these techniques (Pratiwi and Nandiyanto, 2022).

The Lambert-Beer law, which is linked to the concentration of absorbing species in the solution and the path length, forms the measurements with UV spectroscopy. Part of the light energy is absorbed by the solution having a certain thickness, thus weakening the intensity of the light. This absorbance is proportional to the thickness and is expressed by Equation (14.1) (Verma and Mishra, 2018; Guo et al., 2020).

$$A = \varepsilon \times b \times C \tag{14.1}$$

A: the absorbance, is the coefficient of molar exclusion (M^{-1} cm^{-1})

B: the thickness of the cuvette (cm)

C: the concentration (M)

The simplicity and use for many purposes, as well as time and cost savings, have made UV spectroscopy frequently preferred (Mazza et al., 2004; Verma and Mishra, 2018).

UV-Vis spectroscopic analysis of anthocyanins uses the molar absorptivity of the dominant anthocyanin. Anthocyanins generally tend to have typical absorption in the visible region, exhibiting maximum absorbance around the 510–550 nm wavelength regions. Depending on the structural nature, a peak can also be observed in the 310–340 nm range. This peak is absent in almost all nonacylated groups. It is also recommended to perform measurements at pH 2 to avoid interfering substances. A calibration graph constructed from the concentration versus standard anthocyanin absorbance curve can be used to determine the concentration of anthocyanin present in the sample (Mazza et al., 2004; Saha et al., 2020). Table 14.4 shows some of the studies performed with UV-Vis spectrometry to determine the total anthocyanin content.

TABLE 14.4

The Research that Conducted with UV-Vis Spectrometer to Total Anthocyanin Content Determination

Reference	Material	Solution of Extraction	Extraction Method	Conclusion
Agustin et al. (2021)	Red rice	0.1% HCl in methanol	Maceration	The highest total anthocyanin level among red rice varieties of 10.87 mg/g was found in Aek sibundong red rice.
Taghavi et al. (2022)	Fresh, frozen, and lyophilized strawberry	1. chloroform-methanol (2: 1 v/v, acidified with 0.1% HCl) 2. methanol (acidified with 0.1% HCl) 3. methanol:water:concentrated HCl (80:20:1) 4. chloroform-methanol (3:4 v/v) 5. pH differential method 6. 80:20:1 (methanol:water:HCl)	Extraction of six method	The 2, 3, and 4. The method showed the highest total anthocyanin content. The researchers stated that water caused the chloroform phase to separate during the centrifugation and anthocyanins stayed in the methanol:water phase. Due to lower anthocyanin yield and longer processing time, method 6 was not found suitable for the extraction of anthocyanins. The freeze-dried samples exhibited significantly higher amounts of anthocyanin.
Johnson et al. (2020)	Six plum varieties (*Prunus domestica*)	Methanolic extraction (90% v/v aqueous methanol) Ethanolic extraction (25 ml of 50% v/v aqueous ethanol)	Extraction	For the four varieties of plum, the anthocyanin content of the methanolic extract was found higher than the ethanolic extracts.

14.3.2 Near-Infrared (NIR) Spectroscopy

Near-infrared (NIR) spectroscopy is performed in the wavelength region 780–2500 nm, and the absorption bands correspond to the vibrations of chemical bonds such as O–H, C–H, C–O, and N–H in the structure. These molecular bonds undergo vibrational energy changes and as a result, energy absorption occurs in the NIR region of organic molecules when the molecules vibrate. However, in food analysis, a calibration model of NIR spectroscopy is required to confirm less precise reference analyses, which usually means more time. Furthermore, the method is often used in the analysis of major components; the sensitivity is reduced for minor components and some quality attributes (Osborne, 2006; Ertugay and Başlar, 2011; Le, 2020).

While NIR spectroscopy has some key advantages, such as minimal sample preparation requirements, suitability for interrogating moist samples, potential for accurate quantitative analysis, its simplicity and rapid (between 15 and 90 seconds) analysis, and the fact that it can be implemented online are the main reasons that make this technology stand out (Osborne, 2006; Beć et al., 2020).

NIR spectroscopy has been successfully executed in the assay of anthocyanins with the advantage of rapid, in situ (with portable NIR), and non-destruction (Gales et al., 2021). Some examples of studies conducted with this technology are shown in Table 14.5.

14.3.3 Raman Spectroscopy

RS is based on determining the interactions created by photons sent from outside in the molecule. In other words, it produces results with the help of the interaction of light with chemical bonds within a material. Raman spectrometer, which uses the visible or near-infrared monochromatic laser as a light source, provides measurements in the wavelength range of 750–164 nm (Şahin, 2021; Tombuloğlu et al., 2023).

When a light-induced photon encounters a molecule, it produces a scattered photon, which in turn causes inelastic light scattering from the sample. This energy shift of the outgoing photon is measured. The change in wavelength of the scattered light depends on the chemical composition of the molecules responsible for scattering. Two types of scattering can be mentioned here. The first is Stokes Raman scattering, which occurs when the scattered light is a photon with a lower frequency than the original photon. If the scattered light has a higher frequency than the original photon, this is known as anti-Stokes Raman scattering (Das and Agrawal, 2011; Rostron et al., 2016). Each peak obtained in the Raman spectrum corresponds to the organic compound or functional group in the sample. The spectral pattern is called a chemical fingerprint, which allows identification of the compound, while the intensity of the bands can be used to calculate its concentration in the analyzed sample (Yaseen et al., 2017; Saletnik et al., 2021).

The researchers emphasized that RS is a reliable, fast, specific, and sensitive identification method, can be easily and quickly set up, and is a portable analyzing system (Petersen et al., 2021; Qu et al., 2022). Moreover, it has a high sensitivity to even small structural changes, which enable comparative tests (Saletnik et al., 2021).

It should be noted that water can mask other characteristic peaks due to its strong infrared absorption properties. This makes it difficult to use infrared spectroscopy on samples/foods with high moisture content. In contrast, RS only enables the detection of Raman active material and water has no characteristic Raman shift. Therefore, RS has a unique ability to provide chemical and molecular information (Duan et al., 2024). The researcher used smartphone-based RS to measure anthocyanins in grapes, grape

TABLE 14.5

NIR Application to Determination of Anthocyanin Content

Non-Destructive Methods

Reference	Material	Equipment Property	Analyze	Measurement	Conclusion
Amanah et al. (2020)	Black soybean	FT-NIR: indium–gallium–arsenide (InGaAs) detector, region of 10,000–4000 cm^{-1} at 4 cm^{-1} intervals FT-IR: deuterated triglycine sulfate (DTGS) detector, and KBr beam splitter, region of 4000–650 cm^{-1}	Total anthocyanin content and the content of two major individual anthocyanins	Fourier Transform Near-Infrared (FT-NIR) and Fourier Transform Infrared (FT-IR)	The partial least squares regression (PLSR) models for the FT-NIR spectra indicated slightly better R^2 and standard error of prediction (SEP) for the chemical components, compared to FT-IR. The data showed that both techniques could be practical tools for rapid and non-destructive measurement of anthocyanin content in a single seed.
Stuppner et al. (2020)	Wild elderberry (*Sambucus fructus*)	InGaAs detector Spectra region of 10,000–4000 cm^{-1}	Total anthocyanin content	NIR	The total anthocyanin content ranged between 632.87 and 4342.01 mg/kg in *Sambucus nigra*. It was found that NIR spectroscopy is a fast, noninvasive, and cost-efficient alternative for screening TAC compared to pH-differential and UHPLC-MWD-UHR-TOF-MS methods.
Rouxinol et al. (2022)	Red wine grapes	InGaAs detector wavelength range of 1100–2300 nm	Total anthocyanin content	NIR	The total anthocyanin content of four grape variety were ranged between 0.18 and 4.25 mg/mL. The researchers demonstrated the ability of NIR spectroscopy, in combination with a portable device, to quantify entire grapes.

skins, and grape juice in situ, quickly and without destruction. They observed characteristic peaks at 540, 630, 730, 1238, 1334, 1470, 1530, and 1608 cm in the Raman spectra of standard samples by the Raman activity of anthocyanin. Moreover, the intensities of these characteristic peaks were positively correlated with the concentration of anthocyanin solutions. These results showed that rapid determination of anthocyanin content in grape skins and grape juice by Raman technology is possible and eliminates the need for sample pretreatment (Gao et al., 2024).

14.3.4 Nuclear Magnetic Resonance (NMR) Spectroscopy

Nuclear Magnetic Resonance spectroscopy benefits from the interaction between the magnetic properties of atoms and compounds within a sample and the applied magnetic field. When the molecule is exposed to a magnetic field, rotation is observed in the nuclei with magnetic momentum, which are excited (Hassoun et al., 2020; Mishra, 2022).

Removal of the applied magnetic field triggers the atoms to back to their original state into the lower energy levels and this phenomenon causes the releasing of energy at frequency. Thus, the NMR spectrum of that nucleus is obtained (Sahoo et al., 2020). Electromagnetic pulse trains excite and change the spin direction of NMR active nuclei which is commonly used for this purpose, especially ^{31}P, ^{15}N, ^{1}H, ^{17}O, and ^{13}C. When the pulses are turned off, the nuclei return to their base spin state. It has been specific for the molecule and its environment that the time required for return to the fundamental state (Hassoun et al., 2020; Li et al., 2021).

Researchers have emphasized that NMR can be useful for identifying compounds with the same mass and determining the structures of unknown compounds (Sahoo et al., 2020). Compared to mainstream analytical methods, NMR offers strong penetration capability for heterogeneous and complex food systems. It is considered useful for mixture analysis due to the simultaneous determination of multiple analytes (Cao et al., 2021). NMR technology allows working on solid or liquid matrices without altering the sample and generating hazardous waste. Despite its advantages and wide range of applications, NMR spectroscopy is reported by researchers to be far from the desired sensitivity. Furthermore, further development is needed to improve detection limits (Marcone et al., 2013; Emwas et al., 2019).

The researchers have stated that both NMR and UV spectroscopies demonstrate more successful results, especially in anthocyanins analysis of acidic solutions. This is mainly due to the measurement being carried out only in the presence of the flavylium ion, avoiding to copresence of various colored or colorless compounds that occur at higher pH values (Bruni et al., 2023).

14.3.5 Fluorescence Spectroscopy

Fluorescence is characterized as the absorption of light energy at one wavelength and almost instantaneous re-absorption at another wavelength, usually longer than the one absorbed. Fluorescence spectra enable the identification of the compound, which is unique to each compound and is called a fluorescence signature or fingerprint. Fluorescence components may exhibit characteristic spectra that reflect the wavelength and amount of light absorbed (excitation spectrum) and reflect the wavelength and amount of light emitted (emission spectrum) (Naresh, 2014).

Compounds that absorb visible and/or UV light are called chromophores, while compounds that emit light are called fluorophores (Zacharioudaki et al., 2022). In other words, fluorophores have the feature of releasing the light they absorb at a certain wavelength (UV or visible region) at a higher wavelength (Fan and Su, 2022). Due to the high sensitivity of fluorophores to environment, this analysis quite sensitive to other spectroscopic methods (Shaikh and O'Donnell, 2017).

Fluorescence has become an irreplaceable analytical tool with fast and reliable detection properties. Being non-destructive, sensitivity and cost effective are the other reasons for popular use in many disciplines. Despite there may be opportunity, it has been stated that fluorescence spectroscopy lack of major compound analyses due to too low accuracy for large compounds. At the same time, the fact that not all elements and compounds have fluorescence properties also limits their usage area (Bose et al., 2018; Sultangaziyev and Bukasov, 2020; Dos Santos et al., 2022).

References

Agustin, A. T., Safitri, A., & Fatchiyah, F. (2021). Java red rice (*Oryza sativa* L.) nutritional value and anthocyanin profiles and its potential role as antioxidant and anti-diabetic. *Indonesian Journal of Chemistry, 21*(4), 968–978.

Amanah, H. Z., Joshi, R., Masithoh, R. E., Choung, M. G., Kim, K. H., Kim, G., & Cho, B. K. (2020). Nondestructive measurement of anthocyanin in intact soybean seed using Fourier transform near-infrared (FT-NIR) and Fourier transform infrared (FT-IR) spectroscopy. *Infrared Physics & Technology, 111*, 103477.

Amulya, P. R., & ul Islam, R. (2023). Optimization of enzyme-assisted extraction of anthocyanins from eggplant (*Solanum melongena* L.) peel. *Food Chemistry: X, 18*, 100643.

Andrade, T. A., Hamerski, F., Fetzer, D. E. L., Roda-Serrat, M. C., Corazza, M. L., Norddahl, B., & Errico, M. (2021). Ultrasound-assisted pressurized liquid extraction of anthocyanins from *Aronia melanocarpa* pomace. *Separation and Purification Technology, 276*, 119290.

Baeten, V., & Dardenne, P. (2002). Spectroscopy: Developments in instrumentation and analysis. *Grasas y aceites, 53*(1), 45–63.

Ball, D. W. (2001). *The basics of spectroscopy* (Vol. 49). Spie press.

Beć, K. B., Grabska, J., & Huck, C. W. (2020). Near-infrared spectroscopy in bio-applications. *Molecules, 25*(12), 2948.

Bose, A., Thomas, I., & Abraham, E. (2018). Fluorescence spectroscopy and its applications: A review. *International Journal of Advanced Pharmaceutical Sciences, 8*(1), 1–8.

Bruni, S., Longoni, M., Minzoni, C., Basili, M., Zocca, I., Pieraccini, S., & Sironi, M. (2023). Resonance Raman and visible micro-spectroscopy for the in-vivo and in-vitro characterization of anthocyanin-based pigments in blue and violet flowers: A comparison with HPLC-ESI-MS analysis of the extracts. *Molecules, 28*(4), 1709.

Cai, D., Li, X., Chen, J., Jiang, X., Ma, X., Sun, J., …, & Bai, W. (2022). A comprehensive review on innovative and advanced stabilization approaches of anthocyanin by modifying structure and controlling environmental factors. *Food Chemistry, 366*, 130611.

Cao, R., Liu, X., Liu, Y., Zhai, X., Cao, T., Wang, A., & Qiu, J. (2021). Applications of nuclear magnetic resonance spectroscopy to the evaluation of complex food constituents. *Food Chemistry, 342*, 128258.

Cappellini, F., Marinelli, A., Toccaceli, M., Tonelli, C., & Petroni, K. (2021). Anthocyanins: From mechanisms of regulation in plants to health benefits in foods. *Frontiers in Plant Science, 12*, 748049.

Chaves, J. O., De Souza, M. C., Da Silva, L. C., Lachos-Perez, D., Torres-Mayanga, P. C., Machado, A. P. D., & Rostagno, F. (2020). Extraction of flavonoids from natural sources using modern techniques. *Frontiers in Chemistry, 8*, 507887.

Cheah, E. L., Heng, P. W., and Chan, L. W. (2010). Optimization of supercritical fluid extraction and pressurized liquid extraction of active principles from *Magnolia officinalis* using the Taguchi design. *Separation and Purification Technology, 71*(3), 293–301.

Chemat, F., Vian, M. A., Fabiano-Tixier, A. S., Nutrizio, M., Jambrak, A. R., Munekata, P. E., …, & Cravotto, G. (2020). A review of sustainable and intensified techniques for extraction of food and natural products. *Green Chemistry, 22*(8), 2325–2353.

Choton, S., Bandral, J. D., Sood, M., Gupta, N., Singh, J., Langeh, A., & Choudhary, A. N. (2023). Green extraction technology and its application in food industry. *Chemical Science Review and Letters, 12* (46), 79–88. DOI:10.37273/chesci.cs205402576.

Das, R. S., & Agrawal, Y. K. (2011). Raman spectroscopy: Recent advancements, techniques and applications. *Vibrational Spectroscopy, 57*(2), 163–176.

Dini, C., Zaro, M. J., Rolny, N., Caputo, M., Boido, E., Dellacassa, E., & Viña, S. Z. (2020). Characterization and stability analysis of anthocyanins from *Pachyrhizus ahipa* (Wedd) Parodi roots. *Food Bioscience, 34*, 100534.

Dos Santos, I., Bosman, G., Aleixandre-Tudo, J. L., & du Toit, W. (2022). Direct quantification of red wine phenolics using fluorescence spectroscopy with chemometrics. *Talanta, 236*, 122857.

Duan, C., Xiao, X., Yu, Y., Xu, M., Zhang, Y., Liu, X., … & Wang, J. (2024). In situ Raman characterization of the stability of blueberry anthocyanins in aqueous solutions under perturbations in temperature, UV, pH. *Food Chemistry, 431*, 137155.

Emwas, A. H., Roy, R., McKay, R. T., Tenori, L., Saccenti, E., Gowda, G. N., & Wishart, D. S. (2019). NMR spectroscopy for metabolomics research. *Metabolites, 9*(7), 123.

Enaru, B., Drețcanu, G., Pop, T. D., Stănilă, A., & Diaconeasa, Z. (2021). Anthocyanins: Factors affecting their stability and degradation. *Antioxidants, 10*(12), 1967.

Ertugay, M. F., & Başlar, M. (2011). Gıdaların kalite özelliklerinin belirlenmesinde yakın kızılötesi (NIR) spektroskopisi. *Gıda, 36*(1), 49–54.

Esclapez, M. D., García-Pérez, J. V., Mulet, A., & Cárcel, J. A. (2011). Ultrasound-assisted extraction of natural products. *Food Engineering Reviews, 3*, 108–120.

Fan, K. J., & Su, W. H. (2022). Applications of fluorescence spectroscopy, RGB-and multispectral imaging for quality determinations of white meat: A review. *Biosensors, 12*(2), 76.

Farooq, S., Shah, M. A., Siddiqui, M. W., Dar, B. N., Mir, S. A., & Ali, A. (2020). Recent trends in extraction techniques of anthocyanins from plant materials. *Journal of Food Measurement and Characterization, 14*, 3508–3519.

Gales, O., Rodemann, T., Jones, J., & Swarts, N. (2021). Application of near infra-red spectroscopy as an instantaneous and simultaneous prediction tool for anthocyanins and sugar in whole fresh raspberry. *Journal of the Science of Food and Agriculture, 101*(6), 2449–2454.

Gao, Z., Yang, G., Zhao, X., Jiao, L., Wen, X., Liu, Y., & Dong, D. (2024). Rapid measurement of anthocyanin content in grape and grape juice: Raman spectroscopy provides non-destructive, rapid methods. *Computers and Electronics in Agriculture, 222*, 109048.

Geow, C. H., Tan, M. C., Yeap, S. P., & Chin, N. L. (2021). A review on extraction techniques and its future applications in industry. *European Journal of Lipid Science and Technology, 123*(4), 2000302.

Gonçalves, A. C., Nunes, A. R., Falcão, A., Alves, G., & Silva, L. R. (2021). Dietary effects of anthocyanins in human health: A comprehensive review. *Pharmaceuticals, 14*(7), 690.

González, M. J. A., Carrera, C., Barbero, G. F., & Palma, M. (2022). A comparison study between ultrasound–assisted and enzyme–assisted extraction of anthocyanins from blackcurrant (*Ribes nigrum* L.). *Food Chemistry: X, 13*, 100192.

Guo, Y., Liu, C., Ye, R., & Duan, Q. (2020). Advances on water quality detection by UV-Vis spectroscopy. *Applied Sciences, 10*(19), 6874.

Hassoun, A., Carpena, M., Prieto, M. A., Simal-Gandara, J., Özogul, F., Özogul, Y., & Regenstein, J. M. (2020). Use of spectroscopic techniques to monitor changes in food quality during application of natural preservatives: A review. *Antioxidants, 9*(9), 882.

Houghton, A., Appelhagen, I., & Martin, C. (2021). Natural blues: Structure meets function in anthocyanins. *Plants, 10*(4), 726.

Johnson, J., Collins, T., Walsh, K., & Naiker, M. (2020). Solvent extractions and spectrophotometric protocols for measuring the total anthocyanin, phenols and antioxidant content in plums. *Chemical Papers, 74*(12), 4481–4492.

Kharbach, M., Alaoui Mansouri, M., Taabouz, M., & Yu, H. (2023). Current application of advancing spectroscopy techniques in food analysis: Data handling with chemometric approaches. *Foods, 12*(14), 2753.

Le, B. T. (2020). Application of deep learning and near infrared spectroscopy in cereal analysis. *Vibrational Spectroscopy, 106,* 103009.

Li, M., Xu, W., & Su, Y. (2021). Solid-state NMR spectroscopy in pharmaceutical sciences. *TrAC Trends in Analytical Chemistry, 135,* 116152.

Liu, J., Zhou, H., Song, L., Yang, Z., Qiu, M., Wang, J., & Shi, S. (2021). Anthocyanins: Promising natural products with diverse pharmacological activities. *Molecules, 26*(13), 3807.

Lončarić, A., Celeiro, M., Jozinović, A., Jelinić, J., Kovač, T., Jokić, S., & Lores, M. (2020). Green extraction methods for extraction of polyphenolic compounds from blueberry pomace. *Foods, 9*(11), 1521.

Marcone, M. F., Wang, S., Albabish, W., Nie, S., Somnarain, D., & Hill, A. (2013). Diverse food-based applications of nuclear magnetic resonance (NMR) technology. *Food Research International, 51*(2), 729–747.

Mattioli, R., Francioso, A., Mosca, L., & Silva, P. (2020). Anthocyanins: A comprehensive review of their chemical properties and health effects on cardiovascular and neurodegenerative diseases. *Molecules, 25*(17), 3809.

Mazza, G., Cacace, J. E., & Kay, C. D. (2004). Methods of analysis for anthocyanins in plants and biological fluids. *Journal of AOAC International, 87*(1), 129–145.

Mishra, M. (2022). Spectroscopic techniques for the analysis of food quality, chemistry, and function.

Naresh, K. (2014). Applications of fluorescence spectroscopy. *Journal of Chemical and Pharmaceutical Sciences, 974,* 2115.

Nawrocka, A., & Lamorska, J.. (2013). Determination of food quality by using spectroscopic methods. In *Advances in Agrophysical Research*. IntechOpen.

Nelson, D. L. (2017). Introduction to spectroscopy. In *Spectroscopic methods in food analysis* (pp. 3–34). CRC Press.

Osborne, B. G. (2006). Near-infrared spectroscopy in food analysis. *Encyclopedia of Analytical Chemistry: Applications, Theory and Instrumentation,* 1–13.

Oteiza, P. I., Cremonini, E., & Fraga, C. G. (2023). Anthocyanin actions at the gastrointestinal tract: Relevance to their health benefits. *Molecular Aspects of Medicine, 89,* 101156.

Penner, M. H. (2017). Basic principles of spectroscopy. *Food Analysis,* 79–88.

Perez-Vazquez, A., Carpena, M., Barciela, P., Cassani, L., Simal-Gandara, J., & Prieto, M. A. (2023). Pressurized liquid extraction for the recovery of bioactive compounds from seaweeds for food industry application: A review. *Antioxidants, 12*(3), 612.

Petersen, M., Yu, Z., & Lu, X. (2021). Application of Raman spectroscopic methods in food safety: A review. *Biosensors, 11*(6), 187.

Pratiwi, R. A., & Nandiyanto, A. B. D. (2022). How to read and interpret UV-VIS spectrophotometric results in determining the structure of chemical compounds. *Indonesian Journal of Educational Research and Technology, 2*(1), 1–20.

Qu, C., Li, Y., Du, S., Geng, Y., Su, M., & Liu, H. (2022). Raman spectroscopy for rapid fingerprint analysis of meat quality and security: Principles, progress and prospects. *Food Research International, 161,* 111805.

Rostron, P., Gaber, S., & Gaber, D. (2016). Raman spectroscopy, review. *Laser, 21,* 24.

Rouxinol, M. I., Martins, M. R., Murta, G. C., Mota Barroso, J., & Rato, A. E. (2022). Quality assessment of red wine grapes through NIR spectroscopy. *Agronomy, 12*(3), 637.

Saha, S., Singh, J., Paul, A., Sarkar, R., Khan, Z., & Banerjee, K. (2020). Anthocyanin profiling using UV-vis spectroscopy and liquid chromatography mass spectrometry. *Journal of AOAC International, 103*(1), 23–39.

Şahin, Y. (2021). Raman spectroscopy and its applications. *The World of Biomedical Technology, 1*(1), 15–18.

Sahoo, N. K., Tejaswini, G., Sahu, M., & Muralikrishna, K. S. (2020). An overview on NMR spectroscopy-based metabolomics. *International Journal of Pharmaceutical Sciences and Developmental Research, 6*(1), 016–020.

Saletnik, A., Saletnik, B., & Puchalski, C. (2021). Overview of popular techniques of Raman spectroscopy and their potential in the study of plant tissues. *Molecules, 26*(6), 1537.

Shaikh, S., & O'Donnell, C. (2017). Applications of fluorescence spectroscopy in dairy processing: A review. *Current Opinion in Food Science, 17*, 16–24.

Stuppner, S., Mayr, S., Beganovic, A., Beć, K., Grabska, J., Aufschnaiter, U., & Huck, C. W. (2020). Near-infrared spectroscopy as a rapid screening method for the determination of total anthocyanin content in *Sambucus fructus*. *Sensors, 20*(17), 4983.

Sultangaziyev, A., & Bukasov, R. (2020). Applications of surface-enhanced fluorescence (SEF) spectroscopy in bio-detection and biosensing. *Sensing and Bio-Sensing Research, 30*, 100382.

Taghavi, T., Patel, H., Akande, O. E., & Galam, D. C. A. (2022). Total anthocyanin content of strawberry and the profile changes by extraction methods and sample processing. *Foods, 11*(8), 1072.

Tan, J., Han, Y., Han, B., Qi, X., Cai, X., Ge, S., & Xue, H. (2022). Extraction and purification of anthocyanins: A review. *Journal of Agriculture and Food Research, 8*, 100306.

Tena, N., & Asuero, A. G. (2022). Up-to-date analysis of the extraction methods for anthocyanins: Principles of the techniques, optimization, technical progress, and industrial application. *Antioxidants, 11*(2), 286.

Tombuloğlu, H., Öz, F., & Turhan, S. (2023). Raman spektroskopisinin et kalitesi ve güvenliğinin belirlenmesinde kullanimi. *Gıda, 48*(6), 1379–1393.

Verma, G., & Mishra, M. (2018). Development and optimization of UV-Vis spectroscopy—A review. *World J. Pharm. Res., 7*(11), 1170–1180.

Višnjevec, A. M., Barp, L., Lucci, P., & Moret, S.. (2024). Pressurized liquid extraction for the determination of bioactive compounds in plants with emphasis on phenolics. *TrAC Trends in Analytical Chemistry*, 117620.

Wallace, T. C., & Giusti, M. M. (2019). Anthocyanins—Nature's bold, beautiful, and health-promoting colors. *Foods, 8*(11), 550.

Yaseen, T., Sun, D. W., & Cheng, J. H. (2017). Raman imaging for food quality and safety evaluation: Fundamentals and applications. *Trends in Food Science & Technology, 62*, 177–189.

Yousuf, B., Gul, K., Wani, A. A., & Singh, P. (2016). Health benefits of anthocyanins and their encapsulation for potential use in food systems: A review. *Critical Reviews in Food Science and Nutrition, 56*(13), 2223–2230.

Zacharioudaki, D. E., Fitilis, I., & Kotti, M. (2022). Review of fluorescence spectroscopy in environmental quality applications. *Molecules, 27*(15), 4801.

Zang, Z., Tang, S., Li, Z., Chou, S., Shu, C., Chen, Y., & Li, B. (2022). An updated review on the stability of anthocyanins regarding the interaction with food proteins and polysaccharides. *Comprehensive Reviews in Food Science and Food Safety, 21*(5), 4378–4401.

15

GC/MS Analysis of Anthocyanins

Leo M.L. Nollet

15.1 Introduction

Gas chromatography (GC) is a powerful analytical technique used to separate, identify, and quantify compounds in a mixture. However, to analyze the water-soluble pigments, anthocyanins, GC is not the ideal method. High-performance liquid chromatography (HPLC) or ultra-performance liquid chromatography (UPLC) are the methods of choice to analyze, due to their polar nature and thermal instability.

The analysis of anthocyanins consists of different steps:

- Sample Preparation

  Anthocyanins are usually extracted from plant materials using solvents like methanol, ethanol, or a mixture of solvents often acidified (e.g., with formic acid or hydrochloric acid) to stabilize the anthocyanins.

  Next the extract may be filtered to remove solid particles and then concentrated to remove excess solvent.

- Chromatographic Analysis

  HPLC or UPLC is typically employed for separating anthocyanins. Reverse-phase columns (C18) are often used along with a gradient of water and an organic solvent both of which are acidified.

- Detection

  Detection is typically done using a UV-Vis detector, at around 520 nm, where anthocyanins exhibit strong absorbance. Mass spectrometry (MS) can also be coupled with Liquid Chromatography (LC) for more detailed structural analysis.

GC can theoretically be used if the anthocyanins are derivatized to make them more volatile and stable at higher temperatures. Derivatization involves silylation, acylation, or methylation to make anthocyanins volatile.

The derivatized anthocyanins can then be analyzed using GC, often with a flame ionization detector (FID) or mass spectrometry (GC-MS) for identification and quantification.

The literature on GC analysis methods of anthocyanins is scarce and the publications are rather old.

DOI: 10.1201/9781003453260-17

15.2 Gas Chromatographic Methods

The extraction of anthocyanin from Bokbunja (*Rubus coreanus* Miq.) marc generated during traditional wine processing was optimized using response surface methodology (RSM).

One step in this analysis is silylation performed at 50°C for 30 min. The trimethylsilyl (TMS) derivatives were analyzed using a GC -mass spectrometer and a GC (1).

The TMS derivatives of the aqueous fraction eluted from a C18 cartridge were analyzed using a Shimadzu GC-17A. One microliter of each sample was injected into a GC equipped with a FID. Separation of the compounds was achieved using a SPB-5 silica fused capillary column (30 m × 0.25 mm I.D., 0.25 μm film thickness, Supelco, USA). The oven temperature was elevated from 150°C to 300°C at a rate of 5°C/min, and then held at 300°C for 10 min. The helium gas flow rate was 1 mL/min, the split ratio was adjusted to 10, and the FID and injection port temperatures were set at 280°C and 250°C, respectively.

The GC-MS parameters were similar to those used for the GC. An Agilent 6890 N GC/5973N MSD was used to obtain the mass fragmentation pattern of the compounds using a mass range of m/z 50–600 and electron ionization at 70 eV. Identification and quantification of compounds was performed by comparison with authentic standards. Tartaric acid and methyl-3-a-glucopyranoside were used as internal standards for organic acid and sugar, respectively. The specific fragment (m/z 207) of fructose was used for quantitation in the single ion monitoring (SIM) mode of the GC-MS because the fructose and citric acid chromatogram produced using the GC were overlapping. The citric acid content was then calculated by subtracting the fructose content determined by the SIM mode calibration from the total content of the overlapping peak.

The results of HPLC-PDA-MS/MS analysis of the anthocyanins extracted from Bokbunja marc revealed the presence of six anthocyanin components, which were tentatively identified as cyanidin 3-O-sambubioside, cyanidin 3-O-xylosyl-rutino, cyanidin 3-O-rutinoside, pelargonidin 3-O-rutinoside, delphinidin 3-O-rutinoside-delphinidin 3-O-rutinoside, and delphinidin 3-O-glucuronide.

The anthocyanins responsible for the red color of red kiwifruit were extracted in acidified ethanol and isolated by solid phase extraction (SPE) followed by preparative HPLC (2). Five anthocyanins were obtained and subsequently identified as delphinidin 3-[2-(xylosyl)galactoside], delphinidin 3-galactoside, cyanidin 3-[2-(xylosyl)galactoside], cyanidin 3-galactoside, and cyanidin 3-glucoside by a combination of LC-MS/MS, GC-MS, and 2D NMR. Delphinidin 3-[2-(xylosyl)galactoside] and delphinidin 3-galactoside have not previously been reported in the genus *Actinidia*.

Shimadzu QP5050A, the separation column was a 30 m × 0.25 mm I.D. × 0.25 μm film thickness Phenomenex ZB-5 ms (Phenomenex, Torrance,CA). The temperature was programed from 40°C (held for 1 min) to 300°C at 7°C/min (held for 4 min). The derivatized sugar residues were identified by comparison of retention times and fragmentation patterns with authentic standards. Hexoses (glucose or galactose) *m/z* (rel. intensity) 319(36), 205(26), 147(30), 73(100); pentose (xylose) *m/z* (rel. intensity) 307(19), 217(39), 147(20), 103(53), 73(100).

The quantitative and qualitative evolution of the anthocyanins and volatile compounds of four raspberry cultivars (cvs. Heritage, Autumn Bliss, Zeva, and Rubi) growing in Spain were analyzed raw, just frozen, and during long-term frozen storage at –20°C for a one year period (3). HS-SPME coupled with GC-MS and HPLC techniques were employed to study the evolution of the volatile compounds and the individual anthocyanins, respectively. The volatile aroma composition changes produced by the freezing process and long-term frozen storage were minimal. Only a significant increase in extraction capacity was obtained for R-ionone (27%) and for caryophyllene (67%) in Heritage at 12 months of storage. The stability of anthocyanins to freezing and frozen storage depends on the seasonal period of harvest. Heritage and Autumn Bliss (early cultivars) were less affected by processing and long-term frozen storage (one year), and the total pigment extracted showed the tendency to increase 17 and 5%, respectively. Rubi and Zeva (late cultivars) suffered a decreased trend on the total anthocyanin content of 4% for Rubi and 17.5% for Zeva. Cyanidin 3-glucoside most easily suffered the degradative reactions that take place during processing and the storage period.

The GC-MS Volatile Compound Analysis was as follows. An SPME holder (Supelco) was used to perform the experiments. A fused silica fiber coated with a 100 ím layer of dimethylpolysiloxane was chosen to extract the volatile components of the raspberries. One gram of the fruit was cut in four pieces and placed in a 20 mL vial which was capped with plastic film. Extraction was performed by exposing the fiber to the headspace of the sample for 30 min at 30°C.

A Perkin-Elmer model 8500 gas chromatograph equipped with a PTV injector and an FID detector was used to perform the analysis. The system was coupled to a model 2600 chromatography software system (Perkin-Elmer Nelson Analytical). A 50 m × 0.25 mm I.D. fused-silica capillary column (Chrompack) coated with a 0.25 μm layer of CP-Sil-5 CB was used. Helium was the carrier gas. Thermal desorption of the compounds in the fiber took place in the GC injector at 200°C for 15 min in splitless mode for 5 min. The detector operated at 250°C. The oven temperature was programed from 50°C (3 min constant temperature) to 250°C at 5°C/min. The final temperature was maintained for 17 min.

GC-MS analysis was carried out by coupling the gas chromatograph described above to a Perkin-Elmer ITD-50 ion trap detector (EI, 70 eV). The capillary column and chromatographic program used were as mentioned previously. Compounds were identified by comparison of the spectra with those in a general-purpose library. Moreover, the identity of the components was confirmed by matching their mass spectrometric data with those obtained from the same equipment and corresponding to authentic reference compounds.

Metabolite profiling of three blueberry species (*Vaccinium bracteatum* Thunb., *Vaccinium oldhamii* Miquel., and *Vaccinium corymbosum* L.) was performed using GC–time-of-flight-mass spectrometry (GC-TOF-MS) and UPLC–quadrupole-time-of-flight-mass spectrometry (UPLC-Q-TOF-MS) combined multivariate analysis (4). Partial least-squares discriminant analysis clearly showed metabolic differences among species. GC-TOF-MS analysis revealed significant differences in amino acids, organic acids, fatty acids, sugars, and phenolic acids among the three blueberry species. UPLC-Q-TOF-MS analysis indicated that anthocyanins were the major metabolites distinguishing *V. bracteatum* from *V. oldhamii*. The contents of anthocyanins such as glycosides of cyanidin were high in *V. bracteatum*, while glycosides of delphinidin, petunidin, and malvidin were high in *V. oldhamii*. Antioxidant activities assessed using ABTS **2,2′-azino-bis(3ethylbenzothiazoline-6-sulfonic acid** (ABTS) and 2,2-Diphenyl-1-picrylhydrazyl (DPPH) assays showed the greatest activity in *V. oldhamii* and revealed the highest correlation with total phenolic, total flavonoid, and total anthocyanin contents and their metabolites.

Various metabolites act as plant defense molecules due to their antioxidant abilities. The study of five aimed to investigate the influence of UVB irradiation on the accumulation of metabolites, including primary metabolites (sugar, sugar alcohols, amino acids, organic acids, and an amine) and secondary metabolites (anthocyanins, fatty acids, and phenolic acids), and its synergistic antioxidant ability, in purple kohlrabi sprouts (5). Metabolite analyses revealed a total of 92 metabolites in the sprouts. Specifically, the levels of most amino acids increased after 24 h of UVB treatment, and then slightly decreased in the kohlrabi sprouts. The levels of most sugars and sugar alcohols increased after 24 h of UVB treatment and then decreased. The levels of Tricarboxylic Acid Cycle (TCA) cycle intermediates and phenolic acids gradually increased during the UVB treatment. Furthermore, the levels of some fatty acids gradually increased during the UVB treatment, and the levels of the other fatty acids increased after 6 h of UVB treatment and then decreased. In particular, the levels of most anthocyanins, known to be strong antioxidants, gradually increased after 24 h of UVB treatment. In the in vitro ABTS scavenging assay, UVB-treated purple kohlrabi sprouts showed increased scavenging ability. This may be attributed to the increased accumulation of metabolites acting as antioxidants, in response to UVB treatment. This study confirmed that UVB irradiation induced the alteration of primary and secondary metabolism in the kohlrabi sprouts.

GC was performed with an Agilent 6890N GC main frame (Agilent) and an Agilent 7820A GC equipped with 5977E MSD for GC-MS. Each chromatography experiment used DB-5MS (60 m × 0.32 mm I.D. × 0.25 μm film thickness, Agilent). For samples that were only trans-methylated, 1 μL aliquots were injected into the GC via splitless injection, and the GC oven temperature program was set as follows: start at 50°C, increase at a rate of 4°C/min to 190°C, decrease at a rate of 0.5°C/min to 180°C, hold for 5 min, increase at a rate of 4°C/min to 280°C, increase at a rate of 10°C/min to 300°C, and hold at this temperature for 5 min. Ionization energy of 70 eV was used for GC-MS. After performing the total ion chromatography (TIC) of each sample, identification was conducted by comparing retention time using the Chemstation software library.

In the work of de Souza Dias et al. (6) a fast and simple methodology has been applied for the determination of gallic acid, resveratrol, catechin, and malvidin in Brazilian wines by GC–mass spectrometry. The procedure included a stage of ultrasound-assisted liquid–liquid extraction and subsequent derivatization with N,O-bis(trimethylsilyl)trifluoroacetamide (BSTFA) and GC-MS analysis. The limit of detection varied from 0.41 to 1.18 mg/L in all the analytes. The relative standard deviations calculated for 8.0 and 20 mg/L were 1.90 and 0.82% for gallic acid, 3.08 and 1.22% for catechin, 1.30 and 0.44% for malvidin, 1.50 and 0.53% for resveratrol, and 1.41 and 0.61% for quercetin. The developed methodology was applied for the analysis of red wine samples collected in the São Francisco region, Bahia state, Brazil. Quercetin concentration varied from 2.4 to 3.0 mg/L, gallic acid 21.4–56.3 mg/L, resveratrol 1.5–5.9 mg/L, malvidin 15.3–32.2 mg/L, and catechin 11.71–18.2 mg/L. The obtained concentrations are in agreement with those reported in the literature.

The GC-MS analysis was carried out with a Perkin-Elmer apparatus, Model Clarus 500, and the used temperature program were the following: initial temperature of 80°C, for 1 min, then from 80°C to 250°C at a rate of 20°C/min, and held for 1 min; next, it was augmented at 6°C/min to 300°C, held for 2 min, and finally increased at 20°C/min– to 320°C, and held for 24 min.

References

1. Ku, C. S. & Mun, S. P. (2008). Optimization of the extraction of anthocyanin from Bokbunja (*Rubus coreanus* Miq.) marc produced during traditional wine processing and characterization of the extracts. Bioresource Technology, 99 (17), 8325–8330.

2. Comeskey, D. J., Montefiori, M., Edwards, P. J., McGhie, T. K. (2009). Isolation and structural identification of the anthocyanin components of red kiwifruit. Journal of Agricultural and Food Chemistry, 57 (5), 2035–2039.

3. de Ancos, B., Ibanez, E., Reglero, G., & Cano, M. P. (2000). Frozen storage effects on anthocyanins and volatile compounds of raspberry fruit. Journal of Agricultural and Food Chemistry, 48 (3), 873–879.

4. Lee, S., Jung, E. S., Do, S. G., Jung, G. Y., Song, G., Song, J. M., & Lee, C. H. (2014). Correlation between species-specific metabolite profiles and bioactivities of blueberries (*Vaccinium* spp.). Journal of Agricultural and Food Chemistry, 62 (9), 2126–2133.

5. Yeo H. J., Lim S.-Y., Park C. H., Kim C. Y., Sathasivam R., Kim J. K., Park S. U. (2022) Metabolic analyses and evaluation of antioxidant activity in purple kohlrabi sprouts after exposed to UVB radiation. Antioxidants, 11 (8), 1443.

6. de Souza Dias, F., Silva, M. F. & David, J. M. (2013). Determination of quercetin, gallic acid, resveratrol, catechin and malvidin in Brazilian wines elaborated in the Vale do São Francisco using liquid–liquid extraction assisted by ultrasound and GC-MS. Food Analytical Methods, 6, 963–968.

Index

Note: **Bold** page numbers refer to **tables** and *Italic* page numbers refer to *figures*.